de Gruyter Studies in Organization 9

Clark: Anglo-American Innovation

de Gruyter Studies in Organization

An international series by internationally known authors presenting current fields of research in organization.

Organizing and organizations are substantial pre-requisites for the viability and future developments of society. Their study and comprehension are indispensable to the quality of human life. Therefore, the series aims to:

- offer to the specialist work material in form of the most important and current problems, methods and results;
- give interested readers access to different subject areas;
- provide aids for decisions on contemporary problems and stimulate ideas.

The series will include monographs, collections of contributed papers, and handbooks.

Peter A. Clark

Anglo-American Innovation

Walter de Gruyter Berlin · New York 1987

Dr. Peter A. Clark
Work Organization Research Centre
Aston University
Birmingham, Great Britain

Library of Congress Cataloging in Publication Data

Clark, Peter A.
Anglo-American innovation.
(De Gruyter studies in organization ; 9)
Bibliography: p.
Includes index.
1. Technological innovations -- Great Britain -- Management.
2. Technological innovations -- United States -- Management.
I. Title. II. Series.
HD45.C52 1987 338'.06 87-22221
ISBN 0-89925-164-1 (U.S.)

CIP-Kurztitelaufnahme der Deutschen Bibliothek

Clark, Peter A.:
Anglo-American innovation / Peter A. Clark. -
Berlin ; New York : de Gruyter, 1987. -
(De Gruyter studies in organization ; 9)
ISBN 3-11-010572-1
NE: GT

 - Typesetting: Appl, Wemding. - Printing: Gerike GmbH, Berlin. - Binding: Lüderitz & Bauer, Berlin. - Cover Design: Rudolf Hübler, Berlin. - Printed in Germany.

Contents

Part I. Introduction 1

1 The Agenda 3
1.1 Three Core Problems 3
1.2 Organization Studies: Developing a Process Perspective 7
1.3 Cross-Cultural Patterns of Innovation 13
1.3.1 Context: Choice of Technique 13
1.3.2 Britain and America: Convergence or Distinct Trajectories? 17
1.3.3 The Future: Renaissance or Retardation? 19
1.3.4 Typical Variety: The Anglo-American Exemplar 24
1.4 Structure of the Book 25

Part II. Evolvement of Innovations: Shape and Uses 29

2 State of Theory 31
2.1 Introduction 31
2.2 Main Areas 34
2.3 Basic Concepts 34
2.4 Innovations: Multidimensional 38
2.5 Problem Agenda: The Dominance of Economics 40
2.6 Changing Assumptions 46
2.7 Contexts: the Selection of Innovations 49
2.8 Transnational Transfers of Innovations 55

3 Innovation Supply: The Marketing and Imitation Models 57
3.1 Introduction 57
3.2 Economics of Innovation 59
3.2.1 Hagerstrand 59
3.2.2 Mansfield 63
3.3 Sociology of Innovation 65
3.3.1 Rogers I 65
3.3.2 Rogers II 70
3.3.3 Initiation and Implementation 77
3.3.4 Implementation: Current Normative Model 79
3.4 Marketing and Infrastructure Model 83
3.4.1 L. A. Brown 83
3.4.2 Von Hippel 86

4 Technology as Process: Trajectories and Life Cycles 89
4.1 Introduction 89
4.2 Technology and Process 91
4.3 Trajectories and the Role of Paradigms 99
4.4 Life Cycle Thesis: Product, Process and Work Organization . 103
4.4.1 Abernathy I 103
4.4.2 Abernathy II 109
4.5 Technology as Knowledge: Locus of Initiative 113

5 The Corporate User: Innovation-Design Capacity 123
5.1 Limits of the Supply Model 123
5.2 Multiple, Simultaneous and Diverse Innovations 125
5.3 Organizing: Communities, Networks and Bodies of Knowledge 129
5.4 Innovation-Design Capacities 140
5.5 Case Illustration: Contested Innovation 144
5.6 Recurrence, Momentum and Inertia 148
5.7 Cultures, Leadership and Adaptation 151
5.8 Appropriation: The Basis for Survival 155

Part III. Anglo-American Patterns of Organizing 159

6 Transatlantic Evolvement I: Americans and the Absorption Gap . . 161
6.1 Introduction: Absorption Gap Illustrated 161
6.2 Case Study: "Teamwork" 165
6.2.1 The Questions 165
6.2.2 Managerial Templates of Organizing 166
6.2.3 The Evolvement of Innovations: Recapitulation 168
6.3 British Predispositions 170
6.3.1 Nineteenth-Century Societal Context 170
6.3.2 Origins 171
6.3.3 The Embryonic Games 173
6.3.4 Framework of Analysis 174
6.3.5 Crystalization and Schism: 1830s-1850s 176
6.3.6 The Rugby Football Union, 1871 177
6.3.7 Diffusion and Further Schism: 1880s and 1890s 178
6.4 American Predispositions 179
6.4.1 Players and Winners 179
6.4.2 Re-Invention: 1876-1886 181
6.4.3 The Coaches 183
6.4.4 The Professional Game 185
6.5 Comparing American Football and Rugby Union 186

7 Economy, Structuration and Region: A Basic Framework 193
7.1 Introduction 193
7.2 Economy: An Allocative Mechanism through Time and Space 199
7.2.1 Introduction 199
7.2.2 Longwave Theories 199
7.2.3 Transaction Costs, Ideology and the State 212
7.3 Structuration 216
7.3.1 Introduction 216
7.3.2 Asymmetrical Power Relations 219
7.3.3 Knowledge Bases and Thinking Practices 221
7.3.4 Institutions of Work: Capital, Management and Labour 227
7.3.5 Structuration and Transitions 228
7.4 Regions: Britain and the USA 230

8 British Systems of Organizing: Contexts and Directions into the First Divide 233
8.1 Introduction 233
8.2 The Netherlands: Patterns of Innovation 236
8.3 Britain: An Offshore Island 239
8.4 Cotton: Clans and Markets 241
8.5 Entering the First Divide: Templates of Organizing 244
8.6 Institutions of Work 250

9 American Systems of Organizing: The Early Foundations 253
9.1 Introduction 253
9.2 The American System of Manufactures Reconsidered 254
9.3 Location and Resources 256
9.4 Founding Cultures 258
9.5 North Atlantic Economy 263
9.6 American System of Manufactures in Context 266

10 The American Market: A Key Base from 1870 to the 1960s 269
10.1 Introduction 269
10.2 An Integrated World-Economy 272
10.2.1 World-Economy 272
10.2.2 Infrastructure: Transport and Information 274
10.3 Design, Administrative Structures and Corporate Education . 275
10.3.1 Design 275
10.3.2 Administrative Sciences 279
10.3.3 Corporate Education and Training 286
10.4 Market Control and the Modern Enterprise 287
10.4.1 Agribusiness 287

10.4.2 Cigarette Industry 1880-1900 288
10.4.3 Electricals, Chemicals and Automobiles 291
10.5 Innovation as a Filiere: Imitation and Rigidities 294

11 British Systems of Organizing: A Case of Incomplete Modernization? 297
11.1 Introduction 297
11.2 Loose-Coupling and Devolvement in Work Organization . . . 299
11.2.1 Introduction 299
11.2.2 Colonies, Trading Companies and Professions 303
11.2.3 Rationalisation and Bureaucracy 305
11.2.4 Forms of Payment and Skill Ownership 308
11.3 Markets and Sectors 312
11.3.1 Commerce and Shipping 312
11.3.2 Cotton and Cigarettes 316
11.3.3 Cars, Chemicals and Electricals 320

12 Transatlantic Evolvement II: Britain and the Appropriation Gap . . 325
12.1 Introduction 325
12.2 Administrative Innovations 329
12.2.1 Taylorism: Work Study 329
12.2.2 Organization Development (OD) 332
12.2.3 The Multidivisional Form (MDF) 338
12.2.4 Plantwide Productivity Schemes 340
12.3 Technological Innovations 341
12.3.1 Automobile Assembly Lines in the 1920s 341
12.3.2 Information Technology: 1970s and 1980s 345
12.4 Management Education 348
12.5 Assessment 349

Part IV. Implications 351

13 Japan and the Pacific Rim: The New Competition 353
13.1 Introduction 353
13.2 Structuration: Framework 354
13.2.1 Geopolitical 354
13.2.2 New World-Economy: JUSA 355
13.2.3 Structuration 356
13.3 Appropriation Rather than Imitation/Rejection 361
13.4 Markets and Knowledge 363
13.4.1 Current Strengths 363
13.4.2 Markets 364
13.4.3 Production Knowledge and Techniques: Process Innovations 366
13.4.4 Production Institutions: Education 369

14 Summary and Implications 371
14.1 The Core Problems 371
14.2 Innovation Evolvement 373
14.3 Anglo-American Patterns and Transfers 376
14.4 New Divides 378
14.5 Renaissance or Retardation? 381

15 References 383

16 Author Index 401

17 Subject Index 403

Part I.
Introduction

1 The Agenda

1.1 Three Core Problems

Anglo-American Innovation examines and integrates three previously separate arrays of problems:

1. The neglect by organization studies of the pervasive influence of innovations, their diffusion and long-term evolvement
2. The blockages to the understanding of the evolvement of innovations created by the dominance of neoclassical economics and by the fragmentation between disciplines
3. The weakness of many current approaches to cross-cultural analysis, especially the neglect of the problems of the transfer of all types of innovations across national boundaries

These problems may be treated as interrelated (see Fig. 1-1)

This book locates the innovation dimension at the centre of the study of organizations and shows how the retheorization of the processes of innovation can illuminate the problem of cross-cultural transfers. The empirical material and its analysis concentrates upon the Anglo-American case, which is used as an exemplar of principles which can be readily transferred to other comparisons of the transfer of innovation (e. g., America-Japan, France-Germany).

In this opening section the three core problem areas are introduced in a concise, summary format.

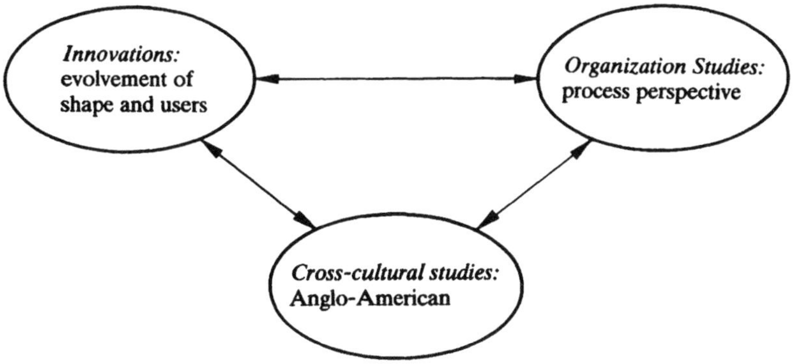

Figure 1-1. Three basic problem areas

Problem 1. Organization studies and innovation

There are five basic problems involved when considering organization studies and innovation.

First, organization studies have largely neglected the phenomenon of innovation and the pervasive role which its occurrence plays in the survival and development of organizations. There are notable exceptions to the neglect of innovation: in the studies of Hage, in the theorizing on population and ecology models by Aldrich, and in the study of seven sectors of the American economy and the analysis of adaptation by Lawrence and Dyer[1]. However, these are exceptions and the mainstream of organization studies[2] has given insufficient attention to the processes of innovation. Mainstream organization studies are still dominated by the problem of bureaucracy, structure and performance inherited from the approaches of postwar sociology. Consequently, the economic dimension, so central for Marx and for Weber, has been overshadowed and translated into the bland, ahistorical concepts of environmental uncertainty. Hence the dimensions of technology and technical change are better analysed by economic historians[3] than by those representing mainstream organization studies. Also, the whole area of the diffusion of innovations is barely mentioned, even in the Organization Development literature.

Second, processual studies of organization lack an adequate theorization of the temporal dimension (Clark, P. A. 1975b; 1985: 35–80). The context of organizations possesses many temporal trajectories (e.g., for technology) whose pervasive, irregular and jagged impacts are ignored. Moreover a processual perspective needs to take account of the significance for structuring of the chronic recursiveness[4] in organization structuring as well as the contingent nature of the periodicities in external signals which organization members attend to as the basis for actions (see Clark, P. A. 1985). Existing thinking fails to recognize the distinction between recurrence and transition (Clark/Starkey 1987). Also, existing models of change have been overlain with the simplistic schemas of social physics introduced in the 1940s by Lewin and these have even influenced the interesting attempts to retheorize transitions in the manner proposed by Miller and Friesen which is too deterministic (Miller/Friesen

---

1 See in particular Aldrich (1979); Hage (1980); Lawrence, and Dyer (1983)

2 For example, Clegg, and Dunkerley (1980); Donaldson (1985); Miles (1980)

3 Relevant examples include the seminal work of Landes (1969); Rosenberg (1976); Rosenberg (1982). Two recent contributions which are relevant to transatlantic comparisons are Elbaum and Lazonick (1986) and Jeremy (1981). On the analysis of management organization in the two countries see Lazonick (1987)

4 The theme of chronic recursiveness is conceptualized in the work of Antony Giddens and applied to a specific historical context by Derek Gregory; see Giddens (1979); Giddens (1985); Gregory (1982)

1982). There is a requirement to give much more attention to macrolevel pulsations of the economy and to the competitive nature of organizational survival[5]. The analyses of phenomena in organizations is too abstracted from its context. The problems are compounded by the methodological device of the taking of samples of organizations which are often so diverse or so narrow in their composition that false assumptions are made about the presence and absence of relationships between variables (Mohr 1982). The result is confusion in the definition of variables – especially technology – and in their causal connections. The difficulties which this creates are very evident in the most highly cited study of technology in the Aston approach (Hickson/Pugh/Pheysey 1969). Fortunately there are more processual approaches to the study of technology and a more temporally relevant alternative is now becoming possible[6].

Third, the study of innovation, largely undertaken by economists, economic historians and economic geographers plus the Rogers (1983) perspective has been largely undertaken without reference to the study of organizing.

Fourth, organization studies have neglected the competition between organizations over time and across national boundaries[7]. That neglect has been hidden by the search for alikeness amongst organizations (Lammers/Hickson 1979).

Fifth, the relevance of the sector and of sector-based concepts such as the sector life course (Clark/Whipp 1985) have rarely been used in organization studies. Also, there has been a neglect of industrial organization and too much concentration on the internal structure (Chandler 1977: 18). There is, therefore, a need to take account of the changing basis of competition and the uncertainty which innovation introduces.

Problem 2. Innovation

First, existing studies suffer from a number of limitations. In general terms the study of innovation is beset with problems, yet the potential for their resolution is highly promising. Innovation studies have been dominated by economists (see Chap. 2), especially by neoclassical economics which has little to

---

5 See especially, Karpik (1978: 15–68) for a valuable treatment of the different kinds of knowledge found amongst managers and staff personnel in medium and large-scale enterprises

6 A more processual treatment of technology is developed in Part II of this book

7 In Britain the Economic and Social Research Council (ESRC) has taken the initiative in promoting studies of Britain's international competitiveness. A major group was established at Aston University with the Work Organization Research Centre (WORC) led by Peter Clark, John Child and Ray Loveridge; the five-year-program was commenced in 1982

offer to policy-related issues (Clark, N. 1985). Too little attention has been given to economic pulsations and their analysis in relationship to the possible bunching of periods of innovation[8]. Overall the study of innovation is highly fragmented and two notions which dominated early thinking still possess a strong influence on common sense reasoning: the pro-innovation bias[9] and the belief that all innovation occurs through imitation (see Part II).

Second, some of the most promising lines of development in the analysis of innovation have arisen from economic historians examining the development of American technology within implicit comparisons to the British case (Elbaum/Lazonick 1986; Landes 1969).

Third, it is now possible to undertake some meaningful disaggregation of topics. For example, the following can be dealt with in a reasonably coherent manner: long-run, fifty-year pulsations in the economy; the bunching of innovations thesis (Clark, N. 1985); diffusion of innovations; re-invention, rejection and imitation; innovation trajectories.

Problem 3. Cross cultural studies

Organization studies have been hooked on the alikeness of organizations across different societal settings. The notion of an Anglo-Saxon or Anglo-American model of best practice is still powerful despite considerable evidence to the contrary (Lammers/Hickson 1979, Chaps. 1, 22, 23). The main developments in organization studies since the early 1960s have been directed towards discovering common structures and common processes irrespective of time and place (Donaldson 1985). Considerable progress has been made, yet there has also been considerable confusion and a displacement of emphasis so that organization design is falsely theorized. Alikeness has been a preoccupation which has shaped the search for simple scalable definitions of key variables such as structure and technology. Alikeness has also been pursued within a theoretical assumption that variables move frictionlessly and in a linear fashion (Mohr 1982; Clark, P. A. 1975).

Alikeness has dominated cross-cultural comparisons until very recently. An early attempt at an alternative cultural paradigm by Crozier (1964; 1973) was largely neglected until the more recent French studies associated with Maurice and with Dubois. There were and are certain problems with Crozier's original formulation and with its modified application to the USA (Crozier 1984), yet it must be regarded as the major forerunner of more recent and

[8] The Science Policy Research Unit (SPRU) at Sussex University, England has been a major international centre for such studies and these are selectively included in Part II and Chap. 7.

[9] The limitations of the pro-innovation bias are dealt with by Rogers (1983: 92–102)

more relevant attempts[10]. Even so, unlikeness still remains undeveloped within organization studies.

The interest in cross-cultural studies has been fired by the economic rise of Japan and by the exploration of American-Japanese unlikeness and likeness (see Chap. 13 in this book). That research has implicitly raised the two further questions:

1. How is the diffusion and appropriation of innovations across national boundaries accomplished?
2. Is the USA experiencing a slow decline and retardation comparable to that experienced by Britain since the 1880s or is this, as some suggest, a period of American adjustment and renaissance?

One way in which cross-cultural studies can be enlightening *is to examine what happens when the same innovation crosses national boundaries.* The original and incorrect assumption was that diffusion occurs through imitation. That assumption is still strong in the American literature on innovation. A recent study by di Maggio, for example, suggests that imitation, or, isomorphism is typical both within a culture and across cultures (DiMaggio/Powell 1983). Moreover, that latter point lies behind the seminal studies of the export of American structures such as the multidivisional form (MDF) by Chandler (Chandler/Daems 1980). The assumption of diffusion by imitation is challenged by the studies of economic historians, especially those specializing in Anglo-American comparison and analysis. Their studies indicate sharp and significant differences in the contexts and processes by which problems and solutions are undertaken in the USA and Britain. Those studies provide a key framework for exploring the focus of this book.

This short introductory section has overviewed the three problem areas to indicate that ongoing developments in theorizing, research and policy-making provide an excellent grounding for the construction of a *new paradigm* in which the facts and problems of innovation constitute the core of investigating the problems and around which a deeper and wider understanding can develop.

1.2 Organization Studies: Developing a Process Perspective

As with any subfield, such as organization studies, it is relevant to penetrate behind the apparent normality and facticity of the contemporary definition of paradigms and to briefly examine which directions were chosen and why. It is

[10] The major European project is the Micro Electronics in Service Sector; a list of publications is available from ESRC, WORC, Aston University, England

equally important to examine which routes were neglected and why. In the case of organization studies it is highly relevant to examine what has happened to process perspectives. Once that task is accomplished the opportunities for altering the directions previously taken can then be considered.

The creation of the field of organization studies has been strongly affected by two factors:

First, the discipline has been split into two aspects: micro and macro. The former has been concerned with leadership, individual differences, expectancy theory and job design[11]. Its sharply prescriptive focus is on the organization as a collection of individuals who have to be selected, trained, motivated, promoted and rewarded. The latter macro focus has tended to be concerned with external contexts, the notion of contingency and the structure-performance thesis (Miles 1980; Donaldson 1985). Each approach implicitly tended to adopt equilibrium models of the organization: comparative statics. Hence change was theorized as the frictionless adjustment of variables.

Second, from an early stage the macro approach introduced a distinction between structure and process in which it was claimed that the study of structure should take priority. The construction of the Aston Programme was pivoted on that distinction and its arrays of research instruments reinforced the tendency towards structural studies (Pugh/Hickson 1976). The problematic nature of that split has rarely been confronted and its contradictions have been evident since Hall's attempt to handle the split by writing a basic text in two parts; the result has been confusing (Hall 1972).

The degree of confusion is made very clear in the review by Mohr of the literature on change and innovation (Mohr 1982). Mohr reports that much of the structural-performance approach hinges around the confounded variable of size. There are many confounding results which might be better understood if the organization was analysed as a set of processes some of which are moving in opposite directions and many of which are so intimately tied into configurations that the frictionless adjustment of variables is impossible. For example, Duncan's frequently cited study of innovative structuring shows that different and contrary structural forms are required to cope with the detection of the directions of an adjustment and the implementation of the chosen plans (Duncan 1972; Zaltman/Duncan 1977, Chap. 11). That finding shows how dangerous it can be to analyse organization in ways that suggest that structuring is homogeneous through time and that the movements of configurations of variables are unilinear.

The structural approach does not deal with the issue of human agency, that is, how collections of individuals acting in concert and in opposition shape the

---

[11] Micro-OB is dominant in North America, whilst macro-OB is much more influential in Europe

directions which are taken by organizations (Compare Donaldson 1985 with Whipp/Clark 1986). So far orthodox processual approaches in organization studies have been very static, over-simplistic and have failed to deal with human agency. Indeed their limitations have justified the extensive critiques which have been made of planned organization change (POC) and of organization development (OD). These are essentially heterogeneous collections of prescriptive frameworks which provide a deceptive image of the ease and speed of achieving deliberate organizational change (Clark/Starkey 1987). Yet an alternative approach to process is beginning to emerge. The alternative builds on the theory that structure and agency are a duality. That is, that individuals as active human agents are both influenced by pre-existing forms of structuring, yet are to some degree empowered to interpret what should be done in the future. The theory will be examined more fully in Chap. 7. Here, certain basic points will be introduced. The theory of structure and agency presumes that individuals learn a collective repertoire of cognitions, normative frameworks and behavioural patterns. The repertoire consists of dormant aspects and active components rather in the same way that a theatrical group like the Royal Shakespeare Company or an American football team like the Chicago Bears possess a repertoire. The components which make up the repertoire can be activated when certain members indicate that a future situation will require them.

The concept of a repertoire is a complex one. The following example, which is drawn from Gearing's reconstruction of the repertoire of an eighteenth century Cherokee Indian village, is intended as an illustration of the principle (Gearing 1958). The Cherokee year consisted of certain relatively fixed events whose specific activation depended on the interpretation of many external events in the movement of the climate, the flora and fauna. Those events possessed a periodicity which was – unlike British weather – relatively free from contingency, though the point should not be exaggerated. The year might also contain events, such as wars and murders, whose specific timing was very contingent. Gearing shows that the Cherokee developed a repertoire of what he terms structural poses which were capable of being activated by certain internal signals from the elders of the tribe; Fig. 1-2 illustrates a year. Gearing, for reasons of exposition, tends to emphasize the recursiveness of the situation and the repertoire. The example of the Cherokee repertoire illustrates the extent to which the unfolding of the repertoire is both temporally differentiated and also highly recursive.

When applying the concept of repertoire to organizations it is useful to start by building on the theory of multiple structures[12] which states that any enter-

[12] The notion of dual structures was introduced in the early 1970s. The perspective of multiple structures was introduced in Clark, P. A. (1976: 231–250); also in Clark, P. A. (1985)

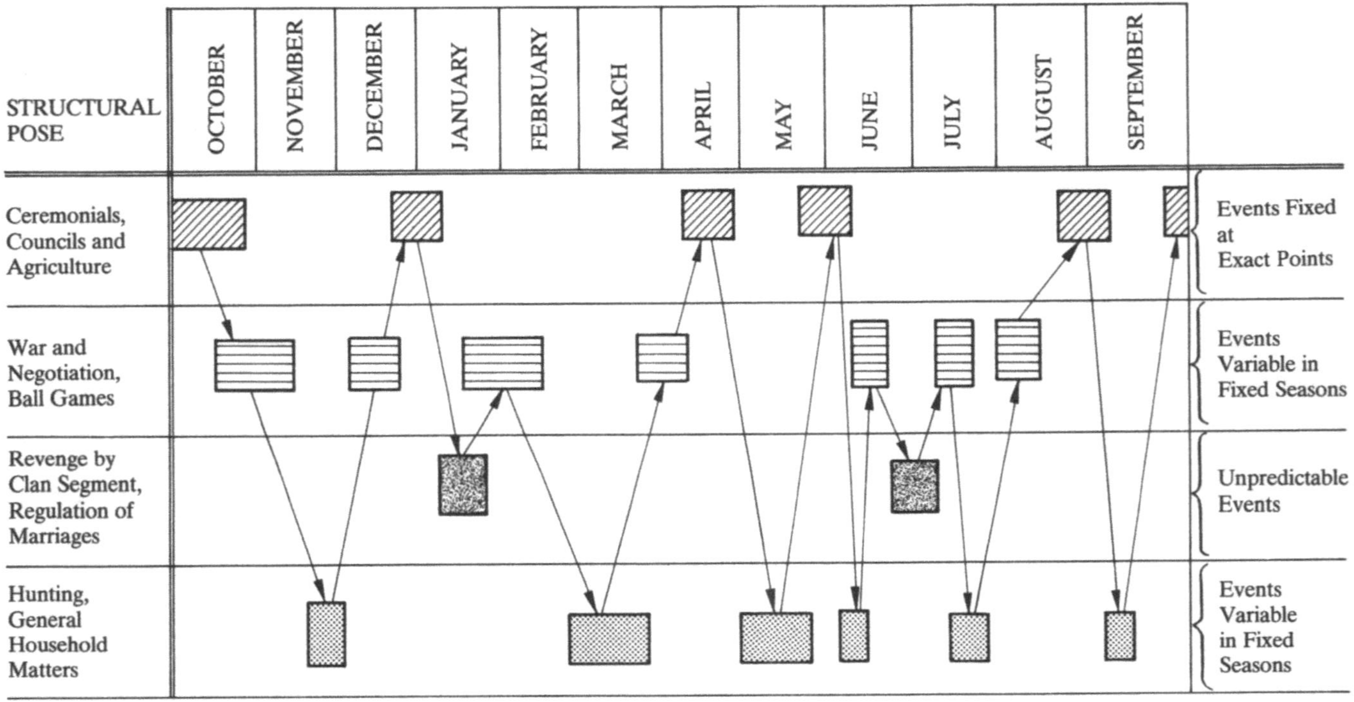

Figure 1-2. Structural pose repertoire: Seventeenth century Cherokee year

prise will consist of a portfolio of productive units which are in different time frames. A basic distinction can be drawn between:

1. Those units which are most heavily involved in planning the future directions of the organization. Such units are usually located in the technostructure of staff specialists. However, only a proportion of the technostructure is involved in the planning of the future, a considerable part is usually involved in executing previously chosen strategies
2. Those units at the operating level (e. g., hospital wards, factories) which are involved in the immediate provision of services and goods

The former are involved in the relations between the future and present, whilst the latter are located in the past into present dimension.

A dual structure distinction can be shown as in Fig. 1-3 where the upper half, commencing in the left-hand corner and moving to the lower right-hand corn-

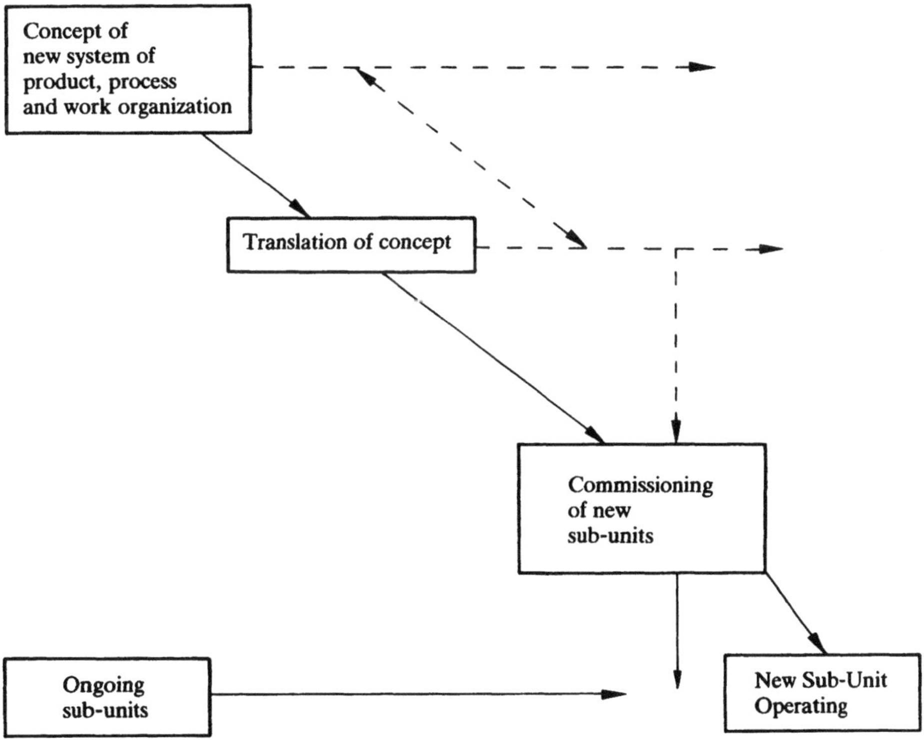

Figure 1-3. Multiple structures: The design-to-operating dimension

er, displays the path followed for the future-oriented staff[13]. It may be noted that their activities are shown in four steps – concept, translation, commissioning and operating. These steps are abstracted from the literature on design and innovation as well as from interviews undertaken in a variety of sectors from the ESRC, WORC[14]. The schematic presentation overstates the coherence of these processes[15]. The lower part of the schema depicts the ongoing operating units and shows that their recursive practices may be a target for those involved in innovation-design. Figure 1-3 is a simplification of many other processes which may be much more chaotic[16]. The notion of dual structuring provides a heuristically useful entry into the analysis of processes.

A further distinction can now be introduced between two kinds of change within organizations[17]. These are:

1. Recursive, recurrent or iterative change
2. Transformational or transitional change

As noted earlier and illustrated in the case of the Cherokee Indians, recursive change has been totally neglected in organization studies. Yet, recursive change – the reproduction of the structural repertoire – is so endemic that recent reviews emphasize the organizational inertia and momentum[18]. It seems likely that individuals and their interstructuring is organized so that sequences of cognitions and behaviours can be unfolded. Hence the repertoire is likely *to become* highly embedded in the expectation that external triggering events will continue to reappear in previously known patterns[19]. Therefore, it must follow logically from this perspective that any transformational change will encompass an alteration in either or both of the structural repertoire and/or the events in the context which trigger its activation. The concept of recurrence is very important in determining the degree of transformational change.

Approaches to transformational change in OD and POC have ignored the investigation of the existing dynamics of recurrence. They have tended to pre-

---

[13] This framework is derived from the research-consultancy study of top executives in the John Players firm during the organization design for a new factory: Clark, P. A. (1972a)

[14] The framework has been used in comparative studies by ESRC, WORC at Aston University of the following eight sectors: speciality cars, construction, coalmining, confectionary, insurance, electricals, hospital design and the provincial press. See, for example, the study of the tendering process in construction firms: Clark, P. A. (1986b: 130-143); Whipp and Clark (1986, Chap. 3)

[15] The iterative and abortive nature of those top level processes is best depicted in the study by Mintzberg, Raisinghani, and Thoret (1976)

[16] Research by David Hickson and colleagues shows that these processes can vary from relatively routine to highly chaotic. See Hickson, Butler, Cray, Mallory, and Wilson (1986)

[17] The distinction is in Clark, P. A. (1975) and developed in Clark and Starkey (1987)

[18] See particularly Miller and Friesen (1984)

[19] Allport (1962). Note the awkward attempt to use Allport's concept of event cycles in Katz, and Kahn (1978)

scribe simplistic schemas for introducing change or reported case studies in which some change is claimed to have occurred, but have not demonstrated that a new repertoire has come into place. The problem is a considerable one. For example, in a study undertaken on Rolls Royce during the development of the original RB 211 engines it was found that the personnel department had devised a strategy for predicting the future sequence of locations in which industrial relations problems were expected to arise[20]. That sequence was based on previous experiences – a decade or more earlier – and on some imaginative extrapolation. Their analysis was substantially confirmed by events. That is, the patterning of cooperation and confrontation followed a similar pattern even though the cycle of events (i.e., engine development) was being spread over many years.

Transformational change must be scrutinized more closely. In particular attention must be given to the historically shaped repertoire of structures which define the inherited capacities of any enterprise to undertake some actions and not others.

A process perspective has essentially to be a political one which examines the power structure of the enterprise, identifies the key factions and their interests, examines their ideologies and investigates their control over the key decision arenas. Previous work on decision arenas, especially the garbage-can models (Cohen/March/Olsen 1972), has emphasized the significance of the agenda of problems and how solutions may search for problems. That is an important contribution, yet the focus of that approach has been on only the first stages of decision-making: the concept stage (see Fig. 1-3). The translation of the concept and its commissioning provide points of entry for further sets of actors and additional factions who can all alter the original intentions. Hence the examination of innovation has to consider the whole array of processes.

1.3 Cross-Cultural Patterns of Innovation

1.3.1 Context: Choice of Technique[21]

The 'choice of technique' is becoming of interest in organization studies and it is clear that the orthodox approach of the structuralists cannot assist in exploring, describing and explaining the differences between societies. Why for example, do the British, Japanese, Americans and Germans all seem to choose such different kinds of technique? Is it because there are inherent cul-

---

[20] Centre for the Utilization of Social Science Research, Loughborough University of Technology, SSRC Program Grant, HR 86

[21] The concept of technique used in this section is based on that proposed by David (1975)

tural predispositions which operate autonomously and which are more powerful than economic and market factors?

In examining this question there is a high degree of convergence in theorizing between the population ecology model and mainstream economic history[22]. They are agreed that there is an interaction between local choices at the level of the organization with the context, especially the extent to which any context provides particular resources and the nature of market competition. Therefore there is choice of technique at the level of the organization, but those choices are highly consequential because the choice may lead to exit from the population of surviving organizations. The assumption is that the evolvement of any technique is the reflection of a changing, transient distribution of production amongst different techniques distributed through a heterogeneous population of organizations. For example, the evolvement of best practice in the world automobile industry shows shifts from European firms to North American, back to Europe and currently to Japan (See Abernathy 1978; Abernathy/Clark 1983).

A key issue is the extent to which the choice of technique is shaped by cultural values and the local noncultural context (see Fig. 1-4). Here the neoclassical economists and cultural historians[23] part company. The former contend that cultural factors may shape choice-making, but do not fully explore the extent to which this may be so, yet they do explore the noncultural context (e.g., resource provision) very thoroughly. One of the best examples of that style of analysis is found in the work of P. A. David (1975) who compares two possible ways in which the choice of technique in any society comes about. One way, which he considers is the least likely, is through the rational, forward-looking attempts by innovators to optimize in response to price information embedded in the market. An alternative, more likely way in which the choice of technique is shaped is through an unintended sequence of events based on the actual experience of learning by doing. That is, from the local interpretations of problems and possibilities. The choice of technique becomes a link through which the prevailing economic climate shapes the unfolding direction of technical change in specific settings. David states that:

The macroscopic character of a society's technological progress may be conceived in terms of the direction and rate of change in the global constraints on resource use performed by the available processes of production. (David 1975: 4)

---

22 The basic approach to population ecology models was introduced by Hannan, and Freeman (1977). Both Howard Aldrich and Bill McKelvey have developed the approach; see for example, Aldrich, McKelvey and Ulrich (1984)

23 Cultural historians are a rare breed. An interesting example is the work of T. C. Cochran (1981). The most frequently cited cultural analysis of Britain is Wiener (1981)

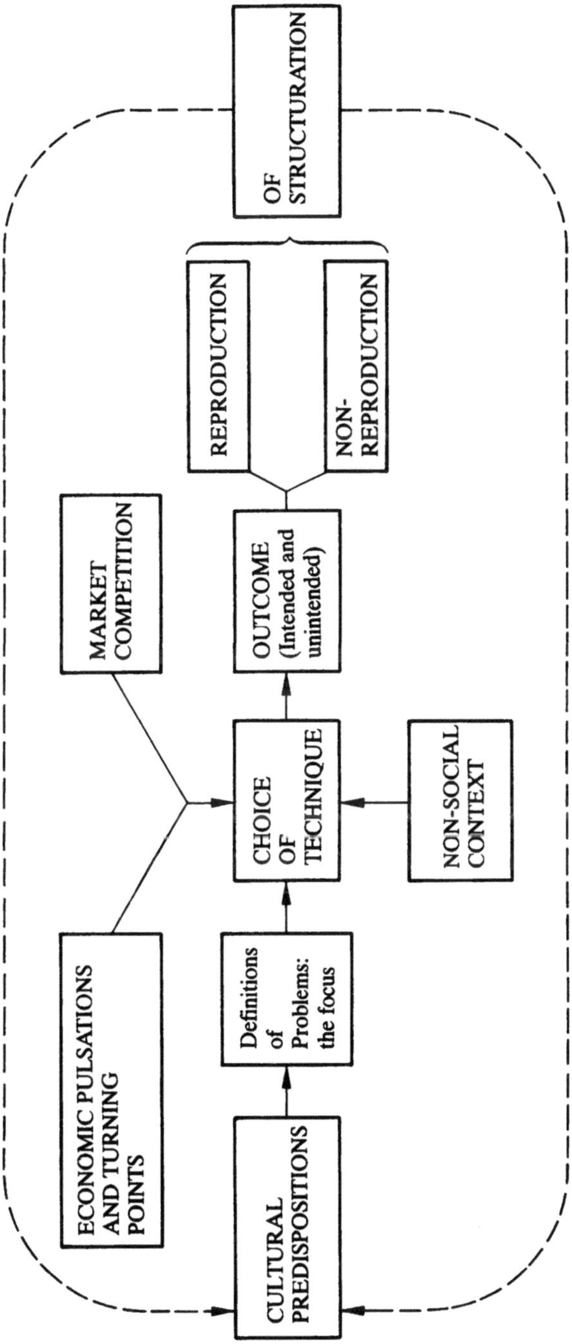

Figure 1-4. Factors influencing the choice of technique

In other words it is previous myopic decisions which guide the direction taken in the choice of technique by a specific set of people, either at the societal level and/or at the level of the organization. Because technological learning is dependent on the accumulation of actual experience in the use of techniques it is the short-sighted framework which shapes and governs what is learned in the future.

For David the context is very important, expecially the extent to which that context provides material resources and the form in which those material resources appear. In a seminal essay he compares the appearance of mechanical reapers in Ohio during the third quarter of the nineteenth century and their rejection by British farmers until much later (David 1975: 246-257). David concludes that the American farmer had large field systems which were firm enough to bear the weight of the new machinery. Also the labour supply was inelastic so that an increase in acreage could only be achieved by the use of equipment. A conjuncture of events led to the small-scale provision of mechanical equipment and to the discovery of a market. The growth in the capital equipment side further reduced the labour supply whilst increasing the volume and cheapness, as well as quality of the equipment. In the USA, the use of the mechanical reaper meant that farmers could increase the size of their acreage very much beyond the level achieved by hand-methods of farming. In other words the increasing size provided economies of scale which raised farms above the breakeven point. Thus, American farmers achieved higher and higher levels of productivity. In the British context, the farmers, who were often tenant farmers, faced small fields whose surfaces had been riddled with fragile manmade drainage systems and similar features. In that context the use of equipment would have been so slow and would have created so many negative side-effects that there was little advantage to the farmers in adopting mechanical equipment until much later.

David's analysis of the context provides a useful antedote to its neglect by much of contemporary organization studies. Yet it is also problematic insofar as the model of the decision-maker remains an acultural one. What David does not do is to separate the two aspects of:

1. Choices about technique made by decision makers
2. The economic processes within which choices face the selection criteria operating in the environment.

The former are socially and culturally shaped whilst the latter are not. Separating these two aspects is essential to the explanation of differences between societies.

1.3.2 Britain and America: Convergence or Distinct Trajectories?

The issue of the alikeness and unlikeness of British and American approaches to innovation in industry and in commerce has been dealt with by economic and cultural historians. They have started with the known data about differences in the relative rates of economic growth between Britain and the USA since the midnineteenth century (Caves 1980; Mathias 1983). Their studies, known as the *retardation debates,* show that there has been a diverging pattern in which American rates of growth have far surpassed those found in Britain. The key debates have been about how the unlikeness in performances might be explained, whether by differences in the relative prices of labour and capital in the two countries and/or by the specific features of national cultures. For example, Wiener's cultural explanation for Britain's relative decline contends that an anti-industrial culture unfolded in the mid-nineteenth century after which pro-industrial values were constrained and emasculated (Wiener 1981). In this book I shall argue that there are crucial and consequential differences between British and American approaches to innovation which have been established in the retardation debates.

The historical differences between the two countries have been ignored by the approach of organization studies because the selected problems are formulated in terms of universal analytic schemas at high levels of abstraction. The limitations in the approach were not apparent during the period when it became established in organization studies as the Anglo-Saxon orthodoxy during the 1960s and 1970s. At that time it seemed that an abstract, utilisable social science could be formulated solely in terms of general laws. Further, it did not seem important to consider the relative rates of survival of firms in either country because a high proportion survived, though a more careful examination of the British scene would have shown that key sectors were already being dominated by North American enterprises: especially in the industrial sectors. In the late 1970s the industrial sectors of each country came under economic pressure and there was massive decline in the British-owned industries of basic raw materials, transport and metal goods. There was little that a structuralist perspective could offer by way of accurate, relevant description or of a sketch for new policies and more effective practices.

A major difficulty lay in the conceptualization of innovation and in distinguishing it from invention. Schumpeter highlighted the importance of the distinction when he identified a considerable gap in time between the invention of knowledge or new items and their commercial exploitation (Schumpeter 1961; 1954; 1939; Penrose 1959). Further, it is also possible for the invention and the innovation to occur in different countries. For example, invention *might* occur in Britain whilst its commercial exploitation could occur in the

USA or in Japan. For that reason it is important to note the extensive buying of American patents by the Japanese (McMillan 1985; Sethi/Namiki/Swanson 1984). Clearly, the distinction between invention and innovation is of great theoretical and practical consequence.

American historians have taken innovation as a serious focus. Furthermore, their conceptualization of innovation has given some regard to social, psychological and organizational innovations. The contrasts between the American and British historians emerge very clearly when comparing Chandler or Cochrane with Hannah, or, in searching for a British equivalent to Rosenberg[24]. Yet, there is an important limitation in much of the excellent American research. That limitation lies in the failure to realize that American forms of innovation are neither the best practice for all time, nor, do they represent the only means by which innovation can be achieved (Piore/Sabel 1985). In that respect American analysts need to be comparative in new ways because the established frame of reference has been with Britain. It has been taken for granted – wrongly – that Britain and the USA are on the same trajectory of economic and organizational evolvement. There is a stronger case for assuming that the two countries are on rather different trajectories. It has also been assumed that Britain was lagging behind the USA. That was a convenient assumption and one which was pleasant for Americans, but its relevance was shattered with the rise of Japan and the increasing importance of the Pacific Rim. Now the Japanese situation has prised open the whole nature of that comparison and in so doing has introduced the requirement for reformulating the actual definition of the problems of innovation.

I shall argue that the perspective of the American historians provides an initial and very fruitful port of entry into the examination of Anglo-American patterns of innovation. Yet, as indicated earlier there are problems with the implicit assumption that America and Britain are on converging trajectories of social, organizational, technical and economic evolvement. That assumption, strongly held and almost hidden, underpins many key comparisons between the two countries. Chandler, for example, assumes that American multinationals discovered and rapidly adopted forms of organization structure which were so superior that multinationals in other countries would be "required" to adopt the same forms (Chandler/Daems 1980) – the multidivisional structure, for instance. Given that assumption, Chandler (and his colleagues) approach the British situation with a particular psychological set and discover similarities with and differences from the USA. They interpret the similarities as

[24] The Business History Unit at the London School of Economics is a major focus for historical studies of twentieth century British enterprise. Its current leader is Leslie Hannah who has provided major overviews of the British situation and also specialised studies of specific sectors. There is little evidence of a business history perspective in British schools of management

imitation and only refer to the differences when explaining why the imitation is not total. The dangers – theoretical and practical – of that approach are manifold. It is an approach which can neither assist the British, nor can the approach show that American practices should and can be found in Britain. An alternative is required.

Fortunately there are several examples of an alternative paradigm in which the diffusion of innovations is examined with reference to specific contexts and to changes in the shape and uses of the innovation. Jeremy examined the diffusion of British textile innovations to America in the period 1780-1840 and demonstrated that the British practices were unsuited to the context[25]. Jeremy shows that Americans modified British equipment which they merged with indigenous devices. It is clear from that account that the American industry began from different founding conditions and unfolded along a distinct trajectory in which there was the use of different equipment, raw materials and end products. Neither of these two examples of transatlantic diffusion could be adequately conceptualized in a framework which presumed that imitation was a required condition.

Our emphasis on Anglo-American differences as contexts for innovation is given added impetus from the re-analysis of the relative positions of Britain and America. First, with respect to Britain the new paradigm emphasizes that Britain was not the site of the Industrial Revolution (Crafts 1984; Maddison 1982). Rather, Britain entered the economic formations of industrial capitalism with forms of work organization and with institutions which were shaped in the periods of agricultural capitalism and commercial capitalism. According to this interpretation it was the USA which became and was the key site for the Industrial Revolution. Further, it may be argued that much of British industrial organization, especially the division of labour within management and the institutional framework embracing capital/labour relations was radically different to the USA (Littler 1982). Second, recent interpretations of American success since 1880 suggest that features which provided the context of that success may be as unsuited to the future beyond the 1980s as were British patterns of industrial innovation over the previous hundred years (Piore/Sabel 1985; Hounshell 1984).

1.3.3 The Future: Renaissance or Retardation?

The retardation thesis refers to the analysis of the extent to which particular societies whose economies once grew at a fast rate enter a long period when rates of growth are low relative to their rivals. The thesis has been applied in

[25] Jeremy (1981); see also the use made of Jeremy's analysis in Chap 6

some detail to two European societies: Spain in the seventeenth century (Braudel 1972) and Britain after 1880 (Elbaum/Lazonick 1986). Recently, parallels have been drawn between the decline of Britain in the twentieth century and the earlier decline of Spain in the seventeenth century.

In the past decade attention has been drawn to apparent parallels between the relative decline of Britain and the static rates of productivity in North America since the late 1950s (Maddison 1982).

Is the USA facing a retardation crisis? The question has been raised and the initial answers have been provided in a series of publications which emphasize the adaptive capacities of American corporations, of American education and of the societal institutions (Abernathy/Clark 1983; Lawrence/Dyer 1983). There is a high degree of optimism in some studies; two may be noted, each of which refers to an American renaissance. Both are from highly regarded academics at the heart of corporate education, namely, Harvard.

First, the late Bill Abernathy and colleagues have produced both a searching indictment of the lack of innovation in the automobile (Abernathy 1978; Abernathy/Hayes 1980; Abernathy/Clark/Kantrow 1981) and also a very optimistic and positive account of how the threats of new forms of international competition can be redressed by readily realizable adaptations of existing practices (Abernathy/Clark/Kantrow 1983).

Abernathy demonstrated that in the US automobile industry the major firms had successfully decoded the structure of styles in cars which the American consumer was prepared to purchase. Following the initiative of GM in creating a central styling department (in 1926-1928) the design policy became one of dealing separately with the visible and invisible parts of the car. The visible parts, the outer skin and the interior, were subjected to creative styling skills aimed at creating a consumer image of evolvement towards high style. Changes in the outer skin were typically of a small-scale, cosmetic nature. Occasionally, say every decade, there would be a major shift (e.g., adding fins). Many of the styling changes could be readily accommodated on the final assembly lines simply by substituting components or adding new items. Beneath the outer skin of the large US cars there was plenty of room for factoring the overall processes of the corporation into distinct subparts each undertaken in very large volume in specialized productive units. This high volume production of basic unseen components could be combined with high economies on a scale much beyond those envisaged by European firms. The engineers in the major car firms soon learned that specialized machines and linking devices could accelerate production and reduce costs. Gradually the working practices and the technology crystalized around highly stable, slow-in-changing forms of work organization. Consequently, the requirement for dramatic alteration of existing practices was greatly reduced. Once learned, these practices formed into 'roadblocks' to innovation. That is, the US companies developed repertoires which were highly appropriate to slow change in existing directions. In practice as long as the market structure remained favourable to the large American roadster and only US firms supplied the consumer preferred choices then the roadblocks were not so serious. However, if those consumer preferences should alter – and they began to in the late 1960s – then either US firms had to follow those changes or face new competition. Abernathy recognized the problem faced by American firms in producing a small car.

Implicitly Abernathy suggests that the US industry tended to sidestep the need for major innovations by designing cars which could be easily restyled whilst retaining basic components beneath the outer skin. His account makes it clear that there were roadblocks to innovation which were inherent in the infrastructure and the superstructure of the industry. By the late 1970s it was becoming evident that the American consumer was choosing an increasing proportion of non-American cars[26].

Abernathy and colleagues examined the cost advantage enjoyed by Japanese car-makers over their American rivals in the *Harvard Business Review* (Abernathy/Hayes 1980; Abernathy/Clark/Kantrow 1981). Two widely read articles crisply documented the differences between the costs of producing a car in Motor City (Detroit) compared with the costs of producing a compact car in Japan and delivering it by ship to the West Coast. The latter could be unloaded in the USA for around 10% less cost than the home-built American competitor. Recent reports from GM suggest that by the mid-1980s Japanese manufacturers had a cost advantage of around two thousand dollars.

Abernathy then sought to construct an explanation of the falling American performance and selected certain targets for change, including the education of management, especially the MBA (Master of Business Administration) (Abernathy/Clark/Kantrow 1983). However, he then assumed that the organizational ferment in the USA which had been evident since 1980 would lead to the deeper changes which were required.

Lawrence and Dyer adopted an interesting and potentially dangerous strategy of enquiry (Lawrence/Dyer 1983). They selected seven diverse sectors of the American economy – hospitals, agriculture, housing construction, cars, steel, coal, and telecommunications – and then examined the extent to which each had adapted its organizational structures over a period of seven decades in the twentieth century. An example of their approach is illustrated in Fig. 1-5 which shows two dimensions of the environment each of which has been trichotomized to create a matrix with nine positions. The vertical dimension ref-

[26] The foreign penetration of the American automobile market rose steadily from 1970 to reach 30% in the mid-1980s

ers to the extent of uncertainty in the informational environment of firms in the auto industry and the horizontal dimension refers to the degree to which basic material resources have been available. It is evident from the profile in Fig. 1-5 that the auto industry has faced change over the seventy-year period. Lawrence and Dyer conclude that the *past ability* of American forms to adapt to changes in uncertainty and in the availability of resources is a reasonable indicator for future adaptability. Further, in a manner similar to Abernathy, they prescribe future requirements and presume that these prescriptions can be incorporated into new policies.

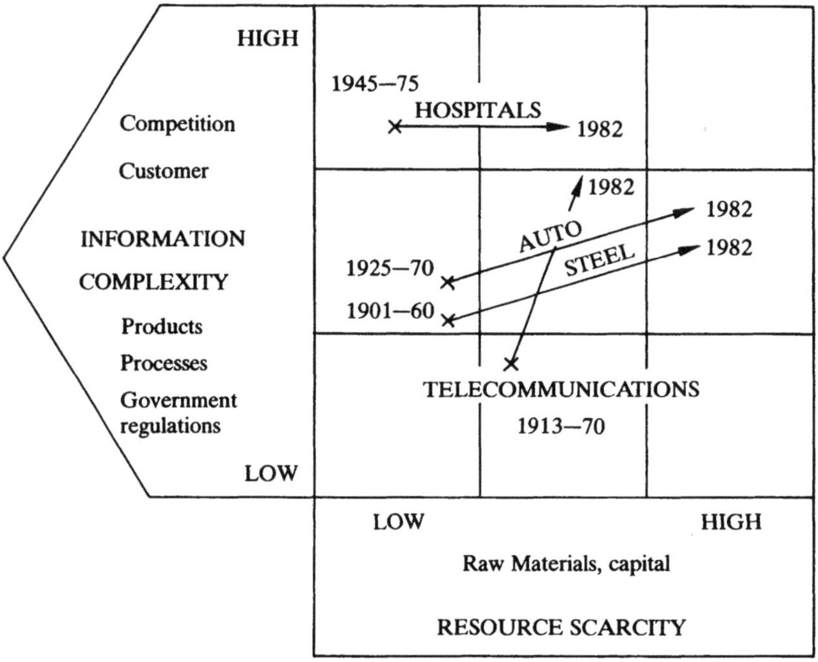

Figure 1-5. Historical tracks taken by four American sectors during the twentieth century (Based on P. R. Lawrence and R. Dyer 1983)

These two analyses assume that when prescriptions are given the American corporations can make the required adjustments. However, that optimistic picture must be qualified in the light of the earlier analysis by Abernathy when attention was focussed on the organizational and ideological infrastructure amongst the interdependent parts of the auto industry. Abernathy does not explain how such structures are altered. Nor does the framework of Lawrence and Dyer explain why and how American organizations can compete better than their rivals (Lawrence/Dyer; Hayes/Weelwright 1985). Now it is relevant to consider the retardation thesis.

The retardation thesis has provoked competing explanations of why economically powerful regions of the world lose their dominance (See Chap. 7). In the case of Britain there is considerable disagreement on causes. Hence, the notion of a retardation debate. However, behind the disagreements there are areas of agreement. One of the most important insights is that the original question requires reformulation. Rather than asking: why did Spain and Britain decline, the question should be one of why a succession of regions in the world become centres of mini world-economies which grow and survive for decades and even centuries, yet eventually lose economic and political centrality and are superceded (Aldcroft 1968; McCloskey 1971; Matthias 1983).

The reformulation of the question is twofold. First, it asks what general and specific factors facilitate the accumulation of wealth? Second, why do specific regions become centres? The focus on regions is quite central because the region is not the same as the state. The closest connection between an economically powerful region and state is that between Amsterdam in South Holland and the Netherlands in the period 1560–1670 (see Chap. 8.2). The Netherlands dominated the North Atlantic economy in this period and displaced the power of the Spanish Empire to create the international rules of trade. In the case of Britain, which eventually succeeded the Dutch by creating a sea-borne commercial and trading empire, there were several key regions including Glasgow and the Manchester-Liverpool corridor. Later, in the case of the USA wealth was built on both the agricultural productivity of the mid-West region and on the automobile industry centred round Detroit. It follows that a large continent like North America might enclose both declining regions, as possibly in the northeastern quadrant facing the Atlantic, as well as rising regions, as in the southwestern quadrant facing the Pacific Rim.

There are two main aspects of the Anglo-American connection. First, from the American perspective the problems of innovation versus retardation is one of the relevance of the existing patterns of societal and organizational learning to the facts of increased competition arising from the pace of innovation in Japan and Europe. Second, from the British perspective the dimensions of the American connection appear to be very different.

Since the appearance in 1851 of the American products at the Crystal Palace Exhibition in London, the British have been unsure about how to **appropriate** American "best practice" to the British context where the market size and its structure of tastes was quite different[27]. In 1853 the British House of Com-

[27] For an examination of the concept of appropriation, see Chap. 6

mons sent a special three-man party of Wallis, Anderson and Whitworth to examine American methods in detail and their report of the following year is a prophetic account of the changing fortunes of the two countries for the next 140 years (Rosenberg 1969). The report to the House of Commons resulted in the establishment of a new arsenal and armoury at Enfield which was intended to follow the template of mass production found in the American state armoury at Springfield. The methods of assembly at Enfield should have been based on the principles of interchangeable parts and standardization which the firm of Colt had been **perceived to be using**[28]. However, the process of learning for the British and the transatlantic transfer of **practices** were so complex that it was several decades before the new principles became firmly established at the level set by American practice in the 1850s.

Since then there has been a mistaken assumption that Britain and America are on the same trajectory in their application of technology, in the choice of raw materials, in their concepts of design, and in their forms of organizing at the shopfloor and board levels. That assumption normally adopts the analogy of a race in which first Britain leads then America challenges and eventually takes over the lead. The analogy is inappropriate simply because the British position is best understood as one of running in a quite different race according to quite different rules.

Unfortunately for the British the "rules of the game" for capitalism have favoured American methods rather than those of the British. The interesting question is whether those rules have now changed for both the Americans and the British. According to the thesis of Piore and Sabel the rules of the game are shifting away from mass production mass distribution at which the Americans have been superior to all, except the Japanese (Piore/Sabel 1985); this issue is further examined in Part IV.

1.3.4 Typical Variety: The Anglo-American Exemplar

One of the most obvious problems in making cross-cultural comparisons concerns the typicality of any statement made about one country. By implication any statement tends to exaggerate the homogeneity of the structures, practices and culture in the society being characterized. How can that problem be minimized? Answer: By taking a range of typical forms which are representative of significant areas in any society.

For example, there has been widespread use of a known continuum of managerial regimes which ranges from Direct Control to Responsible Autonomy. It

[28] The perception by the British of the American methods requires close scrutiny and theoretical examination

would be possible to use that continuum to argue that significant areas of employment in the USA tend towards the Direct Control pole of the continuum, whilst in the British case there is a tendancy towards the Responsible Autonomy pole. It may be expected that there will be numerous readily cited exceptions, yet through the notion of typical variety these can now be labelled and identified as exceptions from the typical variety.

In this book I shall utilize the notion of typical variety and shall refer to the differences between the USA and Britain, but those references will apply to only certain features of action in the two countries. The focus will be on the principal areas of employment and on the ways in which the rules of the game for the key institutions of work and its organization are formulated, sustained and transmitted. Thus, particular attention will be given to managerial authority and decision-making. By implication that focus will also refer to their position within the stratification of each society and in their access to key agencies, such as education.

The concept of typical variety will be applied to the British and American cases to achieve a comparison of the patterning of innovation. Also, the transmission of innovations between the countries will be a focus of analysis. Overall, the Anglo-American case will be used to construct an exemplar which can then be applied to other comparisons.

1.4 Structure of the Book

The objective of this book is to demonstrate the relevance of the evolvement in the shape and uses of innovations to organization studies as a discipline and to show how the transfer of innovations, especially those of interest to organizations, between different societies can be better analysed and how policy-making can become more targetted. Obviously the book focusses on the Anglo-American dimension, yet the ideas and lessons are equally applicable to Anglo-Japanese or to Franco-German transfers or to any other transfers between countries. The major focus is on the case of relevant innovations which are not the prime property of a specific organization.

To reach the objective it is necessary to set out the agenda of problems to be tackled in Chapter 1. Three related problems have been selected: the development of a process perspective in organization studies; the development of a more useful theory about the evolvement of innovations; the problem of analysing alikeness and unlikeness between societies in a manner which facilitates successful policy-making about the opportunities and problems of transferring innovations across national boundaries. Given the agenda of problems the most immediate requirement becomes one of constructing a relevant theory of innovation.

Part II approaches the analysis of innovations through the concept of their evolvement with reference to the shape and uses of innovations of all kinds. The choice of *evolvement* as a concept reflects the determination to strike a balance between the interests of those who supply innovations in their marketing and the concerns of the users in the fruitful absorption of innovations. These two sets of interests have been separated and the interests of the suppliers have dominated those of the users. Thus, Chapter 2 sets out the requirements for a useful theory and Chapter 3 carefully synthesises three supply-side models to demonstrate why evolvement is a more powerful concept to apply to innovations than is the previous concept of diffusion. Moreover, evolvement affects the shape and uses of the innovation, each of which undergoes various transformations. The point is demonstrated in Chapter 4 which takes the technical innovation literature and shows how it can be recast to be of relevance to organization studies and vice versa. Chapter 5 addresses the users of innovations and aims to demonstrate that the prescriptive analytic schemas which have been so popular fail to recognize the position of the user. It is argued that user organizations need to learn to appropriate those innovations which are closest to its core tasks. Yet appropriation involves conflicts with the suppliers. These four chapters form an entity which is then applied to the Anglo-American case.

Part III opens and closes with two chapters (6 and 12) which report cases of the transatlantic evolvement of specific innovations. Chapter 6 introduces the approach of transatlantic evolvement by briefly demonstrating how in the late eighteenth and early nineteenth centuries American textile innovators absorbed British textile practices despite an enormous gap in their relative skills. Moreover, the case explains how absorption became appropriation and how the Americans developed textile manufacture autonomously from the British. The heart of the chapter is an analysis of how a British sporting innovation was introduced into America and then transformed. The choice of sport is deliberate and the case serves several crucial functions in the analysis. Later, Chapter 12 examines the opposite flow of innovations from America to Britain and their evolvement in the British context. The four chapters between demonstrate why the transfer of innovations has to be treated as part of a *filiere,* that is a network of channels which direct actions in some directions and not others. The objective is to demonstrate the different approach to innovation which has arisen in Britain and the USA. The underlying theme is that *in* the USA there did develop a filiere which was very facilitative to innovation in just those areas which were immense sources of capital revenue in the twentieth century. America acquired, partly by chance, a similar unique configuration of benefits to that possessed earlier by the British. However, in the late twentieth century the relevance of the American filiere comes into question. The relevance of the British filiere has already been questioned. Part III treats innovations and organizations as embedded in a societal filiere which shapes the typical variety and scope for innovation, but does not totally preclude novel developments.

To what extent are the Americans and the British meeting the new challenges? Part IV addresses this question by summarizing the Japanese situation in terms of the theory already elaborated in the previous part of the book. It is shown that the Japanese have also absorbed key Western innovations and have so appropriated them that Western societies are faced with the problem of having to become absorbers and appropriators of innovations whose most recent evolvement has taken on a distinctly Japanese dimension. The final chapter examines the claims made by American and British organizations that they have learned the rules for the new bases of competition.

The objective of the book includes taking the evolvement of innovations across national boundaries as an additional means whereby the alikeness and unlikeness of societies can be examined.

Part II.
Evolvement of Innovations: Shape and Uses

2 State of Theory

2.1 Introduction

Part II is a focused synthesis concentrating upon the development of useful theory for the analysis of the evolvement of innovations over time and in different spaces. The objectives of the four chapters are to indicate how cross-cutting connections between different research traditions and perspectives can be achieved to provide an array of concepts and a perspective which can then be applied to the *evolvement of the shape and uses of innovations.*

Existing approaches have provided a fragmentary treatment of innovation from a broad spectrum of perspectives which can, for convenience, be labelled the diffusion approach[1]. The characteristic features of the diffusion approach are its preoccupation with the supply side of innovations and with the tendency to imply that innovations ought eventually to be adopted. That approach contains a number of serious limitations.

First, often insufficient attention has been given to the appropriateness of a specific innovation for all settings. As noted in Sect. 1.3.1 certain innovations may be quite unsuited in some contexts. Also, some innovations are simply equivalent variants of a more generic type of innovation. For example, certain software-based planning systems are largely interchangeable and are functionally similar, so it would not be significant if one of them was only adopted by a small population of users.

Second, innovations have been too often defined in physical terms as with equipment and with seeds or types of building. Too little systematic analysis has been given to the social, economic and technical contexts as a configuration[2]. Consequently innovations have often been stripped from their contexts without attention to which features of their embedding might be crucial for their transfer to another setting[3].

Third, existing perspectives presume that a high degree of imitation ought to take place amongst adopters (see Chap. 3). In practice imitation is both much

---

[1] See especially the themes of Chap. 3 and the authors cited in that chapter

[2] Brooks, and Kelley (1986), The paper refers to the American situation

[3] Nelson, and Winter (1977); Nelson and Winter (1982) The following working papers from CRUSK, University of Michigan at Ann Arbor: Pelz (1982); Pelz, DC (1981); Pelz and Munson (1980)

more difficult to achieve than the orthodox approaches presumed and is also less relevant than they had stated. There are important reasons why the user might want to select only some aspects of an innovation and reject others. Also, there is increasing evidence to suggest that the user should seek to appropriate an innovation by ingesting and accommodating its fundamental features into the evolving stock of specific knowledge inside the user organization. Too little attention has been given to the position of the user (see Chap. 5).

Fourth, to an increasing extent the analysis of innovations treats them as multidimensional (see Sect. 2.3) and recognises that different dimensions unfold in an uneven, often unanticipated fashion. All innovations should be examined as totalities in the first instance. The main dimensions will include knowledge, technology, raw materials, social embedding and economic feasibility (see Sect. 2.4). In addition innovations vary in the extent to which they possess features likely to facilitate their adoption by users (see Sect. 3.3.2).

Fifth, technological innovations have been the focus of more attention than have organizational innovations. However, recent developments in theorizing and research are demonstrating the practical importance of conceptualising technology with reference to knowledge and in treating technological innovations in a process, life-course perspective. These possibilities are investigated in Chap. 4.

Sixth, the position of the users requires detailed consideration. They are the demand side and also, in amongst the population of users will be organizations which take a leading role in the development of sector-specific innovations. The position of the users is dealt with in Chap. 5.

Overall Part II provides a synthesis in which the *evolvement* of innovations is shown to be the normal situation. That synthesis is then applied in Part III where Anglo-American patterns of innovation are addressed.

Previous studies have tended to abstract variables connected to the evolvement of innovations from their relevant contexts and configurations. As Mohr has shown, such an approach has been confusing and creates confusion (Mohr 1982). The approach adopted here treats innovations as parts of a *filiere*[4].

Recent debate about the status of existing knowledge and understanding about innovation has directed attention to the kinds of theorizing which are appropriate. Theorizing can be examined in terms of a two-by-two matrix as shown in Fig. 2-1.

[4] A filiere may be defined as a connected network of channels

	Normative	Descriptive
Variables		
Process configurations		

Figure 2-1. Types of theorising

The distinction between descriptive and normative theorizing is a basic one. In descriptive studies the intention is to reconstruct sets of events as total configurations, giving equal attention to those variables which can be controlled and those which cannot. In normative theories the objective is to work with that minimum array of variables which are both relatively powerful and yet open to manipulation. The study of leadership is an exemplar of the normative approach in American social science because it is presumed that leadership behaviours have a great deal of impact and that the correct behaviours can be taught.

The distinction between theorizing which is oriented to the use of variables rather than to whole processes is also crucial. The management and social sciences in North America have been grounded on the use of variables, particularly on the design of studies utilising an array of variables. The objective has been to reconstruct processes by relying on a small number of variables. The rationale is that it is easier to specify variables than it is to undertake studies based on a consistent usage of process concepts. Thus, the study of organizations and of innovation has been dominated by the normative-variable approach and there has been a dearth of process descriptions. Critics of the former argue for the latter and vice versa. At the present time a synthesis requires several approaches to theory building. Certainly earlier normative-variable studies have created confusing results, yet they were devised to resolve the problems arising from the process approach. Now that the theories constructed from the normative-variable approach require extensive revision it is appropriate to engage in focussed process descriptions. However, it will be equally important for policy-making to put to the test the interpretations which process studies suggest.

2.2 Main Areas

The study of innovation can be arranged into five main areas:

1. The origins of inventions, their patenting and the process involved with bringing them to the point of commercial application
2. The creation of innovations such as for example, float glass (Barker 1977). This includes the topics of radical/incremental innovations: push/pull theories and the role of research and development (R and D) in corporate success (Freeman 1974)
3. The spread, evolvement and diffusion of innovations including the rate of spread, the factors affecting the spread, the competition between existing and new forms, the adoption/rejection process, the marketing of innovations, the role of the user (see esp. Chap. 3)
4. The selection, modification, ingestion and institutionalisation of innovations; the consequences of both innovation and noninnovation
5. The role of societal, political and economic institutions including the labour/capital relations at the level of the sector and the firm

From these five areas this book makes a specific selection from areas (2-5), whilst taking area (1) as an important context. That is, the focus is on the process of recognising innovations, the ability to describe their crucial characteristics, and the capability of undertaking their retention or removal and/or where required their diffusion, selection or rejection including any re-invention and modification.

2.3 Basic Concepts

In the literature on innovations it is important to make certain basic distinctions in three aspects: the magnitude of the scope of the innovation, the area of the innovation, the impacts of the innovation on host users.

The magnitude of the scope of innovations varies in important ways which are indicated by the use of three different concepts: generic innovations, radical innovations and incremental innovations.

First, the concept of generic innovations should be utilised to apply to a broad class such as the automobile, electricity, micro-electronics, commercial sports, assembly lines, management structures, job descriptions, work analysis techniques[5]. The importance of a generic innovation is that its consequences are widely spread. The automobile transformed the nature of distribution of goods and services whilst also altering patterns of housing and changing lifestyles. Likewise, the commercial exploitation of electricity and of electrical motors in the latter part of the nineteenth century altered the whole basis of costs and activities within organizations.

Generic innovations are ones which ought to be adopted *in some form*. Consequently their nonadoption, or, very slow adoption can be highly consequential for the survival of single organizations and even for whole populations of organizations. For example, the more rapid adoption of the electric motor in the USA provided a tremendous stimulus to the major firms supplying equipment (e.g., General Electric, Westinghouse) and to their future international role. Moreover, in North America the electric motor provided a cheap, portable power source which greatly improved the level of output per worker[6].

By implication nongeneric innovations are potentially substitutable by alternatives, or, may not even be required at all by specific organizations.

The distinction between generic and nongeneric should be applied when examining the use of innovations by organizations because many innovations may not be required and so their rejection would be entirely appropriate.

Radical innovations are difficult to disentangle from generic innovations and it is therefore best to use radical and incremental as relative terms whose interpretation is anchored in agreed benchmarks of empirical examples. If the automobile is a generic innovation then it is noticeable that its innovation between 1886 and 1908 was associated with a host of radical innovations in the subsystems (e.g., braking systems, the engine, spring, chassis, power train, electrical systems) (Abernathy 1978). The rate of radical innovations in the product was very high and that rate fell away so that the automatic transmission became one of the few radical innovations in the past half-century. However, research and theorizing shows the dangers of assuming that radical leaps are the only significant forms of innovation. Quite the contrary, research shows that continuous incremental innovation on an annual basis is

---

[5] The idea of generic technologies such as electric power is used by Freeman (1974), who refers to natural technology systems and by Nelson, and Winter (1977; 1982), who adopt a similar conception

[6] The electric motor was diffused very much more rapidly in the USA than in Britain

enormously important[7]. Indeed without incremental improvement an organization is unlikely to survive.

Second, the areas of application will be treated as threefold: the product (or service), the process of its production and delivery, the work organization at all levels. These three areas will be treated as a configuration and it will be assumed that user organizations have to acquire in their design processes a capacity for achieving a balance across the three areas which is compatible with their market contexts.

Third, attention must be directed to the impacts of innovations on existing structures of work organization and authority and on the market linkages of the user. Until very recently impacts in the user organization were neglected in the orthodox diffusion perspective (Rogers 1962); since then they have been given an unbalanced treatment (Rogers 1983: 371-408). The crucial point is to search for impacts, yet not to assume that they will occur. Many innovations are adopted with relatively slight impacts. Indeed the lack of impact might be the main problem because many users continue their old process of organizing in concert with innovations which ought to be asssociated with new, radically different structures of work organization.

The broad point can be made through the schema (Abernathy/Clark 1985) shown in Fig. 2-2. The vertical axis depicts impacts on the market linkages between the user and the market on a scale from low impact to high impact. The axis assumes that some innovations may reinforce existing patterns. The horizontal axis refers to the impacts on the production system and technology on a scale from low impact to high impact. Here also an innovation might lead to the reinforcement of the existing situation.

Impacts should be the focus of an open-minded enquiry designed to examine what actually happened on an empirical basis – that is, not to assume impacts – and to be prepared to interpret the extent of impact within a theoretical framework defining the requirement for impact as part adjustment.

The analysis of impacts has greatly improved in recent studies. These have shown surprising and sometimes opposite impacts of the same innovation in different organizations within the same sector and even in different divisions of the same firm (Boyer 1986). For example, a study of the adoption of the educational strategy of Organization Development in the large British firm, ICI, shows quite different patterns between the major divisions (Pettigrew

[7] Incremental innovation in the American context is examined by David (1975, Chap. 2, 174-191). Also by Rosenberg (1982). The notion of incremental innovation is close to the concept of experience curves used in the literature of business policy. In the business policy literature the impression is sometimes given that incremental learning is virtually automatic, but that suggestion has not been supported in studies of the British paper industry undertaken from SPRU at Sussex University by Scott-Kempes

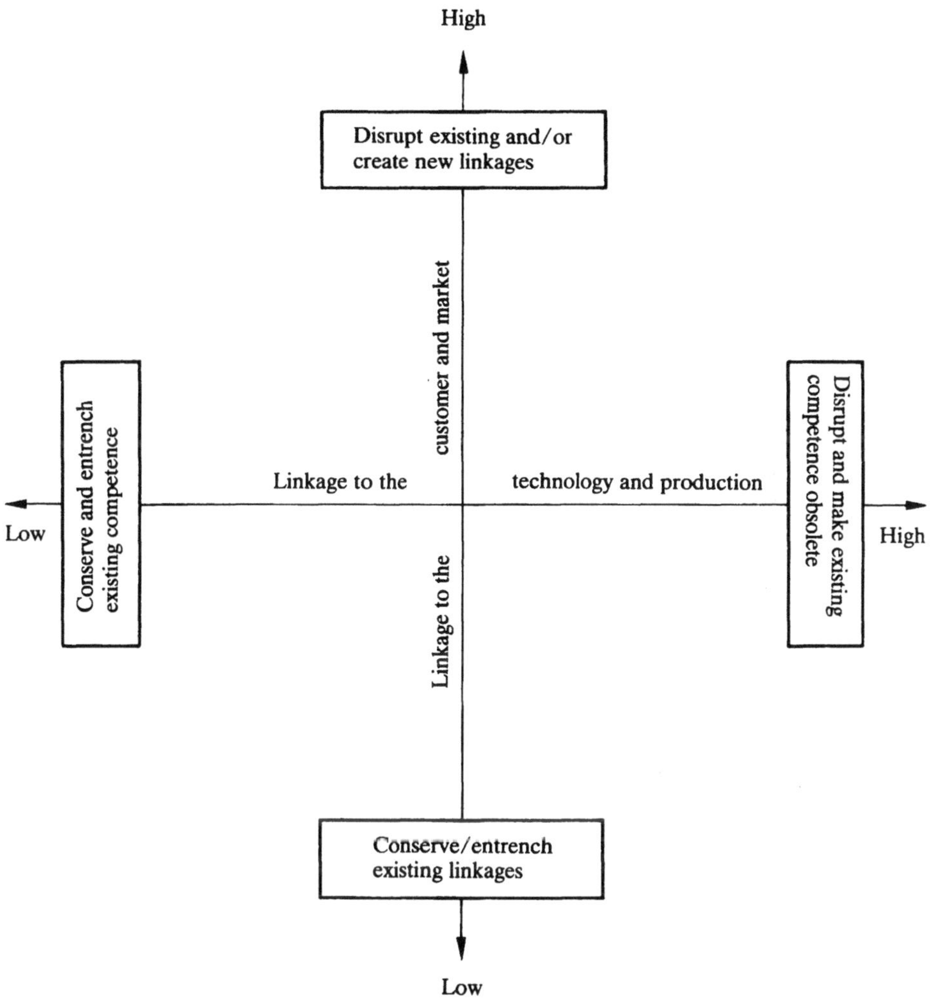

Figure 2-2. Framework for assessing the impacts of innovations on the market and on the production system (Based on W. J. Abernathy and K. B. Clark, 1983)

1985); likewise in the adoption of new technologies in the British sectors of health and banking[8].

These basic, simple conceptual clarifications underpin the following sections.

[8] See the program of research within ESRC, WORC at Aston University, known as Micro-electronics in the Service Sector

2.4 Innovations: Multidimensional

Many studies conceptualise an innovation only in terms of its object features as with the studies of the steam engine, the automobile, hybrid corn, electric arc furnaces and washing machines. Such a narrow focus seems to provide precision, yet closer examination reveals a theoretical confusion which will undermine policy-making. I shall argue for two basic points. First, innovations possess at least four dimensions. Second, all innovations should be examined with regard to their embedding contexts (see Sects. 2.7 and 2.8). This section will concentrate on the multidimensionality of innovations.

Innovations may be considered to possess one or more of at least four dimensions:

1. Forms of knowledge[9]
2. Knowledge embodied in raw materials
3. Knowledge embodied in equipment
4. Knowledge embedded and embodied in associated forms of work organization which may be in a competitive relationship over scarce resources

Each dimension requires unpacking.

First, innovations are increasingly constructed and designed for a deliberate purpose through the application of knowledge[10]. That aspect of an innovation is increasingly recognised, especially in the notion of paradigms as developed in Chap. 4. As such, innovations will vary in the degree to which the knowledge dimension is stable, well understood and easily analysed by the user. This dimension may be examined more fully with an illustration later in the section.

Second, many innovations are in the form of refinements to the attributes of existing raw materials and/or the introduction of entirely new raw materials. An example of the refinement of existing raw materials through the application of new knowledge is the modern tomato specially grown for supermarkets[11]. It has been redesigned to withstand diseases, to be machine-picked, to survive against rapid deterioration and look appetising under the artificial lighting of the supermarket. Another example of a refinement to a raw material is the modern form of cement used in construction (Clark, P. A. 1987). Here scientific and craft knowledge have been utilised to devise cements

---

[9] Throughout this book the concept of knowledge occupies a central place. It is convenient and useful to take the definition of social technology as knowledge advanced by Perrow (1967)

[10] This point is central to the analysis of the economics of innovation by Freeman (1974, Part I)

[11] The design of food for supermarkets has fundamentally altered the shape and texture of vegetables as well as their growing, extraction and processing. Tomatoes are an excellent example and so is the grape. On the latter see Hinkle (1979). Note the development of new large wine-producing enterprises in Italy

with known rates of viscosity and with defined rates of setting. Their usage has greatly accelerated the process of construction and increased the predictability of the construction processes. New raw materials are continuously appearing in a variety of fields: pharmaceuticals, electronics, carpet-making.

Third, knowledge may be embodied in new equipment such as tomato-picking machines or automatic cement-pouring machines; see Chap. 4.

Fourth, innovations may simply consist of new forms of work organization combined with old forms of raw materials and equipment. Far too little attention has been given to organizational innovations and, as is becoming increasingly recognised, object innovations (e.g., the supermarket) are highly dependent upon innovations in lifestyles and life trajectories (see Chap. 3). It is useful to digress for a short examination of organizational innovations.

One of the most systematic attempts to develop a focus on organizational innovations has been the study of managerial activity initiated by Chandler and colleagues with respect to their investigation of changing forms of work organization amongst American management (Chandler 1962; 1977). Yet, organizational innovations are streaming through modern society. Two brief examples will illustrate.

First, in England and much of its former Commonwealth a game known as cricket is played. In its major competitive format the game was played over three to five days, and even then 50% of games ended in a draw, yet until the 1960s large crowds would attend intracountry matches as well as the international matches. However, during the 1950s and particularly in the 1960s the size of the crowd declined rapidly. The result was that the financial basis of the game entered a perilous state. Then in 1969 a combination of commercial interests and the officials of the game designed a restricted variant of the original game which left all the raw materials and the equipment unaltered, but introduced various new organizational parameters which meant that a result was always possible. The new shortened game of seven hours (or less) provided a much-needed financial boost at all levels including the creation of an entirely new format of international match.

Second, the Master of Business Administration (MBA) is an interesting example of an organizational innovation which eventually required alterations in the raw materials of knowledge (e.g., special books) and in equipment (e.g., the lecture theatre). As such the MBA was an innovation because previous master's degrees had been vertical extensions which took undergraduate knowledge to a more advanced state (i.e., the M.Sc. and M.S.W. degrees). By comparison the MBA was an horizontal degree made up of a designed collation from selected disciplines and selected functions with the basis of design centring on a core of commonly occurring managerial problems which were used as the focus of pooling and organizing knowledge.

Cricket is a distinctly English game and the MBA is a distinctly American product. Americans have found considerable difficulty in decoding the knowledge which is *embodied* in the game of cricket. Likewise, as will be shown in Part III the British have found decoding the knowledge embodied in the MBA equally difficult.

Innovations can take many forms combining these dimensions.

2.5 Problem Agenda: The Dominance of Economics[12]

Until the mid-1970s the study of innovation was characterised by three features:

1. Neglect
2. The relative dominance of economic definitions of the problems
3. A high degree of analytic fragmentation, both within economics and between economics and other significant contributions from the remainder of the social sciences and from history

The dominance of economics is surprising when one considers how slight is the attention which has been given to innovation by that discipline relative to other problems. Fragmentation is evident in the absence of cross referencing, until the Venice Symposium of 1986[13], between the writings of Abernathy, Rosenberg, Rogers, Hagerstrand, Freeman, P.A. David or between the subdisciplines of Organization Development, Operations Management, Organization Design, Industrial Organization and Business Policy[14].

Within the past decade the level and scope of interest in innovation as a problem area has increased. Previous neglect is being transformed into high interest at three levels: the state, the corporation, and the academic community. In the academic community the agenda of problems considered to be relevant for understanding innovation has been largely shaped by the theorizing and researches of economists and economic historians. The renewed interest in innovation, however, has provided a challenge and opportunity for a new relationship with the social sciences, *especially with organization studies*.

The increase in interest amongst economists led initially to the rereading of seminal though neglected texts, most notably the classics from Marx,

---

[12] See for example the discussion in Clark, N (1985)

[13] See the papers presented at the Conference on Innovation Diffusion, Venice, March, 1986 in the symposium organized jointly from the Center for Economic Policy Research, Stanford University with IUAV, Venice

[14] The potential for cross-fertilization is indicated in Part II of this book. So far the extent of cross-referencing has been limited, yet there are clear indications of increasing connectedness

Schumpeter and Kondratiev, whose theories addressed the underlying dynamics and pulsations of capitalism[15]. A key question has become: does the accumulation of wealth under capitalism ebb and flow in regular pulsations of, for example, 50 years? If so, is the period from 1940/1945 to the 1990s an example of an upswing to 1970 followed by a downswing which could persist into the 1990s? Attempts to distill patterns are frustrated by the absence of data. Yet, during the 1970s the notion of regular pulsations – longwaves – proved to be a very fruitful heuristic in refocussing attention onto the longer term and into the re-examination of theories explaining the relative growth of different economies. Also, renewed attention was given to the possibility that there were shifts of the core areas of capital accumulation (Braudel 1984; Wallerstein 1974; 1980) as well as geographical shifts in regional centres of innovative activity (Keeble/Owens/Thompson 1982: 62-75).

Neoclassical economics has concentrated upon the equilibrium models of the economy; this is known as comparative statics. However, the events of the 1970s stimulated a revival of interest in Marx who had claimed that there were regular short cycles of business activity (Marx 1954). Marx contended that the inherent nature of competition between firms was a constant vector which tended to depress the average rate of accumulation. The only escape – a temporary one – was to recombine the factors of production (i. e., capital, labour and land) in novel, more intensive ways. The choice of a strategy by the owners of capital was open, though Marx suggested that there were likely to be two basic tendencies:

1. Substituting new equipment for expensive labour
2. Intensifying the process of work organization by eliminating those spaces and times in the production process when the rates of accumulation were lowest

Marx anticipated that the processes of competition between firms would generally lead to the increasing scale of operations. That is, the survival of a small number of very large employers whose practices of work organization were inherently contradictory. These firms and their civil society would provide the means for extracting surplus value from labour whilst simultaneously homogenising labour into a class with revolutionary potential (Edwards 1979).

The importance in economics of business cycles as vehicles for innovation was developed by two critics of Marx who admired the systemic theorizing whilst rejecting crucial details of the dynamics, Kondratiev and Schumpeter. The former collated data on price movements which seemed to suggest that

[15] These aspects are dealt with more fully in Chap. 7

the 7- to 10-year industrial cycle discovered by Marx was simply a component of a much larger pulsation consisting of an upswing and a downswing each of around 23 years. These pulsations, named longwaves, could be used to show that recessionary periods for capital would be followed by periods of expansion. If so, then the functionalist logics of Marx would seem to be inapplicable. Schumpeter carried the same themes further by contending that capitalism was based on a continuing disequilibrium in which there were regular periods when a whole array of new types of products and services were simultaneously launched through a bunching of innovative activity. Schumpeter appropriated the idea of longwaves from Kondratiev and contended that each wave is based on a new set of generic technologies as summarised in Fig. 2-3. For example, railways were based on innovations in engine power as well as a plurality of innovations ranging from the construction of exact timetables and time zones to a new kind of process for smelting. Figure 2-3 contains a schematic history based on the notion of longwaves. It is illustrative of the longwave thesis which is examined further in Chap. 7.

In the postwar period the historical concerns of Marx, Kondratiev and Schumpeter were not on the agenda of key problems because attention was being given to comparisons in relative rates of innovation. For instance, Landes in a seminal study explored the general European scene from the eighteenth to the early twentieth century in an attempt to trace the causes of

Span*	Longwave	Key Carrier Sectors
1785–1840s	First	Cotton, Iron
1840–1895	Second	Railways, Steel
1895–1940	Third	Electricity, Chemicals, Automobiles and lorries
1940–1995	Fourth	Aerospace, Electronics
1995–	Fifth	Information technology, Microprocessors, Bio-engineering

* From the low point to the low point
The peak years are at the mid-point of each span

Figure 2-3. Longwave thesis and generic innovations

early British patterns of successful innovation and their apparent subsequent decline (Landes 1969).

The treatment and recognition given to the topic of innovation in economics, whilst never totally central, led to certain debates over why there were differences in the apparent rates of innovation within a society between various sectors, as well as differences between societies for the same sector (e.g., automobile production)[16]. It was those debates which provided the paradigmatic inputs to the economic and historical analyses of innovation in Britain. In attempting to unravel the causes of Anglo-American differences much, but not all the attention was given to the examination of relative factor prices in Britain and America. Consequently there was a tendency to neglect the processes within firms and the location of firms within very specific sectoral institutions and national institutions. Fortunately the neglect was not total.

Economic analyses of innovation separates into three broad and different frameworks of explanation, each of which tends to compete with the other two. First, one school focuses on the relative costs to the entrepreneur of the factors of production: capital (including equipment), land and labour. Assuming a more or less perfect market, various theories explore the extent to which one factor forms a high proportion of the total costs. The theories assume that rational entrepreneurs will be led to innovate by either the relative cost of specific factors, or, by the high proportion of total cost assumed by any one factor[17].

Second, after Abramovitz discovered that only a small proportion of the total gains in output could be explained by the accumulation of the factors of production attention turned onto the discovery of new forms of knowledge (Abramovitz 1966). That focus, sometimes referred to as *learning by doing* drew its inspiration from the empirical observation of increased levels of output from the same factors of production. Several studies of stable technologies provided clear data on steadily increasing annual output. In an important though restricted analytic breakthrough, Arrow suggested that this might be explained by the discovery and refinement of new production techniques (Arrow 1962). That is, by new forms of local, recipe knowledge, specific to the entrepreneurial firm in question. Learning by doing became the focus of attention. For example, what kinds of information flows did the entrepreneur utilise and were there societal patterns of knowledge making? Temin suggested that there were societal differences between Britain and America in the construction of knowledge (Temin 1975). However, Rosenberg's com-

---

16 Compare the analysis of the American automobile industry by Abernathy (1978) with that for Britain by Whipp and Clark (1986)

17 See Clark, N. (1985, Chap. 4) for an introductory overview. For an interesting application see Harley (1974), and his later comparison of steamships and sailing ships, Harley (1971)

parison of major eighteenth and nineteenth century civilian construction projects in the USA and Britain seemed to highlight the similarities (Rosenberg/Vincenti 1978).

Learning as a focus was highly developed by P.A. David particularly by his postulate that a learning process begins when entrepreneurs choose one type of production technique rather than another (David 1975). David contended that the initial choice was highly consequential because its specific knowledge would provide the paradigmatic framework for creating an agenda of problems and for choosing specific sets of solutions. In other words the technique which was chosen tended to lead the entrepreneur down a specific pathway or innovation-track.

Third, whilst most economists worked on factor prices, some – a small number – analysed production knowledge and sought to explore the institutional framework of innovation (North 1981). These economists contended that, though the influence of markets was significant in providing the selection criteria by which unsuccessful firms were eliminated and the successful firms were rewarded, the practices of innovation were better understood by looking inside the firm at management and labour.

Some economists, such as E.O. Williamson[18], when looking inside the firm, argued that markets and firms are alternative mechanisms for the allocation of resources. They concluded that the market is the hidden hand whilst inside the firm management was the visible hand. It was reasoned that the market mechanisms had hidden costs previously neglected in market models. Two major areas of costs were highlighted: the cost of searching for the best price; the costs of making a contract, of monitoring the contract and of enforcing the terms of the contract in the event of its being incomplete. These costs were referred to as transaction costs and it was argued that firms existed when the transaction costs of a market system would be lowered by internalising the pricing-to-policing process inside corporations. The transaction costs approach developed most strongly in the USA where its exponents focussed on the role of two sources of authority: the state (e.g., North 1981) and management (e.g., Chandler 1962; 1977).

The development of the third school within economics prised open the agenda of problems to the other social sciences and drew attention to the problems of an *only economic analysis* – however sophisticated! There were too many problems for economics to handle as a single discipline.

---

18 ... contribution to the economic analyses of hierarchies and markets as alternative mechanisms for the Institutional economists have made a major allocation of resources; also to the investigation of innovation. The initial ideas and themes are set out in Williamson (1975)

Consequently correcting fragmentation has to be achieved through a synthesis across several analytic perspectives. For example, the diffusion of innovations is tackled in different yet complementary ways in three major perspectives: by Hagerstrand and L.A. Brown in geography; by Mansfield in economics; by Rogers in sociology (see Chap. 3). Moreover, there are important distinctions in economics between the approach of Mansfield and those of P.A. David, B. Gold and N. Rosenberg[19]. The objectives of this book are to show that recent developments in the study of innovation can be synthesised into a useful theory of innovation which can then provide a much clearer understanding of the different patterns of innovation which are found in Britain and the USA.

Criticism of the usefulness of existing theory about innovation has been stimulated by four sets of events. First, the recessionary decline of the 1970s renewed interest in the macrotheories of economic transformation, most notably those of Schumpeter and of Kondratiev. Those theories emphasised the inevitability of enormous structural changes in the economy and in society as an integral part of capitalism. Such major transformations would include the experience of creative destruction by which existing occupations, techniques and firms would be and are eliminated through the competition between alternative modes of production. Second, the rise of Japan as a major economic power has created a strong interest in the extent to which and the means by which technology and practices of work organization can actually be transferred between different societies (see Chap. 13). Third, the optimism generated in the 1960s about the ease of achieving innovation and about the methods by which it can be achieved has been shattered by widespread evidence of failed innovations and by the growing belief that existing theories of economic change are no longer a useful framework to policy-making and practice (Clark/Starkey 1987; Whipp/Clark 1986). Fourth, there is a growing literature which contends that American success in the century from 1880 to 1980 was based on an array of internal and external factors whose favourable consequences for America have been altered by changes in the structure of global markets and by new methods of organization. Piore and Sabel claim that the criteria and basis of international competition are passing through a period of transition comparable to that which happened between 1850 and 1880 (Piore/Sabel 1985). The former period was labelled the 'first divide' and it is agreed that it was a period in which there were three decades of international economic competition between the existing principles of craft, small-scale operation (e.g., Britain) and large-scale, standardized modes of operation and distribution (e.g., USA). Hence the current three decades since 1970 are being termed the second divide. It is suggested that the

[19] David (1975); Gold (ed) (1975; 1980; 1981); Rosenberg (1982)

new international economic competition is between the established mass production and mass distribution (e.g., USA) versus flexible specialization (e.g., winner unknown!). The second divide thesis has immense consequences for Britain and the USA, though in rather different ways.

2.6 Changing Assumptions

There is widespread agreement that existing theory about innovation is less useful than it ought to be because of the fragmentation between different perspectives, yet the prospects for useful theory are more promising than has been assumed.

The problem with the existing state of theory is that the study of innovation is in transition from one set of assumptions – the orthodoxy of 15 years ago – to another, more complex, contrary set of assumptions. At the present time there is a process of incomplete paradigmatic transformation. Consequently the old assumptions are being used to examine and explain innovation in various areas such as business policy and organizational sociology. The old assumptions have been discarded in a very fragmentary manner and *the new assumptions have not yet been coherently organized.* The result is that existing theory is too weak to guide policy-making and is too confusing to be of assistance to practice.

Looking backwards to the early 1970s it is possible to identify the outlines of an orthodox paradigm and a shared set of assumptions, particularly amongst economists and economic geographers. There was an orthodoxy even though it was never wholly accepted.

The main assumptions of the orthodoxy were:

1. Innovation ought to occur: the pro-innovation bias
2. Any innovation is easily recognised and described by both those involved
3. Innovations possess a stable shape and set of uses
4. The marketers and users of innovations are rational actors possessing sufficient information to make relatively clear choices between out-dated practices and innovations: the rational decision-maker
5. Innovations spread through the relevant populations through a process of contagion or imitation
6. The choosers of an innovation are able to implement their chosen innovation and institutionalise the new procedures; the political systems of the firm are presumed to be neutral, yet there is an integer power centre (Clark 1972b: Chap. 9), while the role of labour is assumed to be passive

Each of these postulates has contributed to the dominant studies undertaken over a longish period of time by, for example, Hagerstrand and Mansfield.

Throughout the 1970s each assumption was challenged.

First, the pro-innovation bias was shown to be dubious (Rogers 1983: 192f.; 1962). The research of the early 1970s showed that at any one time there were many alternative innovations for the same type of problem and that many of these alternatives were functional substitutes for each other. Thus a pro-innovation bias could lead to the dysfunctional replacement of the existing practices. It became evident that certain mixtures of innovations may be incompatible[20]. Therefore an enterprise which pursued a policy of pro-innovation might be in danger of trying to ingest unnecessary and irreconcilable practices and knowledge. Moreover, there is a good deal of evidence which suggests that successful firms may let ambitious rivals try out new practices and equipment whilst closely monitoring the problems and preparing themselves. It is possible that the first adopters may be the eventual losers unless they can cope with the learning processes involved.

Second, it was initially assumed that the potential users of innovations could readily recognise any innovation and that both marketers of innovations and potential users would perceive the innovation in similar ways. The assumption was widely held in economics, but was revised as the influence of psychological and social theories became more salient. The key new influence has been the theories of psychological schemas and bounded rationality (March/Simon 1958). These theories emphasise the ways in which subjects select the visual stimuli to which they respond.

Third, the innovation itself is now known to be evolving rather than stable in its shape and uses. The computer is a prime example. Initially these were massive machines in fixed locations and it was thought that there would be a limited usage for processing restricted types of data sets. It is said that in the 1960s Watson of IBM believed that a few computers would satisfy world demand. Today, computers can be portable and can be located in aircraft. Their uses are multiple, not least in entertainment and in the home. The current assumption is that any innovation (e.g., CNC machines) will continuously evolve[21]. Also it is now assumed that the extent of evolvement influences the length and content of the life course of the innovation. It seems therefore that although the exponents of innovations often present the innovation as a tightly bonded array of elements, the users and secondary agencies of diffusion are inclined to prise open the package and to continuously reconfigure the contents. This new assumption is highly consequential. It means that those studies based on a stable assumption might require revi-

[20] See for example the Working Papers by J.J. Williams from the University of Alberta on the compatability of Zero Based Budgeting Systems with other forms of organizational control

[21] The significance of the uneven evolvement of an innovation over time is shown very clearly in a study of automation in the USA by Noble (1984)

sion. Practically it means that those firms which claim to be ahead of their rivals when adopting a new practice or technology need to continuously modify their new possession in order to simply remain at the same competitive level.

The orthodoxy tended to assume that innovations unfolded through quasimechanical or biological processes which were labelled natural (Nelson/Winter 1977, 192; Freeman 1974). That notion did much to counter the excessive weight placed by historians on particular 'heroic entrepreneurs' by giving more attention to the broader stream of processes in which any innovative act is embroiled. However, the result was the underestimating of the central role of human agency (Giddens 1985). Innovations take particular directions in part because, as Noble shows, their tracks are shaped by a configuration of political forces of promoters and enemies and rivals (Noble 1984). A considerable impetus to this approach was given by the work of Schon who demonstrated that the contextual cultures influenced the tracks taken – or, not taken – by an innovation. In a similar vein Abernathy reconstructed the cultural values which underpinned technological innovation in the American automobile industry (Abernathy 1978).

Fourth, it is now accepted that the diffusion of innovations occurs in an environment of many kinds of uncertainty[22]. The advantages of specific innovations are hardest to unravel when their choice is most fateful. By the time the choices are clear it is sometimes too late. Thus decision-making is made on the basis of incomplete information and must involve a fair degree of luck, chance and unintended outcomes. Moreover the benefits of innovation may be dependent on external agencies over which a focal firm has no influence. For example, the adoption of just-in-time inventory systems in the British automobile industry during the 1960s would have further weakened the power position of the assemblers relative to the component supply industry (Whipp/Clark 1986, Chap. 3). The rational model has to be abandoned in favour of a model which recognises intersubjectively rational political coalitions, bargaining and institutional power relations.

Fifth, numerous studies are showing that the users of innovations are major agents in the process of modifying the innovation. There is now good reason to believe that the successful users are those enterprises which enter a process of re-inventing the innovation in order to ensure that it is appropriate to the context and that its shape and uses are collectively understood. Other studies demonstrate that the users unpick the elements of an innovation and selectively adopt some elements whilst rejecting others. For example, the

[22] The notion of uncertainty is central to organization theory and has been incorporated into the economics of innovation in Nelson and Winter (1977; 1982), and by Gold (1975; 1980; 1981)

spread in Britain of Taylorism and its major variants such as Bedeaux in the 1920s and 1930s is characterised by the selection of some elements and the total rejection of others which the originators considered were quite central to success. It follows that considerable caution must be expressed concerning the claims about diffusion, especially with regard to those studies which imply that adoption has been whole scale. Studies showing imitation may only be addressing the most superficial features of alikeness whilst omitting attention to core, fundamental processes.

Sixth, there is still too little attention to the influence on innovations of the authority structures of enterprise and the institutions of the sector in which it is located.

The level of analysis in the orthodox approach tended to be very global and decontextualised. There was, consequently, a wide range of variables selected for attention hindering the focussing of policy at the appropriate level of the sector or the enterprise. Much economic work provided policy-makers with the illusion of control. Today studies are increasingly targeted. For example, studies of the diffusion of innovations in Canada incorporate a systematic examination of the Province as a key contextual dimension (Martin/Swan/Banks/Barket/Beadry 1979).

These new assumptions reveal the limits of the former orthodoxy and provide a sketch of the requirements of a more useful theory. The orthodox theory was clearly very deficient in its conceptualisation of processes and dynamics, particularly of the cumulative, longitudinal aspects.

The search for a useful theory of innovation has been underway for a decade, yet is incomplete because of the imbalance between disciplines and the variable focus.

2.7 Contexts: The Selection of Innovations

In previous analyses the examination of innovations has often stripped them from key features of their embedding in social and economic institutions and the context has been treated as though it were neutral and therefore unimportant. That limitation requires correction (Nelson/Winter 1977, 1982). In practice the context of an innovation is directly involved in the selection process by which some innovations are diffused and continue to evolve whilst in other contexts the innovation is only slightly adopted and does not evolve. The point may be illustrated by comparing American and British approaches to photography.

The development of the camera in the midnineteenth century and the subsequent development of cinematic techniques was equally balanced between

European and American innovators. However, there is a major difference in the use to which Americans put the new techniques. Americans were fascinated by the capacity of the camera to shrink days into seconds and vice versa and they soon learned to apply the camera to scientific and commercial uses to observe what the eye could not see: for example, in the analysis of the surface of the moon. In the 1870s in a series of famous experiments, Muybridge used 24 cameras triggered off in a sequence to record a horse moving and to demonstrate that when a horse gallops there are moments when all four feet are off the ground[23]. The same ideas were also developed by Maret (in France) to study birds in flight. From that evolvement of the technique it was a relatively short step *for Americans* to apply the technique to the recording of action and its minute analysis by freezing movements on a frame-by-frame basis: the stroboscope. The principle was used for a variety of purposes: to reveal anatomical detail, to construct picture books of "what the butler saw", to analyse sport, to study human movement at work – for example, in the very common task of bricklaying. Time-lapse photography was then used to study the growing of crops. Thus, the multiple and extensive uses, for pleasure, domestic relations and commercially, all provided a strongly favourable market for the array of related products based on the use of the camera. Yet, the extensiveness of the usage, for instance, the applications to sport and work, grew rapidly in the USA and very slowly in Britain. Whereas, a North American college team would record its major football games as a routine matter from the late 1960s onward, that practice was only very occasional in Britain even during the 1980s. So it has been, also with the analysis of work.

Treating the context systematically requires that attention be given to two basic features: first, to the extent of uncertainty which may surround the diffusion and adoption of an innovation; and, second, to the institutional features of specific sectors. These two features differ sharply between different sectors in the same country and in the same sector between two different countries.

Uncertainty as a dimension of the evolvement of innovations was largely ignored in early studies because it was assumed that the market mechanism provided suppliers and potential users with complete information at no cost to either party. Later, when the imperfections in the market model were recognised and retheorized, attention was suddenly focussed on the uncertainties facing potential adopters. In particular Gold, Nelson and Winter emphasised the extent of uncertainties which face users (Nelson/Winter 1977, 1982). Their work makes it clear that the adoption and rejection of innova-

[23] See particularly the work of E. Muybridge, (1830–1904) for early photographic studies of animal and human locomotion in the period 1872–1882

tions may contain a great deal of hope, chance and judgement. Also, there are two stages of uncertainty:

1. The uncertainty facing those who create and construct the innovation, for example, as faced by the creators of the float glass technique (Barker 1977)
2. The profitable dissemination of the innovation, as exemplified in the case of float glass where the discoverers had to devise patents which ensured that only licensed users possessed the new techniques (Barker 1977)

Each of these stages is surrounded in uncertainties. However, not all stages and facets are equally uncertain.

Uncertainty should be treated as a variable, so that degrees of uncertainty can be identified as set out in Fig. 2-4. Uncertainty is in two parts: the degree of uncertainty embodied and embedded within an innovation, and the extent to which users already possess levels of knowledge and skill which enable them to systematically encode the uncertainty and to devise means for handling its level[24]. Each part can be treated as a variable in the format set out in

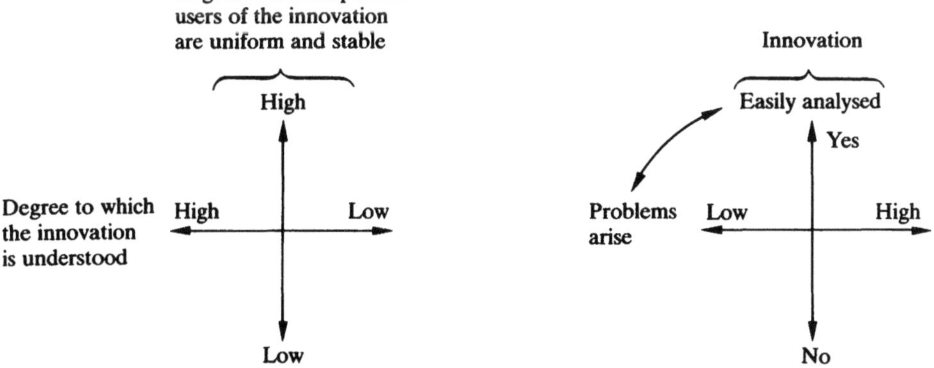

Figure 2-4. Uncertainty and innovation (Derived from C. Perrow, 1967)

[24] See the treatment of uncertainty and the construction of a theory of change in Clark, P. A. (1972)

Fig. 2-4 and therefore it is necessary to ascertain uncertainty with regard to specific situations. Moreover, two firms in the same sector looking at the same innovation may have quite different corporate capacities to handle the uncertainty.

Uncertainty is an important dimension, but it is crucial that it be treated as a variable with differing states, rather than as a single polar magnitude which is either high or low.

The discussion of uncertainty raises the question: what are the necessary conditions for the survival of an innovation? At any one time there is a flow of diverse innovations in varying states of evolvement flowing through any sector of the economy. These innovations encounter varying niches of resources so that some are picked-up and developed, whilst others disappear – though they might be revived later in another format. For example, in the early 1970s it did seem as though organization studies had created the innovation of Organization Design (From Clark, P. A. 1975b). Its features are set out in Fig. 2-5 and, as may be observed, the innovation embodies a high level of uncertainty for suppliers and users. Its image is fuzzy and so users are uncertain of its relevance. Figure 2-6 shows that organization design was more likely to succeed where the users possessed capacities which were complementary to the innovation. In fact that innovation in its original, generic format has largely disappeared and been replaced.

The choice and rejection of an innovation affects its evolvement as well as affecting the users. There is a reciprocal two-way process. The point may be illustrated through the comparison of the evolvement of colour television in the USA and Japan (Rosenbloom/Abernathy 1982). Colour television, which had been pioneered in the USA by RCA and in Europe by N. V. Philips, was the main story in the 1960s and sales took off in the USA so that 80% of sales, worth two billion dollars, were in colour sets. Japanese demand took off in 1967 and grew rapidly until 1973, whilst European demand lagged slightly behind. By 1969 the USA and Japan each held an equal share of three-fourths of the world market. In 1952 Philips licensed Japanese developments and also engaged in joint ventures and in 1960 they also licensed their patents on the audio cassette to Japanese firms. The Japanese absorbed the foreign technologies by developing process innovations and by exploiting their robust domestic market. Later, they adopted export-oriented strategies (see Chap. 13). By 1969 the Japanese investment in the *design and development* (D and D) of technology led to the creation of the Trinitron tube by Sony and a solid-state colour receiver by Hitachi. Those developments, and others provided the basis for the next stage with the development of the home video player.

In this example, it was a combination of Japanese innovations with their acceptance in a domestic market which provided the basis for the success in

Factor	Range	Organ. Design Requires
A. Problem		
1 Crystallization	given/negotiable	Negotiable
2 Urgency	great/small	Small
3 Ownership	closed/open	Open
4 Territories	distrib/integrated	Integrated
B. Competence and power		
1 Orientation	local/cosmopolitan	Cosmopolitan
2 Awareness of Org. Science	low/high	High
3 Temp. horizon	brief/several years	Years
4 Goal setting	specific/general	General
5 Control: ext. envt	low/high	High
6 Control: int. envt	low/high	High
7 Willing to expt.	slight/great	Great
C. Expectations of consultant		
1 Intervention Style(s)	single/multiple	Multiple
2 Input	results/analysis	Analysis

Figure 2-5. Requirements for successful organization design (From Clark 1975)

Factor	Adopt: CCE	Adopt: Distrib. Ltd.	Adopt: Ecu Parish	Reject: Brit. Eng.	Reject: MOD	Reject: Custo-dial
A. Problem						
1 Crystallization	4	2/3	4	2	4	2
2 Urgency	5	3	3	1	4	3
3 Ownership	4	3	4	2	3	2
4 Territories	5	4	5	2	3	1
B. Competence and power						
1 Orientation	5	4	5	3	3	4
2 Awareness of org. studies	3	4	3	1	2	4
3 Temp. horizon	5	4	4	2	4	3
4 Goal setting	5	5	4	3	4	2
5 Control: ext. envt.	5	5	3	2	1/2	1
6 Control: int. envt.	5	4	3	2	4	1
7 Willing to experiment	4	4/5	3	3	2	1
C. Expectations of consultant						
1 Intervention Style(s)	4	3/4	4/5	1	1	2
2 Input: type	3	3	4/5	1	2	2
Ave.	4.4	3.8	3.8	2.0	3.0	2.2

Key
CCE Ltd is a large cigarette manufacturer
Distrib. Ltd is an oil firm
Ecu Parish is an ecumenical parish
Brit. Eng. is a large aero engine firm
MOD is part of the military
Custodial is a remand centre for a large conurbation

Score of 5 is a high fit between requirement and actual

Figure 2-6. Factors affecting the outcome of trials in organization design: The user

consumer innovations. It may also be shown that these contextual features were less favourably present in North America.

The evolvement and survival of innovations is crucially shaped by the contexts in which they are generated. The point applies equally to the transnational transfer of innovations.

2.8 Transnational Transfers of Innovations

Part III will concentrate upon the transfer of the same innovation across national boundaries with particular attention to the Anglo-American case. The Japanese dimension is examined in Chap. 13.

The cross-cultural transfer is a specific case where the analysis of the context becomes crucial. For example, many American innovations in fact embody (or require) American ways of organizing and operating to make them viable. Consequently, their transfer requires an identification of those contextual features which would be crucial in another context. The basic principles are addressed in the following three chapters and then strongly illustrated in Chap. 6, where the transfer of a form of teamwork from Britain to the USA is analysed and it is demonstrated that its appropriation by the American users led to small modifications which in turn led to a total transformation of the original innovation. That example is of an organizational innovation for which there was not a powerful transfer agency. Later, in Part III several examples of the reverse flow from America to Britain are examined. In most cases there were agencies to promote the transfers, yet problems arose.

An important area of cultural difference between the USA and Britain lies in the expectation and capacity of Americans to engage in future-oriented large-scale projects. American innovations are more often presented by their backers to the accompaniment of an all-embracing, future-oriented imagery which is exciting and legitimating. That cultural feature has been evident through the twentieth century and even influenced the American concept of a museum! That cultural propensity is one of creating comprehensive control concepts with a high level of public acceptance – even when opposed as in the case of 'Star Wars'. The search for a workerless factory, commenced in the 1940s inspired the novel *The Player Piano* and has been shown by Noble to have provided an organizing framework which embraced the universities, the military and commerce in the collective development of an array of specialized machinery (Noble 1984; Vonnegut 1955). There was no comparable tendency in Britain. However, the American propensity may also contain the seeds of future problems when the military-industrial complex is utilised because the needs of the military may inhibit commercial innovations. The case of computer numerical control is revealing. American military influence provided the impetus, yet seems to have led to the creation of a supply system which is oriented to expensive machines capable of exacting standards. In contrast the Japanese suppliers faced users who wanted cheap machines to handle less-exacting components. At the present time, it is the Japanese context which has proved the more favourable.

3 Innovation Supply: The Marketing and Imitation Models

3.1 Introduction

The investigation of the diffusion of innovations is in a slow, irregular state of transition away from the dominance of a group of paradigms which presumed that diffusion occurred through a process of imitation[1]. These paradigms assumed that all relevant innovations were marketed through well-organized agencies who possessed a complete and objective understanding of the requirements of the customer. Hence, the orthodox approach denied the relevance of the rejection of innovations or the usefulness of their modification and reinvention by the users[2]. There is little doubt that all analysts tended to assume that imitation and contagion was the typical situation. That presumption, now hotly challenged within the field, still has a grip on nonspecialists[3].

Amongst specialists there has been a tendency to understate the limitations of the imitation paradigm and to give inadequate attention to the situation of the user (see Chap. 5). Instead of making a significant reformulating of the study of the diffusion of innovations by addressing the overall balance between their marketing and their usability, the general approach has been one of introducing slight refinements to established thinking. Three areas of refinement may be highlighted: the re-invention or modification of the original innovation[4], the initiation and implementation model (Duncan 1974), and the partial appropriation model (see Chap. 6).

The orthodox imitation paradigms treated any innovation as a largely static bundle of elements which ought to be adopted as a totality by the user. Thus, we have the early studies by Hagerstrand, Mansfield and Rogers, who all ini-

---

1 An indicative early example of the use of an imitative model is Mansfield (1961). The limitations are reviewed in Clark, P. A. (1983)

2 See the early work of Rogers (1962)

3 The imitation model is implicitly used in a number of recent studies, most notably those of the transatlantic diffusion of work study, the assembly line and the multidivisional form of organization; see Chap. 12

4 The notion of re-invention surfaced in Rogers, E. M. and R. K. Rogers (1976, Chap. 6); Larsen, and Rogers (1977). The re-invention theme is fully developed in Eveland, and Rogers (1980)

tially assumed that the user was a passive selector of an innovation[5]. Later and more recent studies, especially those by von Hippel, have shown that the user may well be capable of making significant revisions to the original innovation[6]. A similar interpretation has been proposed by Eveland and in the recent work by Rogers in their concept of re-invention[7].

It is tempting to juxtapose imitation and re-invention as apposite explanatory paradigms and as contending prescriptive approaches. However, as L.A. Brown shows, that would be misleading and unhelpful because both processes exist and require attention (Brown 1981). This chapter starts by examining how and why three leading analysts of the diffusion of innovation – Hagerstrand, Mansfield and Rogers – have each shifted their earlier position from focusing solely on imitation to giving more equal attention to imitation and re-invention.

The balanced treatment of the supply side and the user side is a necessary step towards the replacement of the one-way linear models, like those of Schumpeter, which suggested that invention and innovation were two very separate activities (Schumpeter 1939). The two may be separated in certain circumstances, yet many studies reveal a continuous iterative and recycling process linking invention and innovation. Further, is is clear that in certain sectors the supplier is the major source of innovative activity (see von Hippel 1978; 1982a, b; 1983). Duchesneau demonstrates that the locus of innovative activity for production techniques in the American shoemaking industry has almost always been with the suppliers of equipment (Duchesneau/Cohn/Dutton 1979). In that case the supplier markets the machinery and the innovation is handled through the agency of the supplier. However, many innovations, especially process innovations of an organizational kind, such as the Japanese just-in-time productive systems or autonomous work groups may not be handled by intermediaries, or, even if they are, their relevant features may, for a variety of reasons, be obscured from the eventual user. Such innovations lack a well-developed infrastructure on the supply side and pose the potential user with problems which require attention.

As indicated earlier the process of imitation is more complex than has been assumed. Yet, many texts contend that imitation is widespread (see Chap. 12). Their claims raise two problems which will be dealt with later in Part II and in Part III. First, in order to establish that imitation has occurred it is important to demonstrate that the crucial features of an innovation have really be transposed from one setting to another. It has to be shown that imitation is not simply a surface cosmetic, but a fundamental implementation

---

[5] For example, see Hagerstrand (1952); Mansfield (1968); Rogers (1962)

[6] Von Hippel (1978; 1982a; 1982b; 1983)

[7] See note 4

(Clark/Tann 1986). Second, several analysts have indicated that the American context is one in which imitation might occur more readily than in Europe (DiMaggio/Powell 1983). These claims require careful consideration and will be dealt with in Part III. For the moment I shall contend that the potential – the capacity – for certain sorts of imitation is more fully articulated and developed in North America than in Europe, especially more so than in Britain.

This chapter is in three further parts, starting with an examination of the development of diffusion models in neoclassical economics and in economic geography. The following section presents the more sociological perspective associated with Rogers. These earlier models emphasised imitation as a process, yet gradually began to explore the processes of adoption within the user corporation. That process was extended by the approaches dealt with in the final section. All these approaches are from the supply side. They view the user as an object rather than as a subject.

The following chapter unravels technological innovations as a special case and it continues the critique of the orthodox paradigms. Chapter 5 completes the critique by examining and highlighting the position of the user.

3.2 Economics of Innovation

3.2.1 Hagerstrand

Hagerstrand's contribution came in two major steps[8]. The early work, in the 1950's, attempted to demonstrate the existence of a deterministic set of processes based on the notion of imitation. The later work, so far less influential, sought to demonstrate that one of the most important factors affecting the adoption or rejection of innovations is the time-space trajectories which individuals create when they organize their daily lives. Hagerstrand has been based at the University of Lund in southwestern Sweden and the school of investigation he has created is often known as the Lund School.

In his early work in the 1950's Hagerstrand modelled the spatial diffusion of manufactured innovations and claimed to find a striking spatial patterning in their adoption. This approach, which was strongly influenced by operations research techniques constructed to prescribe rational location patterns (e.g., Monte Carlo), identified three basic patterns. First, the S-shaped diffusion curve through time as shown in Fig. 3-1 A. The curve indicates that the

[8] The early studies are illustrated by Hagerstrand (1952), but the later studies are best summarised in Parkes, and Thrift (1980); see Chap. 6

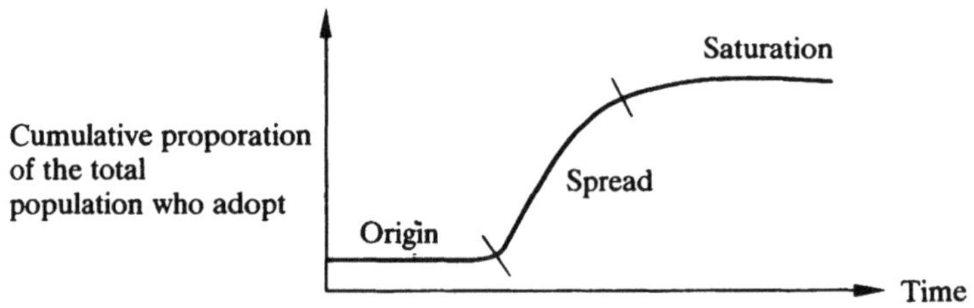

Figure 3-1 A. S-shaped diffusion curve through time

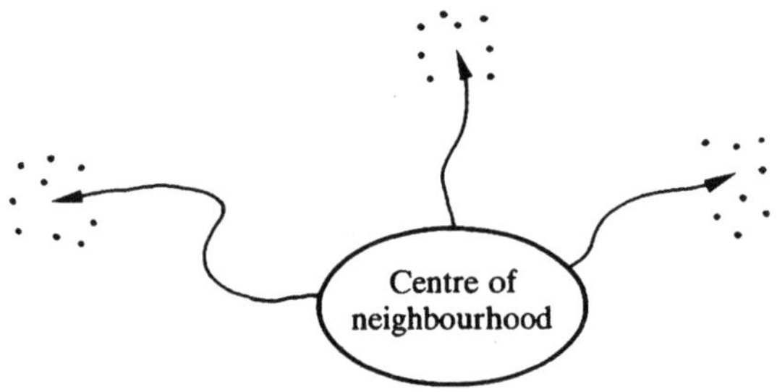

Figure 3-1 B. Neighbourhood effect for diffusion in a region

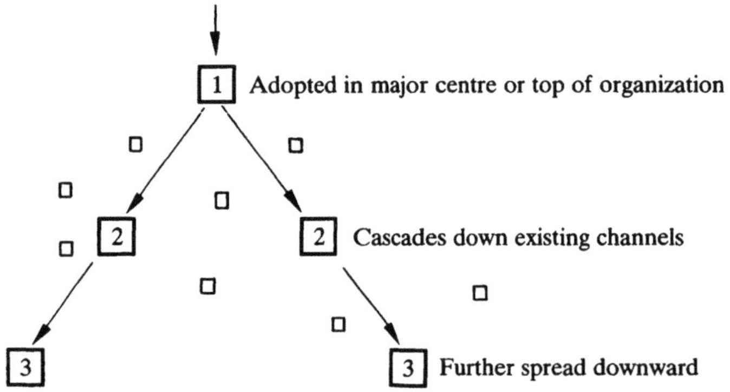

Figure 3-1 C. Hierarchy effect in for diffusion

diffusion of an innovation from its place of generation begins very slowly with a primary stage during which initial centres of diffusion are established. The spread of adoption grows very rapidly leading to a final plateau. The S-shaped diffusion curve has itself been widely diffused into, for example, basic texts on marketing. Second, there is the neighbourhood or contagion effect when diffusion proceeds in a wave-like manner from the initial centres to locations which are further away (Fig. 3-1 B). The early work suggested that this effect could be estimated very simply in terms of measured distance, but more recent studies suggest that the presence of such patterns has to be explained in terms of both the journeys taken by potential adopters and by what they notice and pay attention to when they perceive an innovation. Rogers cites the case of a farmer in Ohio driving past 100 miles of a new strain of cereal without noticing its features (Rogers 1983). Third, the hierarchy effect refers to the tendency for innovations to become established in major metropolitan places and prestigious urban centres, rather than to be diffused into the hinterland (Fig. 3-1 C).

The hierarchy effects and neighbourhood effects play an important part in the S-shaped diffusion curve. Hierarchy effects would be strong in the initial stage, whilst neighbourhood effects would be crucial during the middle, rapid diffusion stage.

These basic concepts became an integral part of the geographical study of diffusion. Hagerstrand presumed that the empirical regularities which the data revealed could be interpreted as showing that imitation was occurring. The assumption was that adoption was the outcome of a simple learning process, based on the rational flow of information. Clearly, the flow of messages about a new product within a population can be affected by physical factors like distance, however, the application of operations research models soon revealed many examples of diffusion where the expected pattern did not occur. There was the discovery of inconsistencies between the model and the observations. In order to explore these inconsistencies Hagerstrand shifted his attention.

The new focus was placed on the 'activity systems' of individuals. It was reasoned that individuals operate at two levels:

1. In the long run, individuals construct projects from their various intentions; these projects affect their distribution of activities through time and space
2. In the short run, the daily lives of individuals are embedded in many recursive routines which are so taken for granted that they are only rarely examined

He demonstrated that the adoption or rejection of innovations was inhibited or facilitated by how the innovation affected the time-packing and spatial trajectories of individuals. The point may be illustrated with reference to possible innovations in forms of transportation, or, in the introduction of flexible-time working schedules. Hagerstrand reconstructed daily patterns for Swedish citizens as illustrated in Fig. 3-2. These depict the day of a wife and her husband. It may be seen from such schemas how existing routines

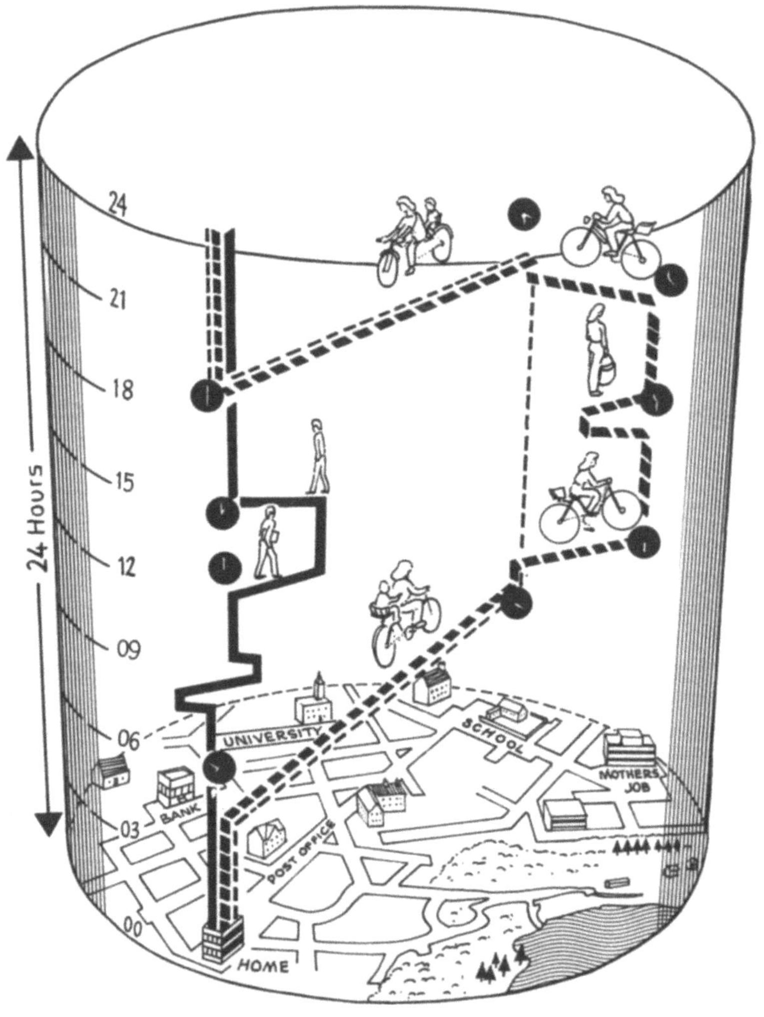

Figure 3-2. Industrial trajectories through time and space in one day

involve certain forms of time packing and how the day is organized around certain given events which act as pegs for shaping the day. The objective is to examine the constraints of existing activities and how they may affect the opportunities for innovation.

He does not really search into the underlying processes which give shape to the day. Moreover, there is a tendency to regard individuals as only physical traces on charts rather than as potentially active agents who exist within contexts of influence and power. So for example, it would be impossible to understand the excellent study by Cullen of how students at London university spend their time without knowing about the university's practices for allocating lecture hours or about student predispositions (Cullen 1972). The timetable, a crucial innovation in the Western world, prescribed the lecture times. Few students anywhere have an influence over their construction. Also the predispositions of students vary. So whilst an identifiable segment used the timetable and its gaps between lectures as a collection of fixed events around which they organized other activities – library, shopping, laundry, courting – other students with different intentions followed different trajectories.

Hagerstrand also studied social innovations and their diffusion, (e. g., rotary clubs). Overall the tendency has been to rely on the initial raw assumption that diffusion occurred through an imitation affect even though that very basic assumption was not fully examined. Hagerstrand unwittingly gave credance to the notion of diffusion as a deterministic process involving the spread of items whose features remained essentially stable.

3.2.2 Mansfield

Mansfield's contribution to the economic analysis of innovation is widely acclaimed for the many empirical studies and for the elaboration of the imitation thesis[9]. It is neither relevant nor possible to do justice to the empirical studies except to draw attention to their extensive coverage from beer to diesel engines, from bituminous coal to CNC machines. Our interest is in the basic principles which he developed and which, for a time, became a crucial part of the intellectual architecture of the economic study of the diffusion of innovations[10].

Mansfield, like Hagerstrand, began with a model of the imitation process, though later that model was revised. His approach has been that of the neoclassical economist using an equilibrium model to argue that market signals are the key source of stimulus to the activities of those who run firms. It was initially assumed that corporate decision makers had relatively complete knowledge about alternative innovations.

Mansfield's research and theorizing focussed on technological innovations. This focus occurred in the period after Solow and others had uncovered the large amount of variance which could not be explained by existing factor-input models (Solow 1957; Leontief 1985). The examination of technology by

---

9 Mansfield (1968 b, Chaps. 1, 4, 9; 1977: 85–122)

10 Compare with the interests of industrial economists in general (e. g., Clark, N. 1985)

Mansfield represented a shift in paradigm and a definite advance. First, it was assumed that technological change, that is, innovation, arose from an advance in knowledge. However, it was assumed that technological change represents embodied capital. Second, technology was distinguished from technique. The latter was a combination of equipment, products and work organization and the ways in which they were used.

The early studies set out to identify and to measure the underlying economic variables which explained the process of diffusion of a variety of technologies (Mansfield 1968b, Chaps. 1, 4, 8). Those studies demonstrated the overall slow rates of diffusion and the wide differentials in adoption rates between innovations. The studies provided a clear image of innovation as a relatively determinate process in the manner of the S-shaped curve. The findings served to refine the imitation thesis rather than to suggest any serious revisions. Mansfield reported that two factors were strong predictors of the rates of adoption: the enormity of the investment required and the actual profitability. Likewise these conclusions seemed to affirm the neoclassical model of the informed decision-maker.

Later, in the 1970s (Mansfield 1977), the earlier assumptions were reviewed. For example, it was recognised that early innovation, the pro-innovation bias, might be unwise:

> The first user is important because, (he), as well as the supplier, often takes considerable risks. (Mansfield 1968b: 83)

In other words there might be real advantages in tracking behind the leaders and learning from their mistakes (Rosenberg 1982).

Mansfield also began to disaggregate the global definitions of technological innovations and to treat them in a more discriminated manner with considerable advantage. For example, in a neat comparison between the diffusion of data processing and computerised numerical counting (CNC) machines, he demonstrated that the diffusion of the latter differed from the former. That was because although numerical counting (NC) machines were known in the 1950s, the state of production techniques in the 1970s meant that they were most appropriate for medium-batch manufacturing rather than for small – or large-batch production. Consequently, NC machines took 15 years to capture a quarter of the market. By comparison, innovations such as the semiconductor captured four-fifths of the vacuum-tube market in that same period. Mansfield therefore concluded that the effects and appropriateness of a new technology will be incomplete and selective.

Mansfield's work has been a model of careful empiricism, yet the underlying tendency has been towards deterministic models and assumptions about processes. The earlier assumptions about imitation have been qualified in detail rather than substantially revised and elaborated.

3.3 Sociology of Innovation

3.3.1 Rogers I

Rogers' contributions, encompassing more than a quarter of a century of work, have been a sustained focus and refinement of the supply side of the diffusion of innovations (Rogers 1962). In addition there has been an important attempt to bridge into the field of organization studies (Rogers/Rogers 1976, Chap. 6). The more recent work extends and refines the basic model of the supply side rather than reformulating an overarching theory to accommodate both the supply of innovations and the position of the corporate users[11]. Rogers' contribution is significant because of its origins in an extensive attempt by the American agricultural extension agency to pool knowledge about the factors which affected the diffusion process. Moreover, that model represents a very clear and graspable account of the imitation model. It is convenient to split Rogers' model into two parts in order to highlight the features detailed.

Rogers' models were heavily anchored in the activities of American institutes which specialised in research and development for the agribusiness sector. Agribusiness has been a major American industry[12]. America has been self-sufficient in agriculture since the early nineteenth century and has been a major exporter of agricultural products since the midnineteenth century. Agribusiness, in the growing of crops and in the raising of livestock, has been a major American concern. At the state level, agricultural practices have been the focus for public investment in research, design and development of all features of the business. For more than a century, the farmer has been a prime target for specially established courses in the land-grant colleges (e.g., Cornell, Purdue) and in the state universities. Research on all aspects of agribusiness has been a sustained and directed program which has incorporated the conscious devising of means to translate research into a detailed, written specification[13]. That specification defines best practice[14]. The aim of agricultural agencies was and is to promote best practice amongst farmers. It is relevant to stress the template[15] of best practice, to indicate its long heritage in American society. The best practice concept has also been

---

[11] Compare Rogers' approach with the position taken in this book in Chap. 5

[12] For an overview of the centrality of agriculture, see later in Chap. 9. More specifically see references to Chandler (1977); David (1975); and Pulos (1983) in Chap. 9

[13] See, for example, the novels by Laura Ingalls Wilder

[14] The template of best practice is a taken-for-granted everyday practice in the USA and to a certain degree also in Canada, but not so in Britain

[15] Template is a collection of social activities for which there are defined and publicly available guidelines; the wedding ceremony is an obvious example

an influence as a template amongst American small capitalists and in major corporations; it predates the Scientific Management Movement of the 1880s and the specific approaches, such as Taylorism (see Chap. 12), which was an embodiment of the generally accepted practice.

During the 1950s the agricultural agencies sought to improve the diffusion of their definitions of best practice. They created special meetings culminating in the decision to pool existing knowledge about how to influence the target audience, that is, the farmer. The practice of pooling knowledge for a practical objective was a taken for granted approach. A committee, meeting around 1955, sought to pool knowledge from the social sciences, especially social psychology with knowledge of farming practice (Rogers 1962). The model, which is shown in Fig. 3-3, focuses on innovations like seeds, equipment and agricultural practices (e.g., planting systems). The target was individual farmers in their communities.

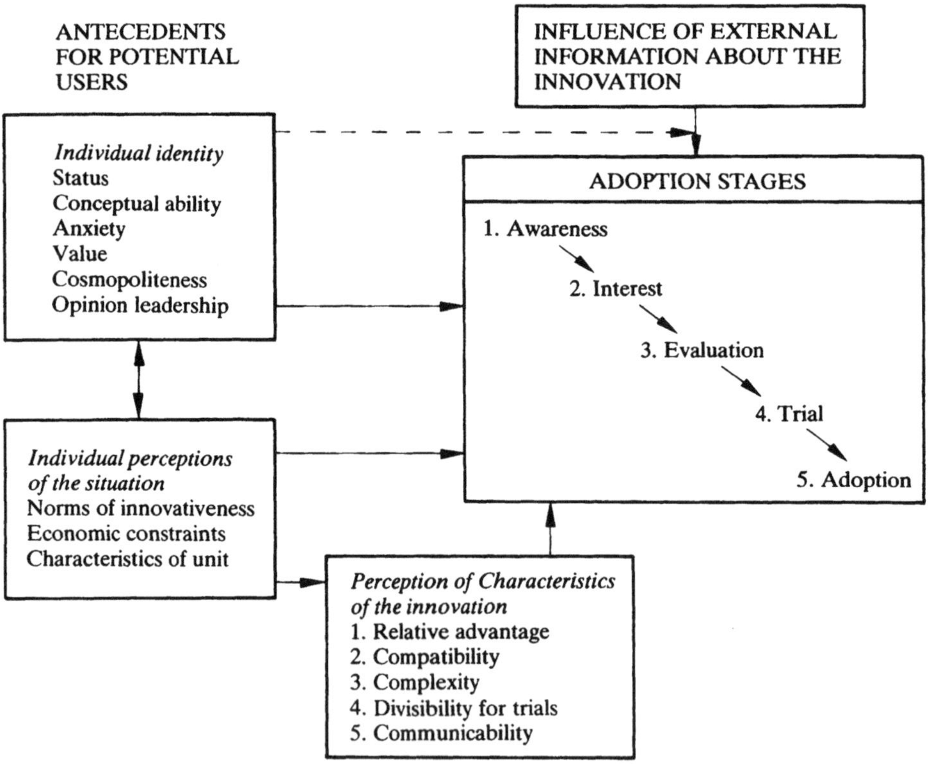

Figure 3-3. Diffusion of innovations framework: Rogers I (Derived from Rogers, 1962)

The model begins from the belief that certain new items are worth diffusing to the users. That is why it has such a strong emphasis on the supply side.

The model will be used by the central planning team of the agency wishing to diffuse a new item. The analytic framework is a sophisticated causal map for arranging their own efforts and for their agricultural extension agents in the field. In other words, the model assumes the prior existence of an agency to market the innovation. The field staff are conceived of as change agents[16] by Rogers, who intends that they will use the framework to plan activities in their locality. The assumption is that extension agents can adapt a general plan to the local context.

Figure 3-3 is the first, original model, *Rogers I.* There are five aspects of the model: the antecedent context of the specific farmer, the characteristics of the innovation, the innovation process, its states, and the results.

First, the antecedent context contains the farmers and their perceptions of the situation. Here it is immediately obvious that this model differs sharply from those used by Hagerstrand and Mansfield. The reason for the difference is that the committee decided to construct a model of the farmer based on psychological and small-group research studies. Hence attention is given to how farmers might construct their own sense of identity and it is theorized that farmers as individuals will vary systematically along the six dimensions set out in the model. Some farmers will be anxious, hold values of individualism, possess high mental and conceptual skills, be of low social status, have a cosmopolitan orientation to events outside their own community and not be opinion leaders. Such farmers may be likely to innovate at an early stage, but their actions are unlikely to influence others in the same community.

The reasoning behind the approach was derived from the early studies of communications which demonstrated that mass advertising is always heavily interpreted at the local level where specific moral codes and normative frameworks evolve. It was assumed that certain individuals were very influential – opinion leaders – in creating a local attitude. Therefore the task of the change agents becomes one of identifying and influencing the opinion leaders who, it was reasoned, would in due course activate an awareness of new practices amongst their local network: the neighbourhood effect noted earlier in the work of Hagerstrand. Consequently it was crucial for a change agent to target certain farmers.

As indicated farmers vary in their social status and in the extent to which their attention is focused onto external events and external sources of information or is tightly bounded by the normative frameworks of the local com-

[16] The roles of change agents are examined in Clark, P.A. (1975a)

munity. The model assumed that the early adopters of innovations would tend to be those of higher status, who were also cosmopolitan in their orientation to sources of information, intelligent and the most emotionally secure.

A model of the farmer as an actor was accompanied by an examination of how a specific group of farmers from a community perceived their situation. For example, communities of farmers will differ in their shared orientation towards innovations and the model reminds the change agent to concentrate upon the most innovative communities at the start of a new program: the hierarchy effect.

Rogers, like Hagerstrand, anticipated that the intention to be innovative would be constrained by economic factors and by the basic characteristics of the farm: its size, profitability, types of activity.

Once the change agent has characterised the pre-existing features of each farming community, the framework can be used to arrange the order of a diffusion program by starting with those communities which are the most innovative; then within those communities to look for those farmers who are opinion leaders. Once the best practice is initially established, then the framework postulates that the effects of hierarchy and neighbourhood will unfold in a more rapid phase of innovation.

Second, the model indicates that the rapidity of the diffusion process can be affected by the design and presentation of the innovation itself. Five dimensions of the innovation are identified in Fig. 3-3.

1. The model expects that an innovation is more likely to be adopted and rapidly diffused when potential users perceive that they will gain one or more forms of advantage from adoption. There are several important sources of advantages: financial, enhanced status, better use of time and space. For instance, new types of sugarbeet seed are situated inside a pellet which protects the seed from diseases whilst being simultaneously simpler to place into holes in the ground during the drilling process. In that case there seem to be several advantages to the user. However, the perceived advantage may or may not be present. The model simply assumes that the perception of advantage is significant and it goes further than the models of neoclassical economics in assuming that sociological aspects such as increased status and prestige may influence the adoption/rejection decision.
2. Innovations whose features and requirements are compatible with existing values and practices are likely to be adopted most rapidly. The importance of compatability may be illustrated from the studies of the diffusion of zero based budgeting (ZBB) by Williams (see Chap. 1, note 19). Studies indicate that ZBB is more likely to be adopted by firms which have previously adopted and ingested management-by-objectives (MBO) because

MBO provides a framework of accountability which ZBB can use as a basis for its own practices. Incompatability between the values of the innovation and the potential adopter hinders acceptance.

3. Complexity, that is perceived complexity, influences adoption. When an innovation is perceived as too complex in its requirements the potential user is more likely to be uncertain of the outcomes and therefore delay adoption or reject the innovation.
4. The more an innovation can be divided into smaller pieces which can be used in trials by the potential user, the more likely is the user to attempt adoption. This feature has certainly influenced the design of a number of organizational innovations in the management field. Perhaps the best example is that of the grid organization development (GOD) devised by Scientific Management Incorporated and variously described by Blake and Mouton (Blake/Mouton 1974). The total application of the GOD program could take about four or five years in a large firm and would cost hundreds of thousands of dollars, depending on the numbers involved. The total package aims to create a major, fundamental transformation in managerial behaviour and action by institutionalising in the firm a specific language for judging one's own performance and that of others. The language is a defined one and is progressively inserted through a subtle variety of training and learning situations. The total package is divided into stages, usually about six, of which the initial stage is short, synoptic and focused on those who will have to sanction the cost. In other words the innovation is deliberately divisible so that users can sample the process.
5. Communicability refers to the ease of describing the results of adopting an innovation. The more communicable the results, the more likely is adoption. As can be imagined an agricultural extension agency can readily provide short, cogent and interesting statements which detail the results from experiments.

The framework concludes that the most acceptable innovations will have a strong relative advantage over existing innovations, be highly compatible, not too complex, able to be divided into stages and the results of their usage will be highly communicable.

Third, the adoption process by the farmer is depicted as a five-stage sequence undertaken by the individual (Coleman/Katz/Menzel 1966). It is assumed that the farmer is empowered to make decisions as a single agent without the hindrance of seeking the support of others.

Those processes form the context of information sources. An individual's encounter with an innovation is then examined in terms of the five steps set out in the centre section of Fig. 3-3. The diffusion agency should start by creating awareness and arousing interest in the innovation. The farmer is targeted through the two-step flow of information: initially through the media of journals, leaflets and the like, followed by personal contacts from the agricultural agent and the informal influence of opinion leaders. Once there is an awareness of an alternative, the next step is the arousal of interest in a target segment of the community. The framework assumes that the farmer then engages in a mental evaluation. (S)he imagines that the innovation is already operating and through that process reflects on the problems and advantages. This mental experiment can be the prelude to an actual trial with a small part of the innovation. Finally, some farmers, satisfied with the outcomes of the trial will proceed to the adoption of the innovation.

At each of the five steps the farmer may withdraw interest and, as the figure shows on the right-hand side, the adoption or rejection decisions may be reversed.

The basic framework provides the diffusion agency and its field staff, the change agents, with an implicit program of action, each aspect of which can be transformed into standard operating procedures.

In *Rogers I,* the framework was treated as being a widely applicable approach. However, as will have been noted there are several important assumptions, most obviously, that (a) innovation is the best policy, (b) the user is a passive agent in the whole process, and (c) the supplier devises the innovation and provides objective, neutral data about the innovation. The model also assumes that the innovation remains largely static, is unmodified by the user and that diffusion occurs through contagion.

3.3.2 Rogers II

The original framework, Rogers I, has been refined and substantially elaborated in a series of studies and several lengthy restatements (Rogers 1983; Rogers/Shoemaker 1971). Revisions have been made to five major aspects: to certain basic assumptions and details of the basic framework; to a limited analysis of the generation of innovations; in the shifting of attention to the organization as the focus for innovations, leading to a reconceptualisation of collective processes of decision-making; to the recognition that innovations are sometimes modified by the user in a process referred to as re-invention; in the examination of the consequences, positive and negative, of adoption. These revisions amount to a substantial extension of the original model whilst retaining a preference for the supply side of the diffusion process.

First, revisions to the basic assumptions commenced with a critique of the earlier pro-innovation bias (Rogers 1983: 92f.; 1986). As indicated in Chap. 2, the pro-innovation bias took the viewpoint of the diffusing agency and presumed that the adoption of the innovation was appropriate for all members of a given population. Later studies showed that new innovations may simply be duplicates or substitutes for existing practices. Also, the pro-innovation bias led to a neglect of the failure of innovations and of their inappropriateness in certain situations. There have been too few systematic studies of failures and the analysis of the failure of an educational innovation – the catalytic role model – by Gross (Gross/Giaquinta/Bernstein 1971) remains one of the best. That study demonstrated that failure arose even when all the teachers were, as individuals, committed to the new innovation. Rogers correctly notes that earlier diffusion models assumed that any failure or lateness in adoption was an individual fault whilst later models give much more attention to outcomes as a conjunction of determinants (Bandura 1986).

These shifts are incorporated in the revised basic framework as shown in Fig. 3-4. It may be noted that the concept of prior conditions has been extended to take into account previous practices and perceived needs. Each of these changes is an implicit recognition of the slowness of innovation in set-

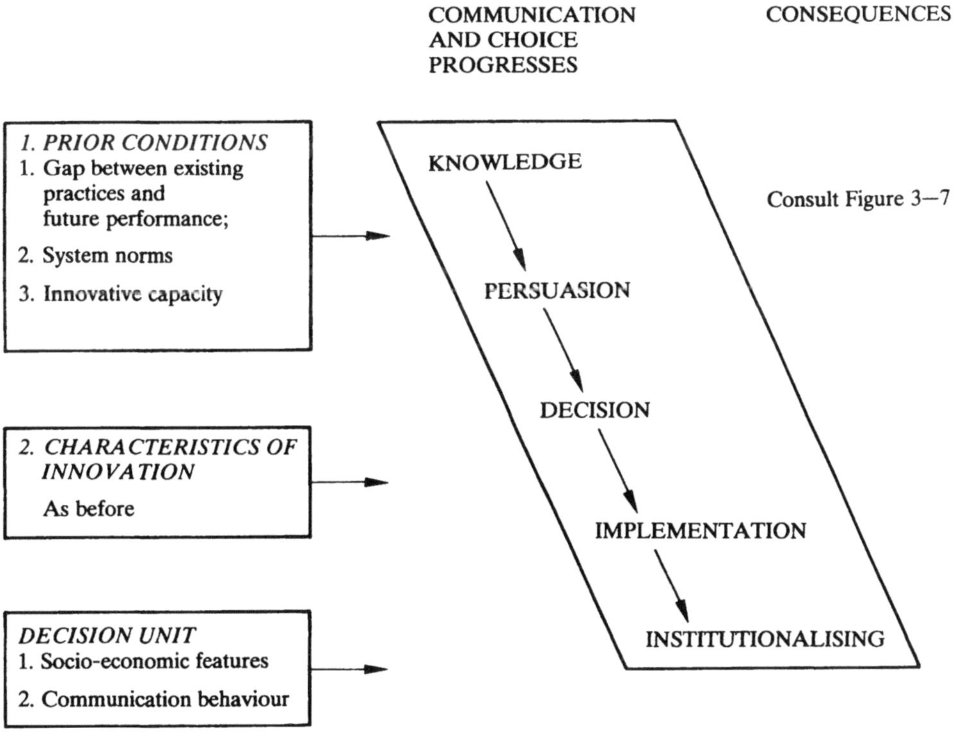

Figure 3-4. Diffusion of innovations framework: Rogers II

tings where the kinds of administrative structure found in the agricultural agency are absent. After *Rogers I,* many studies were undertaken in settings where there was an absence of a diffusion infrastructure to promote the innovation; examples would include the introduction of family planning methods into India (see Rogers/Shoemaker 1971). In effect Rogers' thinking has gradually converged towards that of the later Hagerstrand. Each arrives at the recognition of the difficulties of altering individual trajectories.

Figure 3-4 also indicates a revision in the stages of decision-making. The knowledge stage occurs when a decision-making entity is exposed to the existence of an innovation and becomes slightly aware of the shape and uses of the innovation as they perceive it. Persuasion is when a definite attitude is taken towards the innovation. This may be positive or negative. For example, in large corporations there are likely to be boundary-spanning executives who act as gatekeepers on innovations. Our studies of a British car firm revealed that its key gatekeepers took a very hostile view of the innovative form of assembly process constructed by Volvo at Kalmar. After knowledge and persuasion comes the decision to adopt or reject. As will be noticed Rogers has dropped the mental evaluation stage and replaced it by the persuasion stage. Then come the stages of implementing the innovation and reinforcing the decision. As shown, the revised model permits the decision to adopt or reject to be continuously reversed.

Is this model an improvement on Rogers I? Certainly there are some important extensions. For example, the process is extended to include implementation and that extension provides a very significant bridge into the literature in organization studies (see Sects. 3.3.3 and 3.3.4). Rogers recognises the importance of the bridge and has set out to cross it (Rogers/Rogers 1976, Chap. 6). A major contribution has been made by defining the existence of a bridge, however it cannot yet be said that there is much traffic flowing between innovation studies and organization studies.

Rogers' revision to the stages of the decision process treats mental evaluation in an interesting and possibly inconsistent manner. For example, the continual citation of the work on learning theory by Bandura (1977; 1986) ought to reinforce the importance of mental processes. Also the whole developments in computer technologies of simulation, such as computer aided design (CAD), provide a potential in the persuasion stage which requires further explication (see Part IV).

Recognition of the structural aspects of innovation has drawn attention to diffusion networks, especially to the work of Granovetter (Granovetter 1973) and his analysis of the importance of weak ties between different networks in promoting new ideas. Figure 3-5 shows two separate networks – A and B – each characterised by strong internal relationships, but with only one weak relationship linking A and B. The theory of weak ties suggests, on the basis of various researches, that crucial information may flow down such relationships. It is pertinent to apply that network perspective to the various subfields making up the study of the diffusion of innovations! It has already been noted that until the *Venice Conference on Innovation Diffusion* (1986) there were few examples of cross linkages even with that subfield of enquiry (see Chap. 2, note 13).

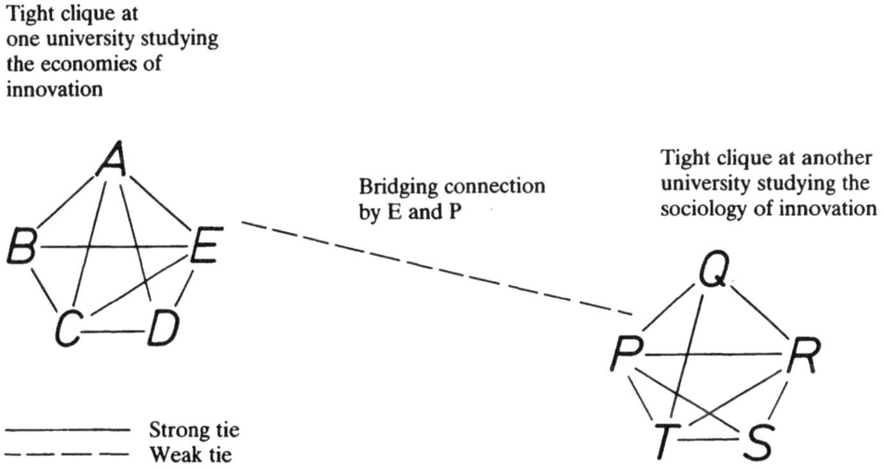

Figure 3-5. Networks and weak ties (Based on Granovetter and Rogers)

Second, Rogers II makes an important extension of the basic framework backwards into the processes by which the innovation is generated (Rogers 1983, Chap. 4). It would be simple to dismiss that contribution as the belated recognition of what economists and business historians have been doing for years. However, such a criticism would wholly restrict the necessary processes of synthesis which Rogers' analysis attempts to sustain.

There is no doubt that the early paradigm took the existence of an innovation for granted because the agricultural agencies which sponsored the development of the paradigm were primarily concerned with diffusion rather than generation. In Rogers II that omission is addressed. However, its treatment is not convincing. Rogers tends to perpetuate the notion that innovation is a simple process driven by needs as shown in Fig. 3-6 (Rogers 1983: 136) when, as will be shown in Chap. 4, the locus of initiative is quite variable and frequently shifts in an iterative manner. Further, it must be argued that Rogers' stages of generation for a radical innovation (Rogers 1983: 142)

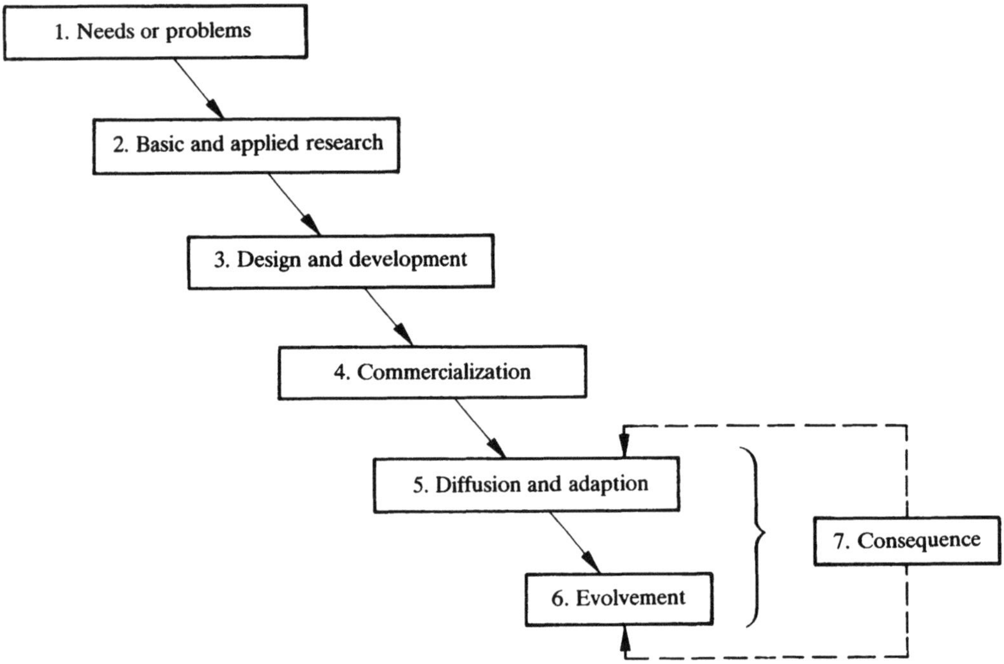

Figure 3-6. Main phases in innovation development (Derived from Rogers E.M., 1983, p 136 and slightly extended)

are an oversimplification. In effect these aspects are more fully dealt with in the established texts of Freeman (Freeman 1974) as well as in the collection of essays on technological innovation by Rosenberg (Rosenberg 1982). Also there are some excellent case studies of generation such as development of float glass by Pilkington (Barker 1977; Quinn 1980).

Third, in an earlier review of Rogers I by Gross and colleagues they criticised the original model for its failure to deal with the implementation processes (Gross et al. 1971: 22). Rogers II, makes an important effort to examine the organization (cf. the individual) as a locus for the adoption of, and implementation of, innovations. Clearly in an organization there is an authority structure and a political system which affect the selection of innovations onto the agenda for consideration, whilst separate and complementary features (e.g., managerial resistance) may influence the implementation of any innovations. Rogers rightly emphasises that the investigation of innovation in organization studies has been beset by a simple model of structure and by cross-sectional studies which consistently produce inconsistent results (see Sect. 1.2) (Mohr 1982). The main thrusts of Rogers new contributions are to treat organizations in terms of network theory (see earlier), and to extend the adoption process as examined in Sects. 3.3.4 and 3.4.

Fourth, *Rogers II* observes that imitation may not occur because the user may alter the innovation or only introduce selected elements (Rogers 1983: 174-184; Eveland 1979). This is a major and significant opening in the original framework and is a development which, when fully developed prises open the whole basis of the imitation paradigm. Rogers refers to the process of the modification of an innovation by the user as re-invention.

Much of the discussion of re-invention by Rogers is directed at noting the limitations of the previous paradigm of imitation, or, in constructing a chronology of the evidence for re-invention. Rogers view is that re-invention is not necessarily bad (Rogers 1983: 178) and that it occurs in the implementation stage because innovations are sometimes too complex or that an innovation may be too unspecific for each application. Also, innovations may be used to handle a raft of problems rather than a single problem. The user's sense of competence and pride might also lead to re-invention. As may be noted these reasons tend to understate the extent to which an innovation may require modification. They also understate the extent to which all innovations are progressively modified through their lifecycle. Even the ubiquitous Ford T – all in black and all alike – was continuously refined throughout its 17-year life course; the alterations are evident in the increasing weight of the car (Abernathy 1978). I shall argue that re-invention has been treated too narrowly and in a too circumspect a manner.

Fifth, the new approach gives much more attention to examining the consequences of adoption, but not of rejection. The adoption of an innovation and its implementation may have unforeseen consequences, as is illustrated by the case of the adoption of wet rice growing on the island of Madagascar. The point is made by reference to the earlier studies of Linton and Kardiner (1952) and is schematically summarised in Fig. 3-7 which shows that over three generations the whole structure of governance, family life and individual life chances were considerably altered. The example is a vivid one, but does tend to attribute too much emphasis to wet rice and too little attention to the facts of former colonial rule and the incorporation of Madagascar into the global economy as a peripheral area (Wallerstein 1980).

To summarise: Rogers II makes an important series of shifts all of which severely qualify the imitation paradigm and several of which provide open linkages to the processual school of organization studies. Yet, overall, the tendency is to retain a good part of the underlying architecture of the original paradigm. The requirement is to break out and to separate imitation from re-invention.

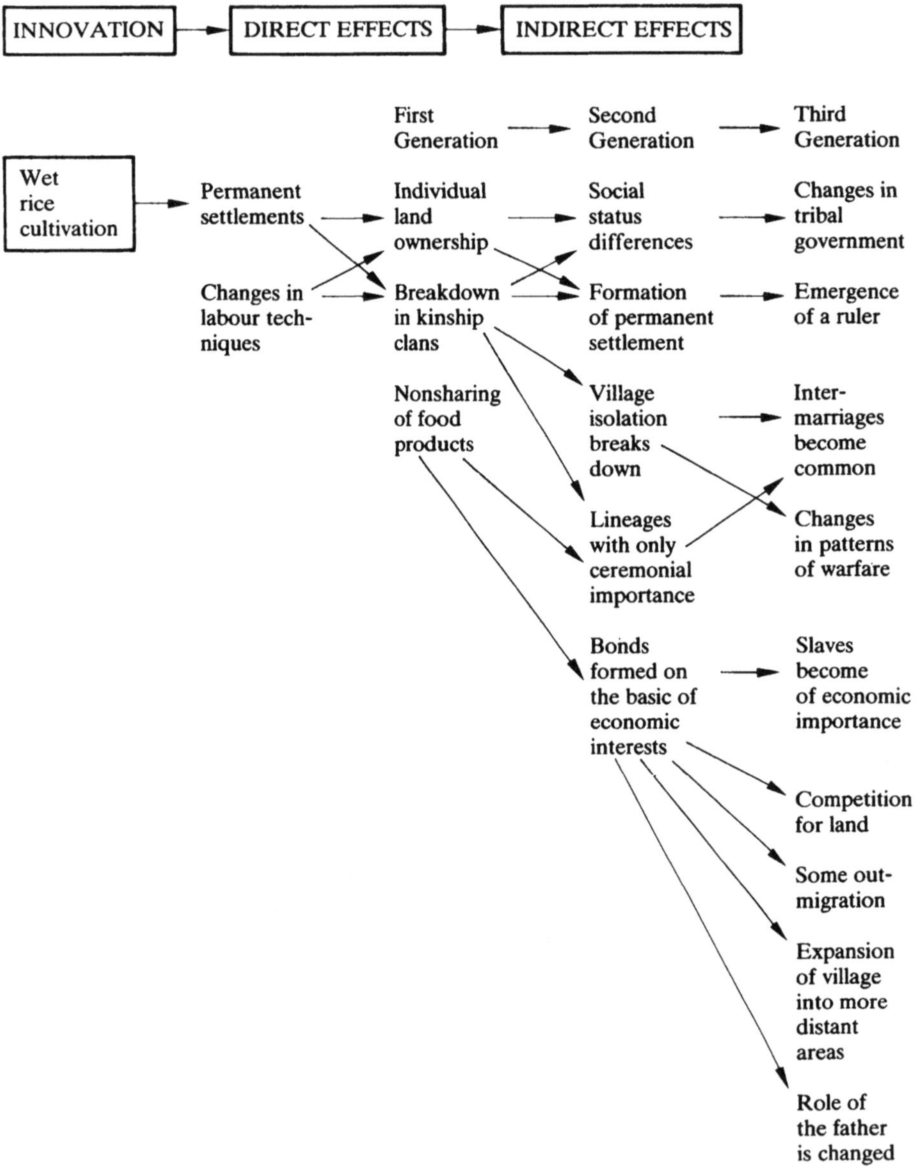

Figure 3-7. Innovation impacts: wet rice in Madagascar

3.3.3 Initiation and Implementation

The processes of initiating an innovation and then implementing the chosen innovation require different forms of structuring according to Zaltman and Duncan (Zaltman/Duncan 1977). Their widely cited interpretation is very structuralist and consists of two stages: initiation and implementation.

The initiation stage itself consists of several substages: awareness, attitude formation and mobilising support. The awareness stage parallels the first part of Mintzberg, with the notion of performance gap (Rogers/Rogers 1976, Chap. 6). Once there is awareness then successful initiation rests on the formation of various aspects of attitudes: the perceived need and openness to change, the perceived potential and perceived capacity in relation to that potential, the degree of commitment. The role of anyone seeking to achieve the innovation is to mobilise energies which are favourable to the proposed solution, possibly by developing comprehensive control concepts which have a wide ideological appeal and legitimacy across a diversity of factions.

In the initiation stage the best results will be achieved when the prevelant form of structuring emphasises decentralization, informal interaction and communication and permits a high complexity of meaning to be encoded and organized. The preferred structuring corresponds to the organic management systems identified by Burns and Stalker. High complexity should result in the involvement of a wide variety of forms of expertise, each of which can provide specific inputs to the overall processes. Yet that level and type of complexity also contains dangers to conflicts and to interpersonal antagonisms. Once the decision to innovate is made, however, Zaltman and Ducan maintain that the structure must alter. Their conclusion suggests that the structuring which is appropriate for initiating should be considered to be one which is collateral or parallel to the forms of structuring which are normally in use.

The implementation stage is likely to contain a shift from a trial use of the new innovation to its actual installation. They argue that implementation requires more direction, so centralization and formalization are increased and the level of complexity is simplified. The formalization is required to provide specific procedures which facilitate the implementation. The whole framework is shown schematically in Fig. 3-8.

The use of different structures presents certain potentials for conflict and therefore it becomes important for adequate conflict-resolution to exist. Also, the structural repertoire should embody widely recognised switching rules permitting the alteration of the structure to occur smoothly. There are few examples of such situations (Clark, P. A. 1985: 56-59).

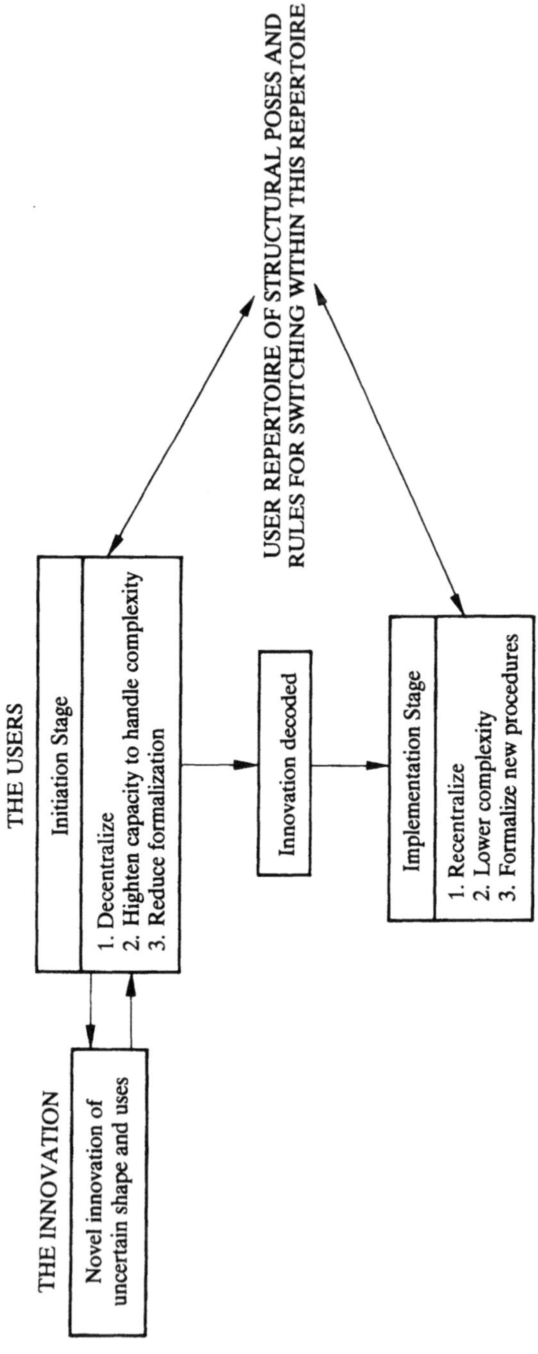

Figure 3-8. Two-stage innovation: initiation and implementation

3.3.4 Implementation: Current Normative Model

The major current model of the corporate user is a normative model of implementation originally constructed by Eveland (Eveland 1979).

The preferred approach starts from the assumption that in the early stages of the diffusion of a new innovation, its shape and uses contain a high degree of ambiguity for the supplier and for the potential users[17]. The choice and implementation of innovations by corporate users is a process of activating the internal regimes of paradigms about technical, material and organizational knowledges (see Chap. 5). There ought to be a well-developed capacity for the absorption of uncertainty. Firms will vary in the extent to which their decision-making can adequately match the equivocality required to encode the shape and uses of the innovation. A comparative study of six enterprises which chose to undertake a complex organizational design project with external consultants demonstrated that all the corporate users perceived the shape and uses of the same 'organizational design' package in a similar way (Clark, P. A. 1975), but that only three enterprises possessed the requisite capacities for successful introduction and implementation. Those capacities were not predicted by earlier theories[18].

According to Eveland the choosing process in successful users will begin with a vague general concept of the innovation rather than a specific blueprint and may lead to a modification of the original innovation as reported later in the hospital-design example. There appear to be several phases in successful implementation. The model by Eveland is summarised in Fig. 3-9. There are two axes. First a flexible time dimension. The reason for this is that Eveland correctly prefers to use a temporal framework in which events embedded in the unfolding process are used as the indicators of timing[19]. It would be unwise, misleading and practically useless to impose a rigid calendar or clock-based system. Along that same axis there is depicted a continuing process of agenda setting. Second, the vertical axis shows the degree of specificity in the innovation from the point of view of the user. The scale runs from low specificity to high. Combining the two axes we see that the whole implementation process commences in the bottom left-hand corner when there is very low specification of the innovation by the user. There are a set of decisions to be made through continuous agenda-setting in four areas: matching the innovation to the user's needs; redefining the innovation so that it can be absorbed, including some possibility of re-inventing; move-

---

[17] Eveland and Rogers (1980), see in particular the discussion of the shape and use aspects of innovations

[18] For example, the earlier theories of Rogers (1962)

[19] The event theory of time is explained in Clark, P. A. (1985)

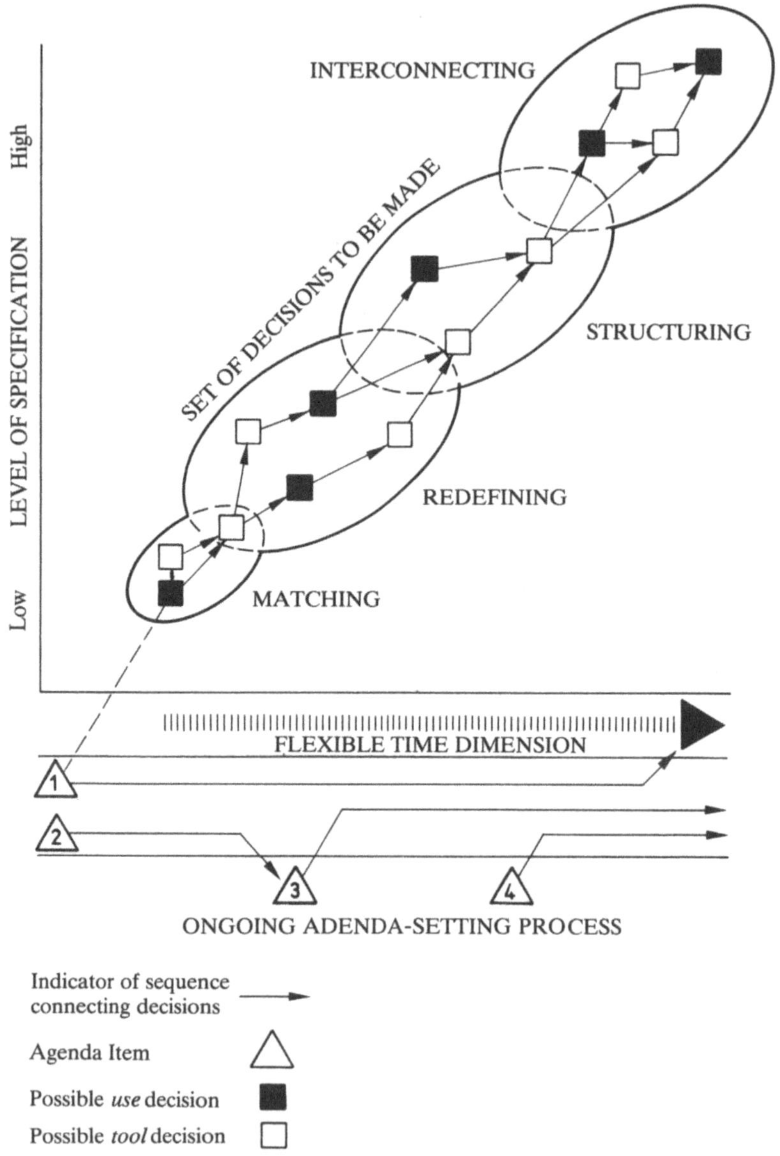

Figure 3-9: Innovation implementation: Stages (Acknowledgement: J. D. Eveland)

ments to structure the innovation into the ongoing processes of the user; and, finally, the interconnecting of the innovation with existing (and altered) control systems and situated practices. That last stage is the institutionalising of the innovation. The figure distinguishes three broad types of decision about the innovation: getting the innovation onto a specific agenda, discussing the uses of the innovation, and discussing its shape. Figure 3-9, which is prescriptive and descriptive, is a heuristic device to aid practice and research.

Agenda-setting is a continuous process of entering different decision arenas and establishing connections between them. The role of senior executives in that early stage probably includes a fair degree of devolvement of details to middle and junior executives. At one time it was thought that these activities were highly fragmentary and broken up into discontinuous episodes, but more recent longitudinal studies show that there are often strong linkages between such diverse activities as short phone calls, corridor meetings, formal scheduled meetings and open debates (Clark, P.A. 1984: 375–383). Their patterning might be as in Fig. 3-10. The agenda-setting stage is a precursor to the next stages.

Matching is when the general problem from the agenda making and possible solutions are linked, albeit in a way which is later recast. One possibility is selected for further attention.

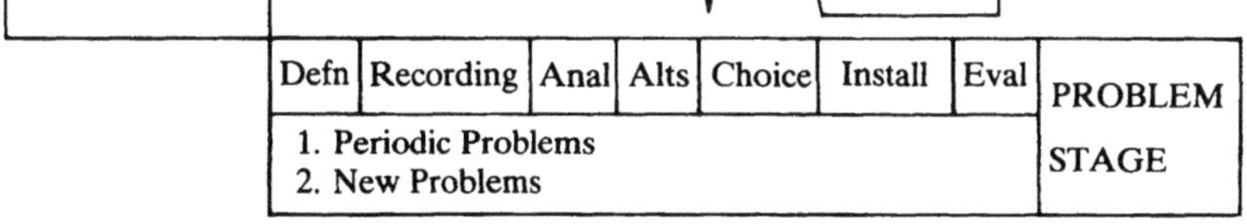

Key: Defn = Problem crystalization
Recording = Collection of relevant data and audits
Analysis = Interpretation of solutions
Alts = Range of alternative couplings of problems and solutions
Choice = Bargained decision
Install = Introduce solution
Eval = Post-audit of decision

Figure 3-10. Managerial activities profiled over time and by level (From Clark P.A. 1984: 380)

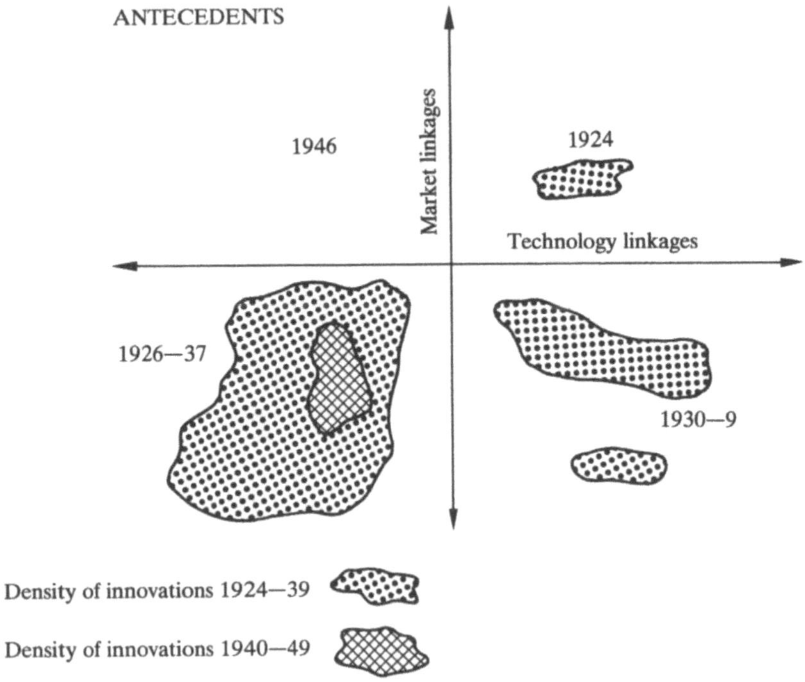

Figure 3-11. Technology ferment: Chrysler 1924–1939 and 1940–1949

Redefining is the occasion when the relevant attributes of the proposed innovation are translated into terms which members of the organization can understand. This stage is a process which may actually leave the innovation intact whilst altering its initial definition of the shape to fit with local values. In many respects it involves the making of meanings which are compatible with the cultures of the user. The relevant forms of organizational language will be invoked and edited to create a commonly shared grammar (Weick 1977, Chap. 8).

Structuring is the stage when members of the user firm engage in a more or less collaborative activity to establish the innovation within the procedures of the firm. The difficulties they face will vary according to the extent to which the innovation reinforces existing practices or requires their significant alteration. That crucial point which emphasises that adoption may only slightly alter pre-existing arrangements can be illustrated from the work of Abernathy and Clark (Clark, K. B. 1985) as shown in Fig. 3-11. The degree of alteration caused by the innovation is depicted on a scale of low to high on each two axis:

1. The linkage between the user and its markets
2. The internal linkages for the user of its production-technology and planning systems

The illustration refers to the situation of the Chrysler company in the 1930s and 1940s when its internal technical innovation was having a very successful impact on its market share. As may be seen, some innovations reinforce existing procedures whilst others create a ferment which has to be absorbed. Structuring, when successful, can be seen in the emergence of new practices.

Interconnecting is the stage when there is a process of making the structures for the innovation fit with those of the whole user organization.

These stages may take a great deal of time. For example, in an early attempt to install the autonomous work group concept in factories and offices, the change agents often underestimated the problems of getting the new innovation "wired" into the pre-existing operating procedures of the user. That process might well take as long as two years[20].

The stages are sequential and yet will involve backtracking. At the end of the process the innovation will be incorporated. That process may have included the re-invention and modification of the innovation, though it may alternatively have left the innovation "intact" whilst changing the users feelings and perceptions of it. It is likely that the five stages will occur differently in each user system. More so when there are no common agencies supplying the innovation.

3.4 Marketing and Infrastructure Model

3.4.1 L. A. Brown

The marketing and infrastructure model complements the models of Rogers by elaborating on the linkages between an innovation and its potential users in either firms and/or households. The key focus of the market and infrastructure model proposed by L. A. Brown and associates (Brown 1981) is twofold:

1. On the absence or presence of a specialised agency for marketing a service or product
2. On the differences between firms and households as targets for agencies

---

[20] Very few studies of the autonomous work group indicate the practical problems which are involved; relevant sociopsychological research is being directed by Toby Wall and colleagues at the MRC Research Unit of the University of Sheffield, England

I shall concentrate upon the role of the agency.

Clearly the Agricultural Extension Agency was an example of a marketing agency, so what Brown contributes is a more intensive examination of the institutions on the supply side of innovations. It is relevant to this book to precede the examination of Brown's analysis with two comments. First, the fact of having to give such attention to the marketing of innovations reinforces the claim of this book that imitation requires assistance and skill. It cannot be taken for granted, especially not in firms. Second, there is considerable evidence and interpretation which will be more fully examined in Part IV to show that in North America the marketing domain has been more highly developed than in Britain, both as a cultural predisposition and as a set of situated practices (see Chaps. 9, 10) which are routinely activated when the opportunity occurs. For example, it was noted by observers at the Second International Conference on Quality of Working Life (QWL) in Toronto[21] that many of the sessions examining new concepts of work design were followed by informal caucusses of organizational consultants who discussed the marketing of the new concepts and even attended to the allocation of territories of influence. Such happenings are not at all unusual at North American conferences, yet they are rarely observed in Britain (or, in continental Europe). In other words, the establishment of agencies has been a distinctly American cultural capacity.

Brown is concerned with two features of agencies. First, they provide a point of distribution for an innovation in a given area. For example, the offices of the Boston Consulting Group in Paris provide a base for the marketing of its consulting concepts to firms and branch-plants in Europe[22]. Second, the temporal sequencing of the establishment of agencies will influence where and how an innovation will be available. For instance, in an important study, Chandler claims that the multidivision format of company organization was discovered by American firms in the 1920s (e.g., du Pont, General Motors) who recognised its appropriateness for firms which were developing portfolios of increasingly unrelated products in diverging geographic areas (Chandler 1962). Presumably American consultancies played an important role as agents of diffusion. Certainly, the later establishment of McKinsey Consultants in London had an identifiable impact on the spread of the diffusion in the British context[23]. McKinsey established procedures which were designed to present the innovation to firms.

---

[21] Quality of Work Life Symposium, 1981, organized by Harvey Kolodney and colleagues from Toronto University; I am indebted to Keith Newton from the Economic Council of Canada for these observations

[22] The experience curve concept is closely associated with the Boston Consulting Group

[23] Channon (1973; 1978); Hannah (1983; 1976)

According to Brown, the majority of innovations are promoted by specific agencies. There are three broad types of agency:

1. The centralized agency which is itself the locus of developments in the innovation and plays a key role in the licensing of new users. These will typically be profit-oriented agencies whose major decisions will include the choices of a sequence of urban centres in which to locate their goods/services (i.e., hierarchy effect). Their decisions will be based on an analysis of alternative centres.
2. The decentralized agency with a coordinating structure of propagation. For example, the spread of Master Charge services through Ohio and West Virginia in 1969–1970 (Brown 1981: 75ff.). These agencies influence the pattern of diffusion whilst leaving the user with some room for small-scale refinements.
3. The loosely linked decentralized agency without a coordinating propagator (e.g., the Planned Parenthood movement in the USA).

These three forms of agency cover a wide range of innovations and confirm Brown's suggestion that many innovations are deliberately marketed. The three points also provide a checklist for refining the existing tendencies and it may be noted that there is an extensive body of knowledge underpinning their operation.

Diffusion agencies have to devise strategies to develop organizational capabilities and to create an infrastructure. Further, they have to package the innovation, set a price, provide promotional material and select favourable market segments. The development of an infrastructure may include the handling of issues of government regulation, both actual and potential. In these ways the channels of diffusion are shaped by the agencies.

The existence of diffusion agencies is likely to affect the five characteristics of the innovation identified by Rogers as affecting diffusion. The skills of the agency can do a great deal to transform otherwise idiosyncratic practices into marketable items. For example, the discovery of just-in-time has led various consultancies (e.g., Cooper Lyband in Britain) to develop an expertise in their matching to specific contexts.

The roles of the agencies raise two points. First, unlike the Agricultural Extension Agencies, the profit-centred agencies will not necessarily provide data on their innovations which is neutral. The agencies may well highlight those features favourable to purchase. Second, there are some innovations which are of interest to firms for which there are no obvious agencies.

3.4.2 Von Hippel

Von Hippel's contribution with respect to industrial innovations (e.g., raw materials, assembly lines) is for corporate users rather than for households and his argument is that there is a requirement for devising methodologies which reveal customer needs and their active role in creating the solution content of future innovations[24]. An earlier study had concluded that the detection of customer needs was a crucial factor in explaining why some suppliers were more successful than others (Rothwell/Freeman 1974). The objectives of von Hippel are to construct strategies and methodologies which build on what is known about how customers actually develop innovations. In other words, his interest in the corporate customer is very much from the marketing point of view. Von Hippel contends that the subject matter of marketing has developed with strong emphasis on consumer marketing to households and so there is a requirement to distinguish between the differences of that kind of target to the industrial corporation and a customer. The corporation differs in at least three ways: buying decisions usually involves a multiplicity of people at different levels and in different functions, the products are often highly systemic (e.g., packaging lines), and there is direct interaction between the customer and the suppliers (von Hippel 1982b: 240–241). Yet, the models of the customer which academics have constructed tend to treat the corporate customer as passive. There is, von Hippel states, a need for a customer active paradigm (CAP). Moreover, there is also a requirement to examine who appropriates the intellectual property rights when the initiatives of the customer are involved (von Hippel 1982a).

Von Hippel observes that, in the conventional supply-side paradigms for industrial goods, it is the manufacturer who is assumed to be active in invention, commercial development and final marketing (von Hippel 1976). But how accurately does that model account for what happens in practice and what might be best for the supplier? By a secondary analysis of already published studies (von Hippel 1982b: 244–251) it can be shown that there is firm evidence of a degree of customer activity which shows that some customers are quite capable of building and marketing products on their own. By examining these cases and by constructing a simple five-stage scale (von Hippel 1978, Fig. 2), it can be shown that the customer is most usually active in the early stage of defining a need and introducing the content of a good solution. Von Hippel therefore concludes that a consumer active paradigm (CAP) should become part of the general approach of the marketing and infrastructure models. He notes that CAP is only applicable when the customer is overtly aware of the need and possible solution and that this is mostly

[24] Von Hippel (1976), see the references under note 6

with new needs (von Hippel 1978: 258). There are various combinations of actions which the sophisticated customer might undertake: design and construct a new system internally, design a new system and commission a manufacturer, commission a design and manufacture internally, and so on. For example, in the carefully documented study of the internal design of a new coal-cutting machine, the British NCB then marketed the system through joint projects with manufacturers. The latter also benefitted from the additional internal developments which colliery workers suggested (Townsend 1980). There are, as has been indicated by von Hippel, more often situations where the manufacturer can exploit the benefits.

A recent study of 34 medical innovations in the state sector of the British National Health Service illustrated the situation where the customer undertakes considerable development and the manufacturer is able to exploit the benefits (Shaw 1986). In that case, the customers were the departments responsible for clinical testing and for trials for new equipment. Their internal specialists worked on the development of basic and major breakthroughs in collaboration with manufacturers. There was considerable joint work: for the development of prototypes there was about 70% joint activity, slightly less for product development (under 60%), less again for joint marketing (under 40%). The nearer the market the more the manufacturer becomes the most active. In that case the manufacturer was able to appropriate many benefits by having the costs of development defrayed onto the funds of the state. There were two ways in which the manufacturer obtained benefits. First, by getting access to clinical trials which were paid for by the NHS, i.e., by getting access to a key peer group whose stamp of approval would open other markets in the medical world. Second, by obtaining expertise, i.e., by getting orders for prototypes, having the research funded by the NHS, and gaining access to various institutional linkages.

To counterbalance that example, there is an example of the customer moving into the commercial exploitation of internally generated innovations by doing his own marketing through a new internal business agency. This example is from another British state-owned firm, British Aerospace (Foxall/Tierney 1984). There the company utilised its design skills and learned from an earlier failure to undertake commercial exploitation (e.g., advanced aluminium-cutting tools) and succeeded later with a number of projects including a self-normalising ultrasonic test probe.

These examples indicate varying degrees of activity within the CAP. Several writers have claimed that their case studies represent extensions to the original model (e.g., CAP1, CAP2, CAP3 and so on). However, the major point made by von Hippel is captured in the original concept. Further elaboration is undertaken in the following chapter.

Another aspect of von Hippel's approach to the customer is in the analysis of customer behaviour as an indicator of future market potentials for a particular industrial product (von Hippel 1983). Three types of customer are defined: the routine customer interested only in minor variations of existing products; the high-need customer who might experiment with major innovations; the 'lead' customer who has direct experience of a new situation and can be used as a source of learning for the manufacturer/supplier. Von Hippel recommends suppliers to segment customers in order to identify the lead customers who can be used to forecast future requirements. For example, fasteners are an important industrial product for the auto industry. However, it is their use in certain sensitive parts of the aeronautics industry which is most likely to provide information of future demand.

The thrust of von Hippel's message is that manufacturers need to adjust themselves so that customers select them to market industrial products. That process can be quite complex, because as has been shown, the definition of internal needs can be very unstable and can change quite radically within short periods of time (von Hippel 1983). Any redefinition of needs will usually lead to changes in raw materials, so tracking the needs is a difficult task.

Von Hippel appears to cross the bridge from the marketing model into the position of the user, but he does so from the position of the supplier, as do the other paradigms discussed in this chapter. They therefore neglect the whole area of internal research and development[25]. Another way of crossing the bridge from the supplier to the user is to take one particular dimension of innovations, the technology, and then to examine the ideas of technological trajectories. This is undertaken in the next chapter, which provides a lead into the fuller examination of the user in Chap. 5.

---

[25] See Work Organization Research Program in Chapter One note 7.

4 Technology as Process: Trajectories and Life Cycles

4.1 Introduction

Technology occupies a central position in any discussion of innovation. How is technology theorized? Is the central position of technology justified? What is the relationship between technology and other dimensions such as production knowledge and production institutions? Where are technologies developed?

The earlier domination of technical change by economists is now undergoing a double shift. First, within neoclassical economics the dominant paradigms are being revised. Second, the development of the behavioural theories of the firm and the general approach of organization studies are both providing a new framework for theorizing.

In the 1960s, during the natural history phase of the study of innovation, there were a lot of quasisurveys and truncated case studies reporting cross-sectional features of technical change. The best known studies were: *Wealth from Knowledge,* Project Sapho, Project Hindsight, Project Tracer, and the Myers and Maguiles review[1]. Their results were often insightful and interesting, yet systematic reviews (e.g., by Rosenberg) highlighted many limitations in conceptualisation (e.g., of the market) and weaknesses in measurement[2]. The natural history phase was followed by attention to the sector, particularly by Pavitt and colleagues at the Science Policy Research Unit (SPRU)[3], and by Abernathy at Harvard. The sector was the focus for a debate between neoclassical economists developing the field of industrial organizations and the school of neo-Schumpeterians. In industrial economics the focus on the market, by for example Kamien and Schwartz, is being complemented by attention to the sector as a key unit of analysis (Kamien/Schwartz 1982). Significantly, the sector has also become the focus for the Work Organization

---

[1] Langrish, Gibbons, Evans, and Jevons (1972); on project SAPHO, see Rothwell, Freeman, Horsley, Jervis, Robertson, and Townsend (1976); Greenberg (1966); TRACES stands for Technology in Retrospect and Critical Events in Science, from Myers and Marquis (1969)

[2] The best review is by Mowery and Rosenberg (1979)

[3] See studies by the Science Policy Research Unit (SPRU), University of Sussex, England, especially those by Freeman, C, Pavitt, K and R. Rothwell

Research Centre in their studies of innovation in technology and work organization[4].

In the later 1970s attention shifted from the sector onto the macro level of longwave pulsations (see Chap. 7). The concern with the notion of fifty-year longwaves embraced both large-scale economic change (e.g., "creative destruction" and de-industrialisation) as well as the interactions between sectors[5]. Consequently, it becomes relevant to distinguish between the intended and unintended consequences of state policy regulations and interventions. More recent thinking, by the French regulation school (e.g., Boyer) suggests that it is no longer possible to have growth in the economy and full employment as was the case in the previous decades (Boyer/Coriat 1986). Thus, the employment consequences of new technologies become significant[6].

Critics of the neoclassical paradigm assert that its core axioms are being revised and the earlier preoccupation with modelling the production function has foundered. They suggest that technical change is now being analysed from the perspective of the adequacy of the theory of the firm. The problem for the neoclassical paradigm lies in the role of markets in its theories. According to the orthodox view, the market provides key signals, but the behavioural theory of the firm suggests that the behaviour of entrepreneurs and nonentrepreneurs alike is based on a very foreshortened collation of signals, only a few of which are derived from the market (Mowery/Rosenberg 1979). Moreover, market signals are anticipated. Some critics of the neoclassical model now contend that the markets are an epiphenomena of decisions taken in organizations. That is a crucial issue and one which influences this book. If the market is significantly an outcome of organizational decisions, then much more attention should be given to the organizational level (see Chapt. 5) and to organization studies as a perspective. Technological opportunities do vary over time and between sectors. If managements recognise the technological opportunities, then it becomes essential to examine the strategy of the firm much more closely.

Despite the claims of the critics of the neoclassical model that its paradigms are foundering, the continuing output in the journals and in the advanced monographs indicates that the neoclassical school still considers that its perspective can be usefully adapted and refined to the task in hand (see e.g. Stoneman 1983). Hence, there is still a strong tendency amongst economists

[4] The focus on the sector now being taken in organization studies is to be welcomed, but it should be observed that the nature of sectors is being radically altered by new technologies and new materials

[5] See Chap. 7 for an examination of longwaves

[6] See, for example, Boyer (1986) and the studies by Chris Freeman and colleagues from SPRU

to view innovations as objects (see Chap. 2) and that point is very evident in the notion of generic technological trajectories as natural phenomena.

In this chapter four issues are examined. First, definitions of technology as process are examined. Second, the notion of trajectories is scrutinised and the suggestion of natural trajectories is replaced. Third, attempts to model the life course of technology and the sector are introduced and extended to accommodate work organization. Finally, attention is given to locations in which new knowledge is most likely to be generated prior to its embodiment in technologies.

This chapter takes technology as a focus for both the suppliers and users. It is presumed that technological innovations are routinely embedded in contexts which include forms of work organization.

4.2 Technology and Process

The theorization of technology has been dominated by three major studies:

First, Bright constructed a schema which considered a single hierarchy of advancements in technology from simple hand-driven tools to highly integrated machines which were self-regulating (Bright 1958); the schema is shown in Fig. 4-1. For each stage on the scale it is possible to obtain some specification of the human skills which are required to operate the technology. The focus is on the individual operative rather than upon the structure of the organization. Bright's concept has been widely adapted, most notably in the sociotechnical framework and in the analysis of deskilling by Braverman (Braverman 1974). It has also been selectively used in hundreds of research projects.

Second, is the production engineering typology of types of production system based on the degree of continuity in throughput of fluids or components (Wild 1971). The scale ranges from the situation of small-batch production (i. e. low continuity) through large- and volume-batch to continuous production. Low-batch production is treated as discontinuous.

When the scale was constructed in the 1950s certain features were obvious, but these have now changed somewhat. For example, small-batch production inevitably meant that very little technology was utilised and continuous production referred primarily to the handling of fluid items (e. g., chemicals) rather than solid items like components. Each of these features have changed so that complex technology can be combined with small- and medium-batch production (Boyer/Coriat 1986). Also, the development of robotics and other self-regulating equipment (e. g., transfer machines) means that continuous production can be achieved with components as well as fluids.

Initiating control source	Type of machine response		Power source	Level number	Level of mechanization
From a variable in the environment	Responds with action	Modifies own action over a wide range of variation	Nonmanual Mechanical	17.	Anticipates action required and adjusts to provide it
				16.	Corrects performance while operating
				15.	Corrects performance after operating
		Selects from a limited range of possible prefixed actions		14.	Identifies and selects appropriate set of actions
				13.	Segregates or rejects according to measurement
				12.	Changes speed, position, direction according to measurement signal
				11.	Records performance
	Responds with signal			10.	Signals preselected values of measurement (includes error detection)
				9.	Measures characteristic of work
				8.	Actuated by introduction of work piece or material
From a control mechanism that directs a predetermined pattern of action	Fixed within the machine			7.	Power-tool system, remote controlled
				6.	Power tool, program control (sequence of fixed functions)
				5.	Power tool, fixed cycle (Single function)
From man	Variable			4.	Power tool, hand control
				3.	Powered hand tool
			Manual	2.	Hand tool
				1.	Hand

Figure 4-1. Technology scale

A similar production engineering scale was used by Woodward (Woodward 1965). Her studies of manufacturing establishments seemed to demonstrate that in the most profitable firms the degree of variability in consumer demand affected various aspects including the following: batch size, the production-planning mode, the technology and the organization structure. Her conclusion is illustrated in Fig. 4-2 where it is emphasised that the technology is simply one facet of an overall configuration of dimensions. The model, which is intentionally prescriptive, posits that best practice is achieved by getting a congruence across the dimensions. It is not a model of technological determinism[7].

Third, is the Aston Programme (Hickson/Pugh/Pheysey 1969) scale of operations technology which consists of a suite of dimensions as shown in Fig. 4-3. The objective behind the approach was to create a single comprehensive scale. It seemed that the objective was achieved on the original developmental application, but that has not been confirmed in subsequent replications[8]. Even so, the scale is the single most frequently cited scale and it has also been widely used in various studies.

These three examples tend to be utilised with cross-sectional studies rather than longitudinal studies of innovation. There is considerable debate and confusion about their value[9].

Longitudinal studies involving technology tend to be either a single case study or to be highly abstracted 'before' and 'after' comparisons.

The former can provide a great deal of insightful material, especially on the consequences for work organization of the introduction of certain new technologies. However, the absence of comparisons is a severe restriction. For example, a detailed study of the introduction of new steel-making techniques into a British firm failed to compare the emergent technology and forms of work organization with competitors in North America from whom the technology had been imported (Scott/Halsey/Lupton/Banks 1956).

The 'before' and 'after' studies tend to overstate the significance of technology because they treat the new technology as the only source of change in the situation[10]. Thus, certain studies of the introduction of electronic point of sale (EPOS) in retail outlets have revealed that many changes in supermarket

---

[7] As may be seen from Fig. 4-2 the market dimension is quite central in what is a normative and prescriptive framework. Woodward's critics constantly fail to recognise this point

[8] In a replication study based in the Canadian textile industry the quality of the scale was shown to be very poor, see Clark, P. A. (1981)

[9] There is an enormous literature on the confusing results from existing scales of technology

[10] The 'before' and 'after' format to research design tends to exaggerate the extent to which changes should be attributed to technology; compare the studies of technical change conducted during the 1960s and more recent analyses of technical change in the service sector

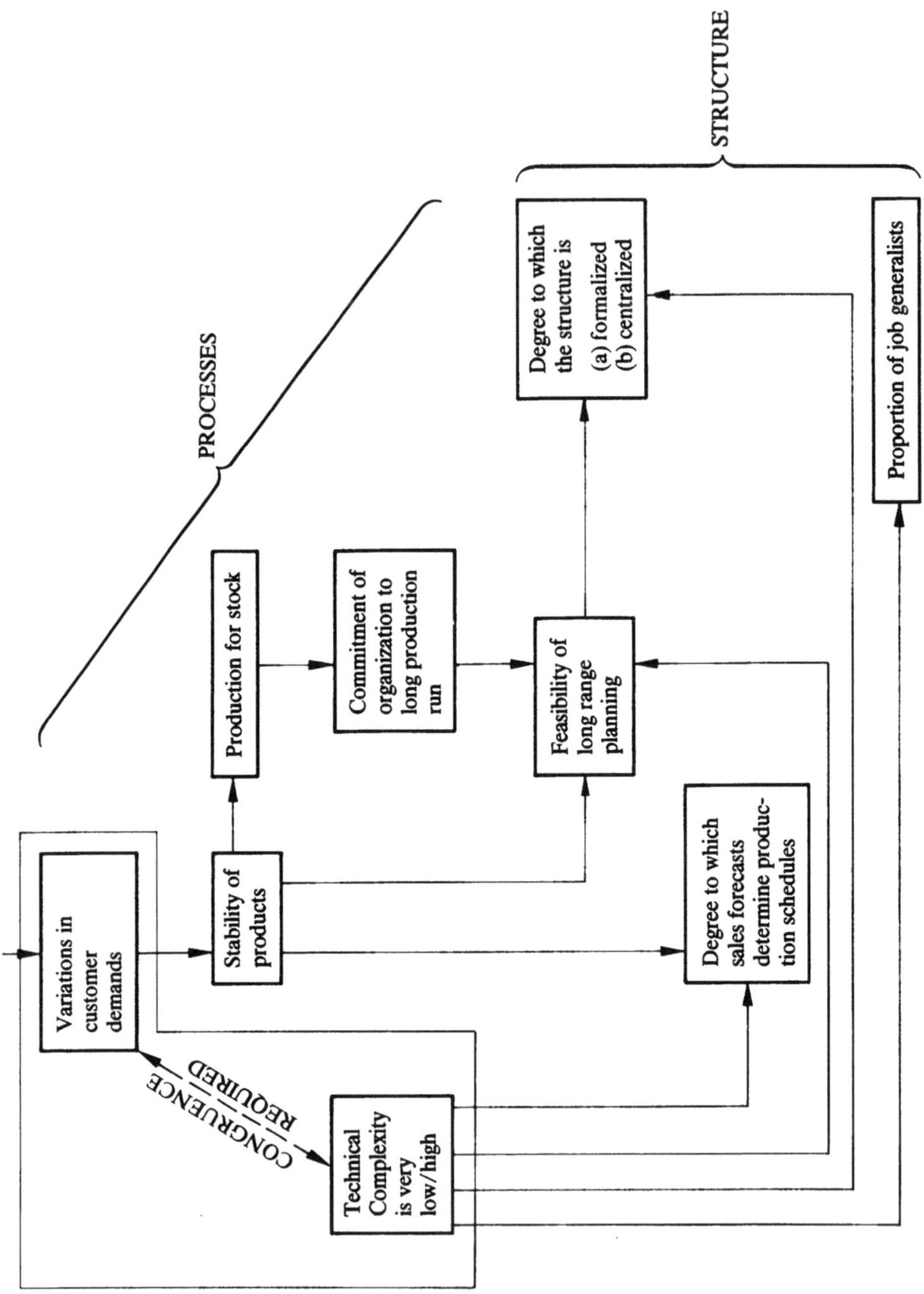

Figure 4-2. Consumer variability and organization structuring (From Woodward)

DIMENSIONS	1	2	3	4	5	6	7	8	9	10	11	12
1. Mode of technology	–											
2. Range of technology	34	–										
3. Evaluation by technology	01	−29	–									
4. Interdependence between segments	30	−04	−06	–								
5. Continuity of workflow	−00	−07	13	−20	–							
6. Equipment single purpose or not	30	−11	06	15	30	–						
7. Sequence of operations fixed	12	−27	20	13	26	31	–					
8. Buffers in the workflow	−07	−07	−04	−18	19	10	22	–				
9. Single source input	−00	−13	24	−31	34	08	21	15	–			
10. Possibility for rerouting	−15	−15	07	03	−04	09	35	14	−16	–		
11. All stops if break down	−07	26	02	07	−06	12	10	24	−04	28	–	
12. Some stops if breakdown	29	−33	28	09	18	17	01	−18	−24	−14	−30	–

* Pearson product moment correlations (N = 47)

Figure 4-3. Technology: Aston framework

operation have altered concurrently with the period within which there were introductions of EPOS. However, few studies provide a detailed causal analysis.

An interesting and promising variant of the before/after framework is the study of the introduction of three similar technologies into a number of different European countries[11]. That research suggests that the use of the technology differs systematically across countries[12].

Many attempts to construct scales for examining technology are highly static and much too one-dimensional. They are therefore not useful for investigating technical change as a process. However, the schema proposed by Bell does provide a fruitful way of constructing a synoptic viewpoint on the evolvement of production technologies (Bell 1972). I shall examine the features of the framework, look briefly at its use in a longitudinal study and provide a sketch of its application to the situation of coal mining in Britain (Townsend 1980).

Bell identified three independent dimensions for examining production technologies:

1. Equipment which transforms a raw material or alters workpieces: *a transforming machine*
2. Equipment which transfers raw materials or workpieces between segments: *transfer machines*
3. *Control machines* which can be applied to either or both of transformation machines or transfer machines

The usefulness of the three dimensions is that it is possible to examine whether their trajectories of evolvement interact with the adoption of specific forms of work organization, types of production knowledge and production institutions[13].

Coombs has used the threefold schema to trace their historical evolvement as shown in Fig. 4-4. Although the detailed patterning is uneven, there are general tendencies. Transforming machines were developed during the early part of the eighteenth century around changing forms of motive power. Initially, with water power, the machines were at a fixed point but with the development of electrification their potential for mobility was greater, though

---

[11] The European program is known as Micro Electronics in the Service Sector (MESS) and is examining three areas: hospitals, banking and retail; the focus is on the introduction of specific technologies such as electronic point of sale systems in supermarkets

[12] I am indebted to John Child and Ray Loveridge for their interpretation of the initial data from the MESS project

[13] Compare Bell (1972) with the work of Touraine as discussed later in the chapter

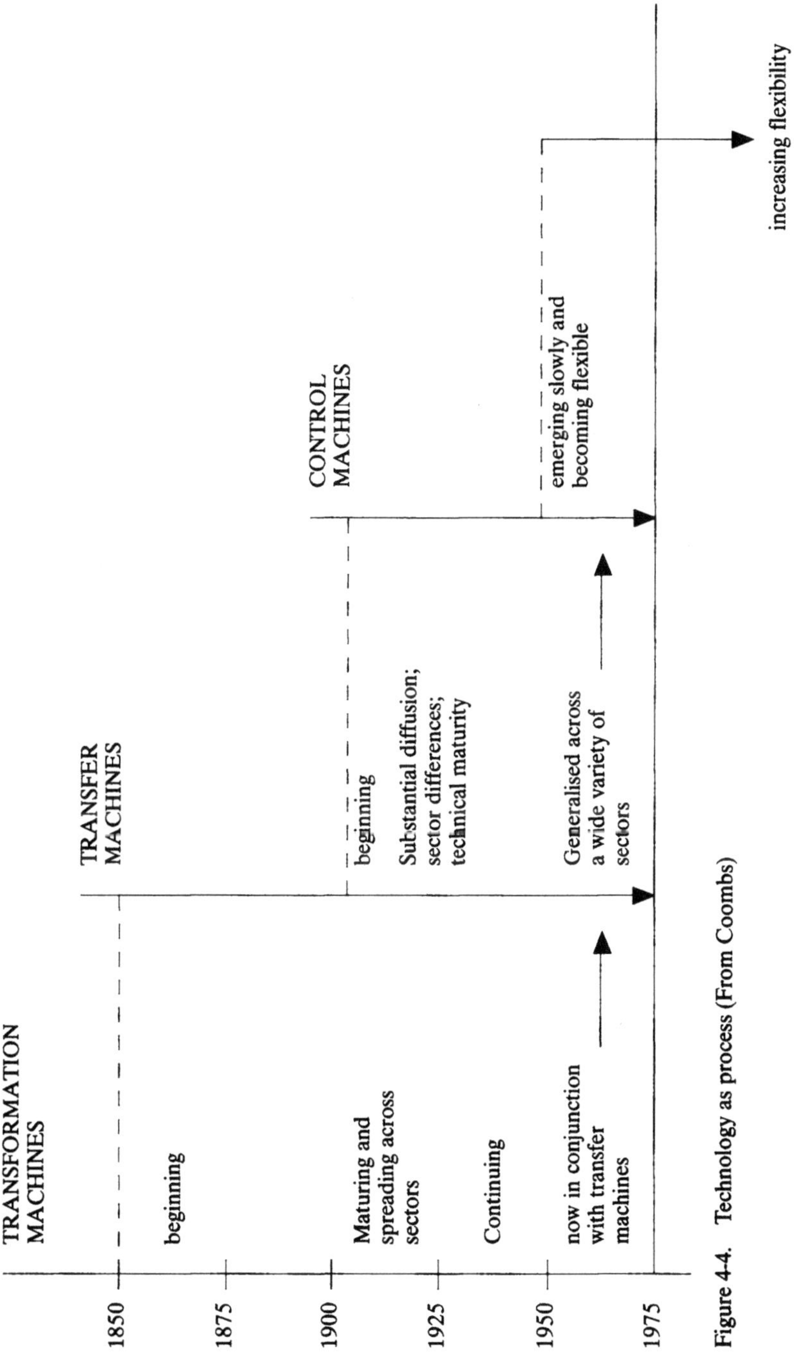

Figure 4-4. Technology as process (From Coombs)

this was not always exploited appropriately, particularly in Britain[14]. Gradually transforming machines increased in their speed and scale. For example, the cigarette-making machine developed by Bonsack in the 1870s vastly increased the rate of output and that tendency has continued[15].

According to Coombs the increasing power of transforming machines drew attention to bottlenecks in the flow of materials between the transforming machines. This same point is also made by Rosenberg and is central to Chandler's analysis of American corporate managements' role as the coordinator of abstract flows of materials, finance and people (Chandler 1977; Rosenberg 1976). Thus, the bottlenecks focused attention and provided the opportunity for the introduction of transfer machines. The major developments came initially in petroleum and sugar refining: the continuous flow industries (Chandler 1977). Shortly afterwards, the assembly line in its powered format was developed at Ford circa 1914 (Hounshell 1984). Transfer machines became a major area of technological evolvement and by the 1940s they were being closely integrated with transforming machines. That specific linkage was facilitated by the third dimension: the development of control machines. Hence the engine transfer lines of the American automobile industry represent an integration of all three dimensions (Abernathy 1978: 204-205).

The usefulness of the threefold dimensions can be further illustrated by examining the case of coal mining in Britain (Townsend 1980). Unlike the USA[16] there has been very little open-cast coal mining because British seams have tended to be deep, narrow and often twisted. The use of transforming machines was slow to occur even in comparison with other countries with similar seams (e.g., Germany). The early machines were essentially for undercutting the coalface so that when an explosion was fired above the undercut, the coal fell downward and could be removed on simple conveyors. That system was more efficient than hand-got methods, but required a complex three-shift system to complete the 24-hour cycle of activity (Trist/Higgin/Murray/Pollock 1963). In the 1950s, after nationalisation of the industry, various inhouse R and D and operations research groups were established. Their long-term project was to create the minerless colliery. An early development was a transforming machine which cut coal directly from the coal seam, therefore eliminating the need for previous machinery and for the 24-hour cycle. That development was extended by machinery which inte-

---

[14] Flexibility in the location of power was greatly facilitated with the development of electric motors after 1900, but that potential was exploited more extensively in the USA than in Britain

[15] See the adoption of the Bonsack cigarette-making machine as discussed in Chaps. 10 and 11

[16] The position of the American coal industry is examined in Lawrence and Dyer (1983)

grated coal cutting with self-adjusting and moving pit props so that further increases in productivity could be obtained. By that stage the main area of low productivity had become the transfer of coal from the seam to the surface and design efforts became concentrated there. The development of coal cutting and transfer machinery was greatly facilitated by the development of self-regulating control apparatus for each of the other two dimensions of technology. The result of the later developments has greatly reduced manning requirements[17].

4.3 Trajectories and the Role of Paradigms

Following the studies by economists in the early 1960s, which concluded that the simple analysis of inputs of factors was insufficient to account for the high long-term rates of capital accumulation (Solow 1957), attention has shifted onto other aspects, mainly the role of knowledge[18].

An important development in the economics of innovation has been the investigation of whether there are generic innovations which possess identifiable trajectories of evolvement, and if so, how such trajectories might be explained (Freeman 1974, Part II; Nelson/Winter 1977). There is a high degree of agreement amongst economists that trajectories can be detected.

In this section, I examine the main themes and introduce certain qualifications to the purely economic analysis. The economic analysis *was* almost entirely focused on equipment. However, to an increasing degree the idea of learning and the notion of technological paradigms have become central. These are useful developments, yet so far insufficient attention has been devoted to the embedding of equipment in the organizational structure or to innovations in organization.

From a policy perspective it might be useful to know whether generic innovations followed an identifiable pattern or trajectory in their evolvement (Clark, N. 1985). If patterns did exist then it might be presumed that future

[17] In Britain there were massive reductions between 1979 and 1986 from 219 collieries to 133, 40% less employees and a reduction in output to 85% of the 1979 levels; from *The Economist*, 2 August 1986, p 23; in 1984-1985 there was the most extensive and lengthy strike by the coalminers since 1926

[18] The focus on knowledge was central to the early work of Langrish et al. (1972) but its new significance is reflected in themes such as the economics of information and in the whole analysis of learning already noted in the approaches of P.A. David, N. Rosenberg and E. von Hippel. That shift of emphasis also reflects recent criticisms within economic history of the emphasis upon technology as an object; I am indebted to the insights of Peter Mathias on this point

patterns were similar to past patterns and this insight could influence policy-making at several levels: the state, the sector and the firm. So far the concept has only been applied to technology and to raw materials. The main attention has been given to technological trajectories. If such trajectories do exist, then it would be important to know whether their directions were shaped – however indirectly – to the market signals of the neoclassical model or whether the trajectories evolved more or less autonomously, that is 'science-push' rather than market pull[19].

The case for technological trajectories has been strongly put by Nelson and Winter (Nelson/Winter 1977). Their view is that when technologies are examined retrospectively it is relatively simple to discover a generic technology or technological system which possessed "a certain inner logic of its own" and was "certainly not fine tuned to the changing demand and cost conditions" (Nelson/Winter 1977: 56–57). Nelson and Winter emphasise the homogeneity within a specific cluster of technologies.

To explain the trajectories they argue that the directions discovered in the trajectory are created by those most centrally involved in R and D and innovation-design within the firms. It is these specialists who actually shape the direction at the level of the firm. This interpretation, which is an important extension to the simplistic learning models proposed by Arrow (Arrow 1962), is both revealing and problematic.

It is revealing because Nelson and Winter provide a fascinating sketch of the cognitive dimensions of the decision-making within R and D. They argue that specific firms develop their own technological regimes of thought. The regimes are defined as broadly conceived sets of cognitions and beliefs about what is feasible and worth attempting, including the potentials, the constraints and the opportunities to be exploited (Nelson/Winter 1977: 57). They indicate that the in house specialists use a variety of heuristics which impel the development of innovations down some tracks whilst precluding other tracks. There are heuristics which shape the search for problems worth working on by setting goals as well as providing procedures which identify and filter potential projects. There are also likely to be heuristics which factor complex problems into parts which can then be allocated within the firm as well as indicating those problems which are best dealt with by other firms or agencies. Nelson and Winter are clearly and neatly elaborating on a theme of the subculture of R and D within the firm and so it is relevant to note the claim by Pavitt (1985), based on an extensive study of British innovations over 40 years, that much of the necessary knowledge came from within the enterprise (about two-thirds).

---

[19] See the discussion of push/pull theories in Rosenberg (1982)

The idea of technological regimes at the level of the firm has also been examined by Dosi and by Lall who suggest that these regimes are similar to the paradigms in Kuhn's theory of normal science (Dosi 1982; Lall 1980). That would suggest, very plausibly, that the directions are driven by the local and specific agendas of problems in the firm and that these agendas define frontiers of potential achievement. In that way the paradigm defines which past solutions were good, which past solutions should be abandoned, and which problems will be worth working on. Dosi concludes that the content of production knowledge can therefore be investigated through the concept of paradigms and through the search for exemplars, i.e., locally defined standard solutions. In practice production knowledge will be much less homogeneous than Nelson and Winter suggest, though there is likely to an implicit architecture of ideas which are specific to the firm and which permit only certain directions (Kuhn 1970). Inside the firm there will be pieces of knowledge, some theorizing, many heuristics and recipes and some procedures, all of which will become a normative framework which is open to certain directions, yet blind to others.

At this point some brief illustration would be useful. Newsome reports the case of a British chemical firm, Albright and Wilson, and the activities of their internal R and D with the design of control mechanisms for their factories (Newsome 1982). Apparently, the chosen trajectory of R and D activity relied on the use of pneumatic controls for the factories. Newsome shows how these controls were inadequate for a pilot plant built in Nova Scotia, yet the R and D regime persisted down the same trajectory even when alternative trajectories for control equipment were available. Many problems arose and these contributed to the eventual takeover of the firm by another company which introduced a different system.

Likewise, in a study of a major British electrical firm supplying components to the automobile industry, it seems probable that the dominance of mechanical engineering thinking inhibited the development of components which were electronic and electromechanical[20]

Also, in a brilliant historical reconstruction, Noble traces the development of the concept of the automatic factory from the early 1940s into the 1980s (Noble 1984). Significantly his account shows that a great many firms and universities contributed to the overall trajectory which can now be discerned. However, the separate contribution of each group shows that many directions were being taken simultaneously, and that suggests that there were

[20] I am indebted to Ray Loveridge for his interpretation of events since 1965 in the West Midlands, multinational firm, Lucas; that study is part of the ESRC, WORC program already noted

many potential lines of direction with considerable potential for branching. Moreover, it seems that a considerable degree of luck and change was involved. Thus, it may be concluded that any regime may well follow a wrong branch for quite some time.

These three brief examples provide strong support for the claim by Nelson, Winter and Dosi to the effect that the cognitive frameworks in particular firms take the form of a regime or a paradigm which influences the directions of enquiry. However, Noble's study also directs attention to the role of the state and to the creation of major projects which gain a high degree of political and social legitimacy as comprehensive control concepts[21]. To that extent the approach is revealing, yet as indicated there are problematic features.

Nelson and Winter claim that the regimes are not entirely free from certain external influences. They suggest that those most centrally involved in the making of agendas will be influenced by cost factors, by economies of scale and by the possible advantages of mechanisation. Also, and following Rosenberg's notion of inducements, it is suggested that attention will be focused on bottlenecks[22]. These interpretations by Nelson and Winter are confusing because they conflate the actions taken at the level of the firm with the consequences of each firm's actions for its own survival. For these reasons it is not useful to infer after the event and without supporting evidence the specific content of regimes. That is a matter of investigation.

The point is an important one because there may be a tendency to confuse typical American regimes with all regimes all over the world. For example, American texts constantly emphasise the importance of 'bottlenecks' whilst failing to recognise that the definition of a bottleneck is specific to the paradigm which guides activity (Rosenberg 1976: 125). Hence, what may seem to be a bottleneck to the North American executive with a first degree in production engineering and a postgraduate MBA may seem to be part of the natural order of life to a British production manager whose educational career was based in only one firm.

The issue which arises is that if there are technological trajectories which can be discerned after the event, then their particular shape can only be explained as the outcome of many parallel actions undertaken in a context of high uncertainty. Given that uncertainty, there is likely to be a strong influence from cultural predispositions (see Part III).

---

21 Van der Pijl (1984) The Star Wars program is an interesting example of comprehensive control concept of an international political kind; through its application, the USA would be enabled to acquire Japanese developments in optics, according to *The Economist*, 2 August 1986, pp 67-68

22 Production bottlenecks are central to Chandler (1977)

4.4 Life Cycle Thesis: Product, Process and Work Organization

4.4.1 Abernathy I

Does the technology embodied in the product and in the processes of its production/distribution follow discernable patterns or trajectories over the life course of a sector such as the automobile industry? According to Abernathy there are clear and recognisable patterns for the complex products like automobiles, and other authors have made similar claims for other products and services (Abernathy 1978). This section examines those claims by focusing on two stages in the development of the model most closely associated with Abernathy and his colleagues[23]. The early model, *Abernathy I,* sets out the case whilst *Abernathy II* may be regarded as a refinement and stretching of the original rich concepts of process.

The objective of sector life-course models is to provide an overall profile of the state of the art technology for a sector by identifying the interactions between the technology embodied in the product (or service), the technology embodied in the processes of creating and distributing the product (or service) and in work organization at the level of the sector. The principal model to be examined is that constructed by Abernathy, Utterbeck and colleagues and refined in subsequent studies (Utterbeck/Abernathy 1975; 1978). Their model uses the notion of the sector life-cycle, yet does so without necessary recourse to biological or mechanical analogies, consequently it is better described as a life-course model. The focus of attention is the sector from the time of its founding onward. The basic postulate is that the profile of technology alters in a patterned way over time so that there is a profile of the state of the art which can be used to evaluate all firms in the sector.

In this section the original model will be revised in four ways. First, for any sector there is likely to be a plurality of state of the art profiles. The exact number of profiles will be dependent on the number of specific niches. There have been, for example, important differences between the life course of technology in North America and Europe. Second, the original model tended to understate the problems of transitions and of inertia for firms in any sector. There was a tendency to concentrate upon the winners and to suggest an almost natural process of unfolding of processes. Later studies have indicated much more variability and that can be readily incorporated into a revised model. Third, in particular *Abernathy I* understated the extent to which its postulate of a mature state of the art of technology would re-

---

[23] Abernathy (1978); Abernathy, and Hayes (1980); Abernathy, Clark, and Kantrow 1981) the late Bill Abernathy was a consultant to our program and I am indebted to him for his positive, insightful guidance and for his contagious enthusiasm

quire revision. Fourth, although Abernathy addressed the organizational dimension in a much richer, more extensive and more synthesised way than any previous attempt at modelling technology as a process, certain revisions are required.

In *Abernathy I* there is a distinction between the pattern for product innovations and for process innovations. In the case of product innovations there is a tendency for the rate of radical innovations accompanying the founding of a sector to be particularly high. After that period the tendency is for the rate to decline except for occasional radical innovations in the major subsystems of the product. The overall pattern is shown in Fig. 4-5. During the founding period there is often little understanding of the new product. The car began as the coupling of a coach body with a mobile power source and at first there was little certainty about the power source. Also, in the founding period neither the suppliers nor the users are certain about the shape and uses of the new product. There will be many suppliers, few of whom will hold large market segments. Then gradually there is a convergence about the main characteristics of the new product (Clark, K. B. 1985). It acquires a dominant design architecture as in the Ford T model. At this stage the basis of competition between suppliers shifts towards factors of reliability and price.

The rate of process innovations commences at a lower level because the new product is being assembled with conventional technologies and with conventional forms of work organization at the shopfloor and in management. However, once a degree of stability occurs in the characteristics of the product, then attention becomes focused on the application of specially designed

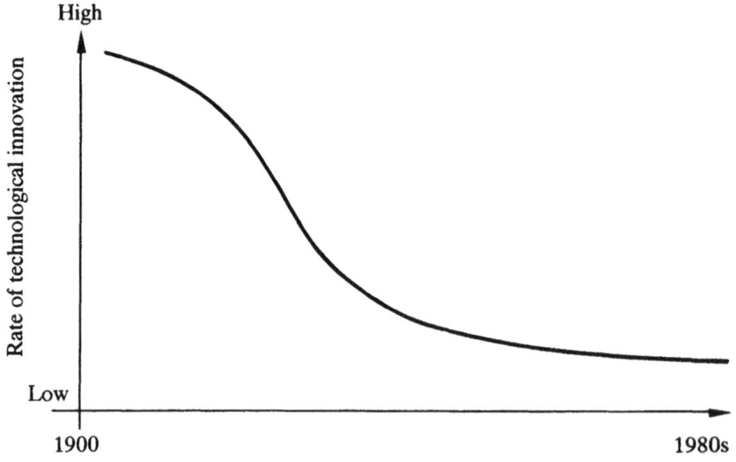

Figure 4-5. Radical innovation in the product

process technologies for its production and there is a sharp increase in the rate of process innovations as shown in Fig. 4-6 which lasts for a period after which the rate declines. The combined profiles for product and process innovations are shown in Fig. 4-7.

In the initial version of the sector and technology life-course model, the focus of analysis was the production unit (Abernathy 1978: 4), that is, the es-

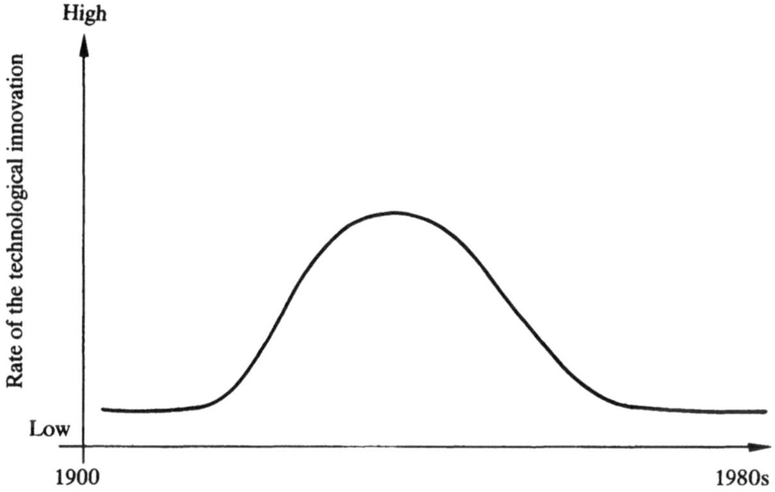

Figure 4-6. Radical innovation in the process

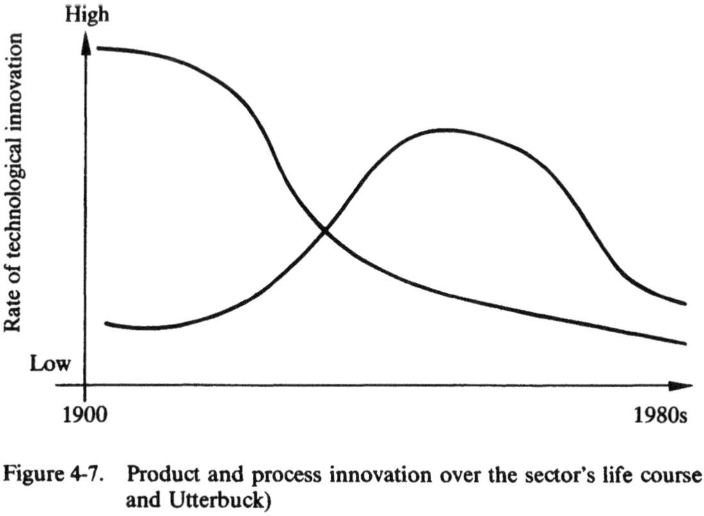

Figure 4-7. Product and process innovation over the sector's life course (Based on Abernathy and Utterbuck)

	INNOVATION	END PRODUCT	FACILITIES/PROCESS			WORK ORGANIZATION AND SKILL
			INPUTS	SIZE AND FLEXIBILITY OF FACILITIES	TECHNOLOGY	
1900/10 Founding era — Fluidity	Frequent and novel changes in the product	Ill-defined product. Cars produced for specific customers in the locality	Common type of components from regional suppliers	Very small in capital, space and employees. Small batches. Flexible units.	Little. Few general purpose machines for isolated tasks	Craft-based manual skills. Long cycle tasks. Piece-work
Specificity → 'Maturity' 1970s	Cost-stimulated incremental innovation. very infrequent radical innovation	Functionally standardized components for the international market	Dedicated components. Integration backward to raw material suppliers	Large batches. Heavy capital involvement	Specialized dedicated and integrated technologies. Process and continuous flows	Monitoring tasks and process maintenance

Figure 4-8. Production units: Fluidity-specificity

tablishment at which a specific activity is undertaken. A large company will possess a portfolio of production units. Over the course of the life cycle there will be a tendency for production units to shift their technology from fluid to specific as shown in Fig. 4-8. That figure contains seven columns covering both the product and the production process as well as the work organization. Its basic principles are similar to those discussed earlier with reference to the theory of Woodward (see Fig. 4-2). Although production units tend to move downward from fluidity to specificity the extent of movement varies between units according to the uncertainty which has to be handled.

This part of the model can be illustrated with reference to the automobile industry. In the founding period, between 1886 and 1908, there were hundreds and hundreds of firms producing cars, many of them producing one a week for specific customers and a small number producing more than 20 cars per week (Bardon/Chanaron/Fridenson/Laux 1982). After 1908, when the Ford T was launched, the number of car firms began to decline rapidly in North America and then also in Europe. Before 1908 the car was the object of scores of radical innovations in various subsystems, after 1908 the tendency was for innovations to be refinements of earlier radical innovations. After 1908 the production process and work organization became major objects of innovation with the development of transfer machines in the production units and of a design capability amongst management as two major developments.

In *Abernathy I* there was an implicit assumption that the technology of the sector would develop through several major stages towards a state of maturity (Abernathy 1978, Figs. 4-1 + 7-1). Those assumptions are shown in Fig. 4-9. The horizontal axis assesses the degree to which technological innovation alters the existing production system whilst the vertical axis measures the impact of technology on the market relationship between the supplier and the customer. Abernathy argued that the sector moved, over time, from the top right-hand quadrant – the architectural phase – to the bottom left-hand quadrant – the routine and mature phase (Abernathy/Clark 1983). In particular it was assumed that the trajectory of the American industry had followed that line. Whipp and Clark contend that the European industry followed the path indicated by the dotted lines (Whipp/Clark 1986). In the mid-1970s it was assumed that the industry was in a state of maturity and that the future depended on the achievement of high volume, mass production through the international division of labour, so that stable, routine components could be produced in the cheapest (Pacific Rim) regions.

Abernathy also addresses the work organization associated with the life course of state of the art technology. The model indicates that in the founding period, the shopfloor work organization tends to retain features of the small-batch and craft forms which are typical of the region in which the new

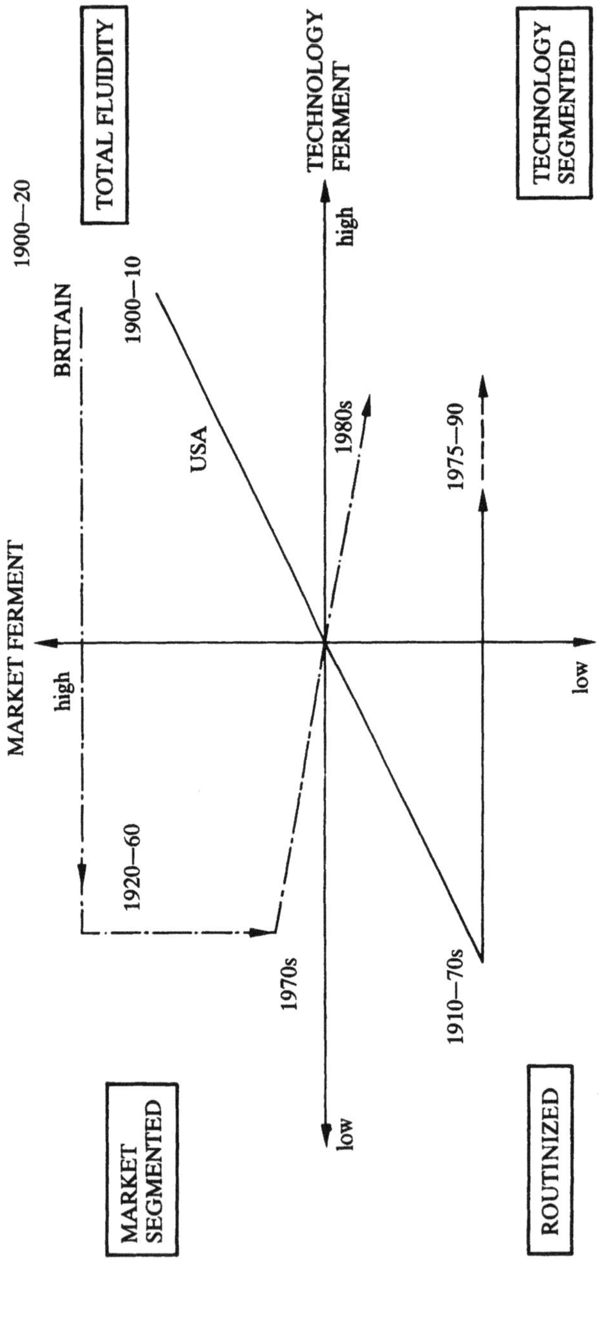

Figure 4-9. Auto industry and maturity

sector is located. Managerial organization tends to be entrepreneurial and autocratic. Most daily decisions are, as Woodward suggested, implicitly policy decisions (Woodward 1965). The managerial system may be relatively organic because there is little differentiation of roles (Burns/Stalker 1961: 126-127). In the next phase, when a dominant design emerges and some firms enter large-batch production the tendency is for new forms of shop-floor organization to be introduced (Abernathy 1978, Chap. 4 + 5). In the automobile industry tasks were subdivided and coordinated through the pacing and structuring of technology which came to embody certain forms of managerial control strategy. At that time managerial structures evolved and whole arrays of new roles emerged, especially in R and D and related areas of design as well as in distribution. The extensiveness of these changes varied between countries. In Britain, for example, the distribution side of automobiles was left largely in the control of the independent distributors and they exerted a significant, possibly fateful, influence on future directions (Whipp/Clark 1986).

Abernathy utilised orthodox organization studies to reconstruct these changes in work organization.

4.2.2 Abernathy II

The original model requires stretching and revision in four ways: the number of profiles which coexist, the problem of inertia and transitions, dematurity and work organization.

First, the original model depicted the technology profile as singular, yet that is an unnecessary limitation. In the case of automobiles, it is recognised that the industry is split into at least two major areas: mass produced and speciality cars. The profile of the latter certainly lagged behind the former from the key period after the founding of the sector through to the late 1960s. In Europe the leading speciality carmakers such as Mercedes, BMW, Volvo, Saab, Jaguar and Rover did not experience major innovations of technology in the production process until the 1960s and 1970s. In other words, at any one time their state of the art technology differed from the mass-produced cars in important respects. The position of speciality cars has been documented in the study of the Rover company and its changing experience of technological innovation. Also, there is an excellent study of the Bombardier snowmobile in Canada (DeBresson/Lampel 1985b; 1985a).

Second, the original model depicted the general bounds within which technological change occurs: an envelope of change. However, the model overstated the smoothness of changes from one state to the next and understated the problems faced by the firms in adapting to transitions and in coping with inertia[24]. The problem of transitions is clearly illustrated in a study of 'burnout' by pioneering firms in several emerging industries which shows that after the founding period, during the shakeout of many small firms, there are a great many who fail to identify the changing market structure and are unable to develop the corporate design capacity (see Chap. 5) which becomes necessary (Olleros 1986). Such firms suffer from organizational inertia, whereby past methods are reproduced rather than altered. The point is that all organizations vary in their capacity to handle the innovations which are required to ensure survival. DeBresson and Lampel contend that all sectors are typified by three models of production, which although they emerge sequentially – custom, batch and semicontinuous – should not be taken to indicate any pre-established direction of change (DeBresson/Lampel 1985a b). Instead reversals and discontinuties are evident. The smooth path would follow the pattern shown in Fig. 4-10 whilst Fig. 4-11 emphasises that there are discontinuities between the three different modes.

Transitions are difficult. In the late 1960s speciality car firms faced the transition from custom to small-batch production. Firms like Jaguar and Rover which then produced about 40000 cars annually failed to make the shift as well as Volvo (100k plus), BMW (400k plus) and Mercedes (400k plus). In fact, the latter two became larger than the British firm of Austin Rover which makes mass-produced cars. The problems arising in Renault in the late 1970s are a classic example of organizational inertia (Clark/Windebank 1985).

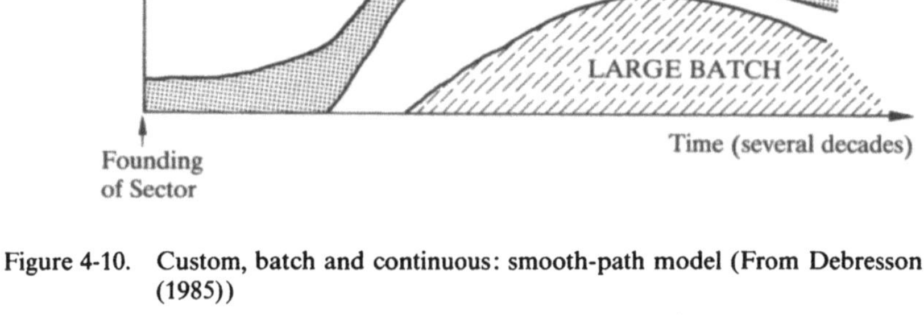

Figure 4-10. Custom, batch and continuous: smooth-path model (From Debresson & Lampel (1985))

[24] See the analysis of transition and inertia in the car firm of Renault, in Clark and Windebank (1985)

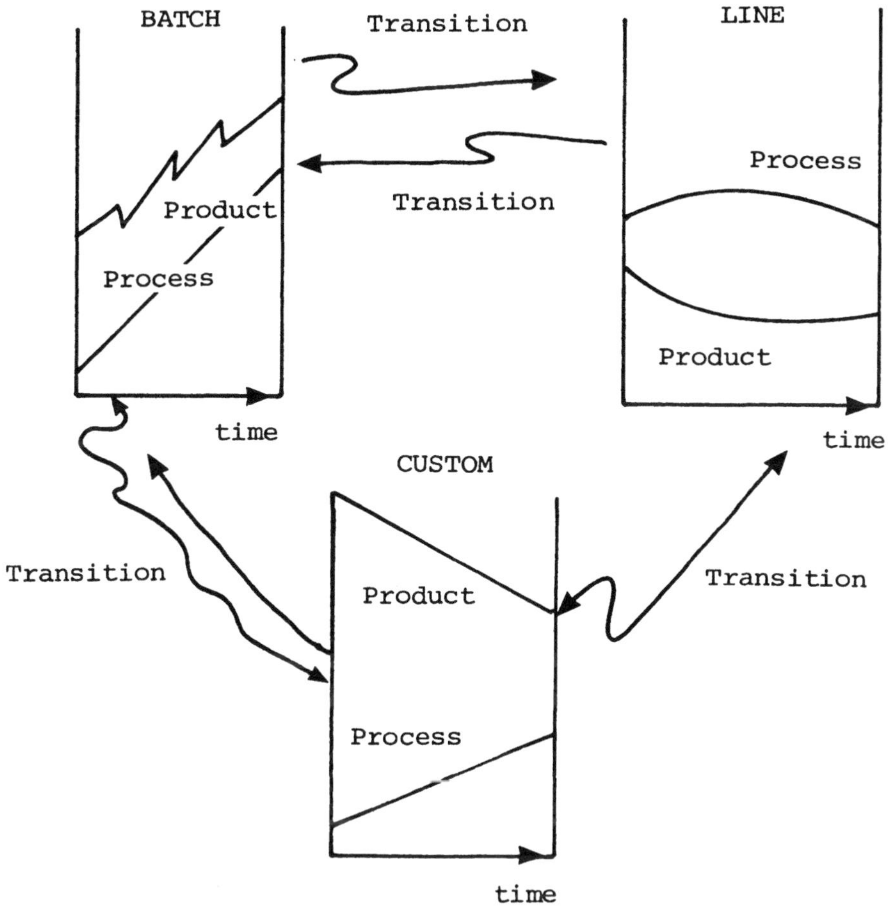

Figure 4-11. Discontinuities and transitions

Third, in his last major study, Abernathy identified the serious limitations of assuming that state of the art technology will enter a stable phase of dematurity and stay there (Abernathy 1978). The concept of dematurity was introduced to emphasize that the automobile industry had, after 1970, shifted increasingly into the lower right-hand quadrant of Fig. 4-9 because the new technology was altering the internal structure of production. This is a crucial revision and emphasizes that the analogy should not be the biological one of birth to maturity, but should be one assuming a life course which is full of reversals, iterations and shifts. For those reasons it is also important to note that the existing Abernathy model may still be insufficiently sensitive to

WORK ORGANIZATION	'A' SYSTEM	'B' SYSTEM	'C' SYSTEM
1. Allocation of Work	Imprecise		Precise
2. Division of labour	Low; polyvalent		Fixed specialist
3. Control of equipment	Self control		Machine watcher
4. Knowledge of signals	Concrete and implicit		Explicit and abstract
5. Contact with raw materials	Direct		Distanced
6. Relation to machinery	Indirect		Direct
7. Adjust of machinery	At each cycle		Occasional
8. Task variety	High		High
9. Cadence of decisions	Rapid		Rapid reactions
10. Worker influence on creation of task knowledge	High		Low; reduced and confined (see 22)
11. Degree of work group autonomy	High		Low
12. Degree of work group integration	High: larger grouping		high: small integrated teams
TECHNOLOGY			
13. Single/multi function equipment	Single and generalised		Multi and specialised
14. Use of handtools	High		Nil
15. Integration of workflow	Low		Very high
MANAGEMENT			
16. Bosses: single/multiple	Single		Multiple
17. Degree of paternalism	High		Low
18. Rule making	Personal		Company rules
19. Rule formalization	Low		high
20. Communications	Network		Horizontal
21. Hierarchisation	Medium		Network
22. Role of work organizers	Low		Fully developed as system designers

D I S C O N T I N U I T Y

Figure 4-12. Changing shopfloor work organization: Touraine's ABC system (Constructed from various publications involving A. Touraine between 1955 and 1968)

changes in the market relationship which have occurred in the last decade (see Chap. 13).

Fourth, work organization is also an integral part of innovation in the sector life course. At the shopfloor levels it might be most fruitful to revive the schema constructed by Touraine in the mid-1950s and shown in Fig. 4-12 (Touraine 1955). The three vertical columns represent the work organization which European and North American managements and labour permitted to evolve in the automobile industry. As may be seen from the final column, headed C, the tendency developed that shopfloor tasks in the successful car firms, following the pattern found in petrochemicals, became such that the operative was expected to be self-regulating within the core values of the employing firm. In practice that was the organizational approach initiated by Volvo (Jonsson 1981; Aguren 1984) and subsequently followed by major Western car firms, but not by the Rover company in Britain (Willman/Winch 1985).

At managerial levels it is important to apply the analysis of technology by Kaplinsky to identify the extent to which managerial tasks are being reconfigured through the introduction of technologies based on common data sets and using computer-based planning and designing systems (Kaplinsky 1984).

The sector life-course model of state of the art technology still requires further development, yet it provides a useful framework to apply to the evolvement of innovations.

4.5 Technology as Knowledge: Locus of Initiative

The examination of the supply paradigm in Chap. 3 introduced the issue of where is the locus of initiative in the creation, development and appropriation of the benefits of technological innovations.

What is the source of key ideas in the development of major innovations? An important attempt to provide a balanced analysis to that issue is provided from the extensive data base of 2000 *significant technological* innovations by Pavitt and colleagues (Pavitt 1985). Figure 4-13 reports the relative importance of three sources of knowledge inputs for more than a score of manufacturing sectors. The results show that on average the employees within the firm are the source for almost 60% of the knowledge required, whilst other firms provide more than one-third. The small remaining element is supplied by the public infrastructure of universities and research institutes. The lowest proportions of internally generated knowledge (approx 45%) are found in leather goods, glass, electrical components, shipbuilding and plastics. Whilst the highest internal contribution is found in machine tools, electronic capital goods, tractors and motor vehicles. The highest contribution of other firms is made in plastics and the lowest in electronic capital goods. Contributions from public infrastructure are as low as zero in shipbuilding, plastics and pharmaceuticals. The highest are in electronic capital goods and computers.

Figure 4-14 indicates that there is some variation in the overall picture. Even so it is not reasonable to assume that there is a general, public stock of knowledge into which all firms can dip because most knowledge is the product of particular investments by firms and is therefore firm-specific and highly differentiated. That means that a very high proportion of R and D investment, perhaps as much as three-quarters, is spent to mobilise local skills in order to commercialise particular products and processes. The central feature is specificity. Clearly this kind of knowledge is not typically in the form of readily transmitted information, so the costs of transmission from one firm to another could be high.

The disaggregation of the previous overall figure directs attention to the relationship between sectors in the development of technological innovations. Various partial surveys and single cases have indicated that there is a certain degree of variety in the location of initiative for innovation. Thus, a survey of American shoe firms demonstrated that the machinery supplier was the developer of the new technologies (Scherer 1982), whilst the case study of the development of float glass by the British firm, Pilkington, has become a minor classic in the business strategy literature (Barker 1977). The large survey reported by Pavitt provides some interesting and important insights into the total situation through an examination of the process innovations used in a sector compared with product innovations exported to other sectors. Approximately 75% of innovations from a sample of 11 diverse sectors were exported to other sectors as product innovations. The level is similar to a study by Scherer (1982), who also demonstrated that manufacturing produces most of the innovations that are diffused to and adopted in other parts of the economy. The high figure of 75% needs to be scrutinised to reveal the variations.

These show that only certain sectors within manufacturing tend to be the source of process innovations for other sectors. These are in instruments, chemicals, mechanical engineering and electrical/electronic engineering, whilst the following sectors are characterised by the predominance of process engineering innovations developed internally and not exported into the economy: textiles, footwear, shipbuilding, metal manufacture, food and drink. Some sectors have been heavily dependent on innovations developed in other sectors, particularly textiles. There are also sectors which create more general innovations than they use in their own sector: chemicals, mechanical engineering, electrical/electronic products. Conversely, the assembly sectors of shipbuilding and vehicles have drawn on innovations developed elsewhere to a much greater extent than the continuous production industries.

Further analysis of the data base reveals the main technological flows reproduced in Fig. 4-15 and Fig. 4-14. The data suggest, very strongly, that when firms make choices about which innovations to develop, their internal R and D departments are highly constrained in their search processes so that technical change becomes a largely cumulative process specific to the firm. In

Sector	Source of knowledge inputs (%)*		
	Intra-firm	Other firm	Public Infrastructure
Food	53.4	44.6	2.0
Pharmaceuticals	62.8	37.2	0
Soap and detergents	60.0	40.0	0
Plastics	40.4	55.2	4.4
Dyestuffs	68.1	30.5	1.4
Iron and Steel	47.7	44.9	7.4
Aluminium	68.0	28.0	4.0
Marine tools	64.1	29.8	6.1
Textile machinery	61.2	36.6	2.2
Coal-mining machinery	52.3	31.6	16.1
Other machinery	59.1	36.6	4.3
Industrial plant	51.6	41.9	6.5
Instruments	61.6	25.2	13.2
Electronic components	48.2	37.1	14.7
Broadcasting equipment	64.4	33.9	1.7
Electronic computers	50.6	33.3	16.1
Electronic capital goods	67.2	9.7	23.0
Other electrical goods	60.8	35.3	3.9
Shipbuilding	47.9	43.8	8.2
Tractors	78.7	21.3	0
Motor-vehicles	69.3	29.7	1.0
Textiles	67.3	32.7	0
Leather goods and footwear	44.4	48.1	7.4
Glass	48.2	44.6	7.1
Cement	62.5	33.3	4.2
Paper and Board	66.7	28.2	5.1
Other plastics	55.8	41.9	2.3
Average	58.6%	34.0%	7.4%

* Each row adds up to 100 percent
Derived from Pavitt K. (1985)

Figure 4-13. Sources of knowledge in significant industrial innovations

Category of firm		Typical core sectors	Determinants of technological trajectories			Techno-logical trajectories	Measured characteristics			
			Sources of technology	Type of user	Means of appropriation		Source of process technology	Relative balance between product and process innovation	Relative size of innovating firms	Intensity and direction of technological diversification
Supplier dominated		Agriculture housing private services traditional manufacture	Suppliers Research extension services: big users	Price sensitive	Non-technical (e.g. trademarks, marketing, advertsing, aesthetic design)	Cost-cutting	Suppliers	Process	Small	Low vertical
Production intensive	Scale intensive	Bulk materials (steel, glass) Assembly (consumer durables & autos)	Prod. Eng. suppliers R & D	Price sensitive	Process secrecy and know-how technical lags: patents: dynamic learning economies:	Cost-cutting (product design)	In-house suppliers	Process	Large	High vertical
	Specialised suppliers	Machinery Instruments	Design and development users	Perform-ance sensitive	design know-how: knowledge of users patents	Product design	In-house: customers	Product	Small	Low concentric
Science based		Electronics/ electrical chemicals	R & D Public science: PE	Mixed	R & D knowhow: patents process secrecy and know-how: dynamic learning economies	Mixed			Large	Low vertical High concentric

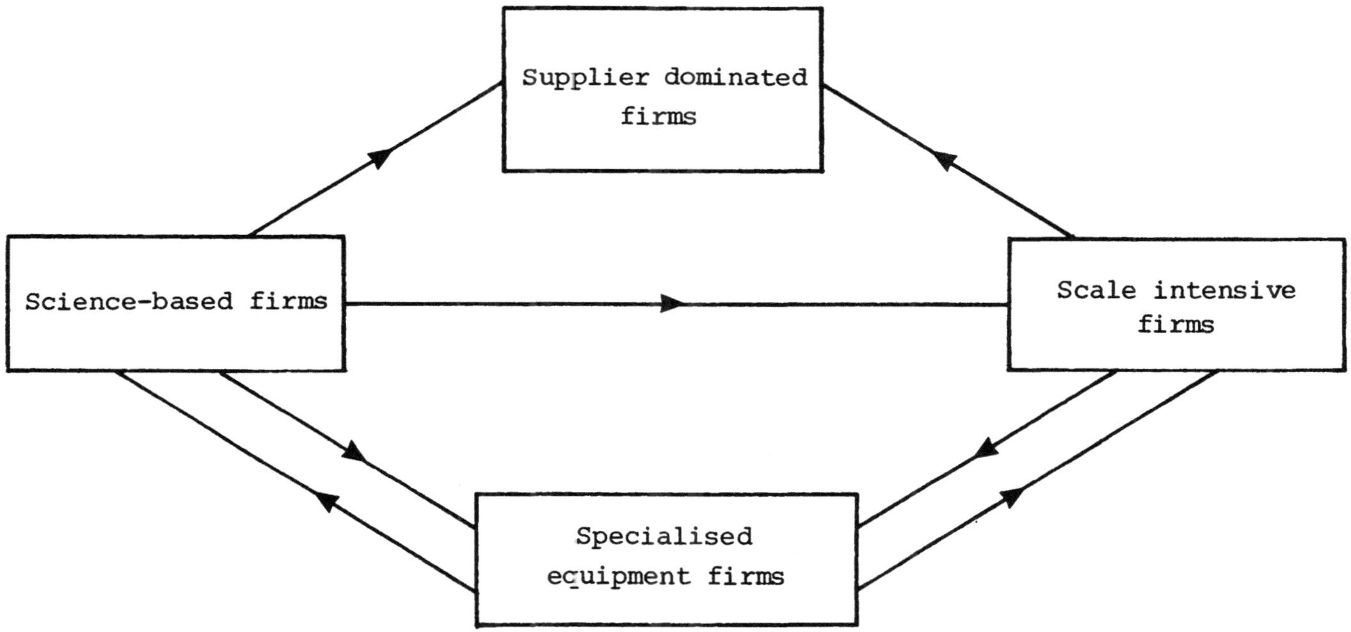

Figure 4-15. Main technological flows between sectors (From Pavitt 1985)

that sense the future can be strongly conditioned by what particular regimes have achieved in the past. There seems to be a high degree of momentum in this key area of organizational learning. When sectors are examined certain regularities begin to appear. There seem to be a whole group of sectors where technical change comes from outside – they are supplier dominated, mainly by suppliers of equipment. Such sectors are the long-established sectors, like textiles and the whole area of nonmanufacturing. Yet other manufacturing sectors do make a considerable contribution to the development of their own process technology (e.g., assembly and continuous process), whilst certain sectors turn out to be creators of technical innovations which can also be transformed into products for sale to other sectors (e.g., chemicals, electrical engineering). Is it possible to produce a taxonomy from these data?

Pavitt offers a taxonomy of four types of situations as set out in Fig. 4-14 and the ingredients of a theory about the interrelationships in the economy as summarised in Fig. 4-15. The taxonomy is based on the search for patterns in the diversity reported in the previous paragraphs. Regularities do emerge in the ways in which technical change occurs. The basic unit of analysis is the innovating firm and there are three features of its situation which are incorporated into the taxonomy:

1. The sources of the innovations
2. The specific needs of the users
3. The extent to which the innovators can appropriate the knowledge through one of several mechanisms (e.g., patents)

First, the sources of technical change may be inside the firms or outside in other firms, or, occasionally in special institutes. Second, the needs of the user vary widely. Those users who require highly standardized structural or mechanical materials assess innovations in terms of their price competitiveness, whilst firms requiring equipment to be slotted into highly integrated systems place a high value on performance and reliability.

The taxonomy and the theory assume that the diversification activities of firms rest quite crucially on their existing technological base and on technological learning which is cumulative and self-generating within the firm. Unlike taxonomies which focus solely on the internal organization of the firm, this one encompasses the linkages with suppliers, customers and additional sources of technical innovations, as well as with the ongoing product and process changes.

In Fig. 4-14 there are four categories of types of firm as shown in column 1 with examples of typical core sectors listed in column 2. The influences on the technological trajectories are itemised in the next three columns (3-5) and the last four columns flesh out the detailed characteristics. Each of the four categories can now be examined.

Supplier-dominated firms make a very minor contribution to either their own technical development or to technical developments which might eventually be sold to other sectors or even to households. Their innovations come very largely from the suppliers of raw materials and equipment. In some cases innovations might actually come from customers. The supplier therefore dominates the user firm's experience of technical innovation even when a high proportion of the innovations which are generated in the sector are destined to be transmitted to other sectors. Examples are to be found in agriculture, housebuilding, the traditional sectors of manufacturing and in the many small service firms, as well as in the household economy. The importance of the latter, and its role in the use of equipment has been demonstrated by Gershuny (1983; Gershuny/Miles 1983). There are other sectors which probably fit this category quite well: textiles; lumber, wood and paper-mill products (Scott-Kempis 1983).

The production-intensive category contains two subgroups: scale-intensive and specialized suppliers. Generally production-intensive firms arise from the increased size of the overall market and from an increasing division of labour amongst firms. These provide the impetus for economies of scale, especially in large-scale fabrication and assembly work. As indicated earlier, in the analysis by Coombs, the technology for exploiting scale economies has developed steadily over the past century (Coombs 1983). Also in continuous processing industries improvements in energy, control systems and input of materials have increased the scale of operations. Attaining the benefits of the scale economies in these firms is often more difficult than economic models have suggested (e.g., the British automobile industry). The requirement for achieving scale economies affects:

1. Firms producing standard materials
2. Firms producing durable goods and vehicles

In these firms the conditions of operating are exacting because of the complex flows and interdependencies between different stages in an overall sequence. Such firms tend to employ internal specialists in production engineering to identify bottlenecks and technical imbalances. Their activities often – though not always or necessarily – lead to cumulative improvements in equipment. For example, in the production of cigarettes, there has been considerable incremental development in the basic manufacturing and packing machines. These internal departments are one potential source of technical innovations for the firm, its sector and other sectors.

Within this larger category there is another subgrouping based around the small, specialized firms which supply instrumentation and equipment to the production-intensive firms. There is a two-way relationship between these suppliers, a small proportion of whom may be quite large, and their customers. The users provide the suppliers with details of the operating experience and with a practical testing of their goods. In return the suppliers provide their learning from several different situations and their capacities to design and build for a variety of different users. Consequently, the supplying firms tend to take different technological trajectories from their customers. Their orientation is towards performance rather than cost-cutting.

Within this large category of production-intensive innovations there are variations in the ways in which the benefits of technical change are appropriated. It seems likely that the large-scale producers with their own internal R and D groups are not very interested in particular inventions. Their skills are in the choice, design, construction and operation of large, integrated systems. Their technological leads over rivals will be embedded in specialized forms of knowledge which are difficult to transmit between firms. Even if the know-how can be carried by bearers the new employer may be unable or unwilling to implement a systemic, core knowledge in a partial way. For example, when the newly merged British Leyland company began manufacturing cars in the 1970s with imported executives from Ford the incoming men found considerable opposition to their know-how from existing managements (Whipp/Clark 1986). For the suppliers of these firms the appropriation position is different. They may be concerned to appropriate the benefits of innovation, but they are less protected by secrecy because they have to retain their competitive advantage by selling that know-how to the next client.

Science-based innovations are the third major category and are an instance of where the major sources of the new technology are within the firms in the R and D activities which are themselves complements to the development of the sciences that underpin these innovations in the universities. The nature of the university linkages is important in this category and it will be necessary to explore the extent to which their coupling differs between Britain and the USA. Science-led innovations are characteristic of the chemical, pharmaceutical and electrical/electronic industries. In these the development of waves of products depends on the previous development of the required knowledge in the basic sciences (Freeman 1974), and that fact has considerable consequences for the structure of power within the firms[25].

The pervasiveness of the linkages between the basic and applied sciences shapes the technological trajectories of the firms. The know-how and knowledge on which this group of sectors is based have underpinned the spectacular growth of certain firms and have limited the capacity of firms to enter certain science sectors, though not all. That said, there have been considerable variations across the science-based sectors in the degree to which they have been active in the exporting of their innovations to other sectors. Firms

[25] For examples in the development of major firms like NV Philips, see Heerding (1986)

in this sector have a variety of mechanisms to facilitate their retention of know-how and the appropriation of its benefits; here, patents are important. Also, the dynamic learning economies of the incremental evolvement of best practice and modifications to the technology are hard to transmit to other firms.

An important feature of science-based knowledge is the constitution, editing and overthrow of particular paradigms. The earlier reference to the case study of Albright and Wilson indicated the difficulties of altering existing paradigms once they gained momentum. Further, their tendency towards autonomous elaboration which all exponents of technological trajectories emphasize, can lead to a corporate imbalance, whereby the hierarchy of paradigms within a collection of laboratories and R and D departments no longer reflects the market potential. That situation of imbalance existed for a period in ICI during the 1970s until the new corporate leadership of Harvey-Jones reconfigured the firm around the growing (in the mid 1980s!) market of speciality chemicals which has been emerging in the USA (Pettigrew 1985).

Chemicals is a particular industry where firms do produce a high proportion of their own process innovations, whilst simultaneously exporting a high proportion to other sectors. The large size of the firms permits them to develop internal structuring for innovation-design and hence to reflectively learn from their practices. For example, in the 1950s, certain groups in the British-based ICI played a leading role in criticising the approaches to productivity of engineering graduates. Those criticisms led them to devise their own interpretation of American thinking and to create work study as an analytic approach (see Chap. 12). Their approach then became widely diffused in Britain, though its characteristics as an innovation were progressively attenuated in the subsequent decades.

Within the science-based sectors, the large diversified firms have tended to make a larger contribution to major innovations than have the suppliers, even the specialized suppliers.

Given these major categories, how are they interconnected, Pavitt suggests the linkages shown in Fig. 4-15. It seems that the supplier-dominated firms get most of their new technology from the science-based firms and from the scale-intensive firms in the production-intensive category. Examples would include power tools, transfer machines and similar systems from the science-based industries, whilst the scale-intensive firms made available new materials like plastics and consumer electronics (e.g., white goods for the household). In turn the science-based sectors are involved both in the transmission of their own innovations and the brokering of specialised technologies from equipment suppliers; likewise for the scale intensive firms. The linkages in the model consist of equipment as well as information and skills.

5 The Corporate User: Innovation-Design Capacity

5.1 Limits of the Supply Model

The understanding of the position of the user has been obscured by the dominance of supply-side models. Now the position of the corporate user must be understood by reference to research and theory which examines how innovation fits into the overall activities of the user. For instance, far too little attention has been given to examining how the pre-existing situation of the user influences the scope for innovation. In particular, attention must be given to the problems of momentum and inertia and the tendency for the innovation-design process to possess a scope which narrows the range of options which will be considered and implemented. An important feature of the user's position is the strategic choices made about the markets for services/products. Following from the previous point, the corporate user must be analysed as a collection of subcultures possessing definite political relationships of competition, conflict and cooperation. Thus, the issue of leadership requires attention.

The position of the user has been emasculated in the existing supply-side models. The supply-side models tend to assume that there is a single linear process running from invention through the development of the innovations to their adoption and implementation. In fact, as already noted, that assumption is not confirmed by the research done on the supply side, yet those researches continue to rely upon a paradigm of innovation which provides a misleading account of the user. That confusion is heightened by the tendency of the supply-side models to depict the user as adopting only a single innovation, or, choosing from a small number of homogeneous innovations. In practice many corporate users (and also households) are simultaneously choosing from diverse sets of innovations in many different areas, e.g., raw materials, equipment, new work practices.

The supply-side approaches reported in the closing section of Chap. 3 illustrate the narrow concern with simple normative schemas rather than with the analysis of the user corporation (Clark/Starkey 1987). Perhaps the most striking, though unintended, slighting of the corporate user is in the tendency of the supply-side models to see how the supplier can appropriate benefits from the contributions of the user. The treatment of the user in von Hippel's model illustrates the dominance of the supply and marketing perspective by turning the user into the marketer's free R and D laboratory[1]. Further, as the examination of the locus of innovative activity has clearly demonstrated (see Sect. 4.5) the role of user in the development of an innovation can vary very greatly from substantial to very slight.

Exponents of the supply side of innovation perspectives have sought to make various refinements to the analysis of the user, such as the re-invention model and the modelling of the internal processes of adoption, but the problem is that these revisions do not go far enough because they fail to see the situation from the user's total perspective. The perspective of the supply-side models tends to treat diffusion as though only one item or a small number of items were being diffused. If that were the case the choosing behaviour of the potential user would be relatively simple.

The position of the user is highly complex because there is a distinct set of problems which users' normally face and these include the decision of whether to develop new ideas internally or to import them. For example, Skinner maintains that Japanese corporations obtain an advantage by developing much of their own requirements for equipment on an *internal basis* (Skinner 1984). There are many examples of British and American firms pursuing similar policies: General Motors with software and the former Imperial Group whose cigarette companies (e.g., John Players) had close financial and working relationships with Molins the supplier of manufacturing and packing equipment (Monopolies Commission 1961).

The user is typically involved in multiple and diverse choices between areas of incremental and radical innovation in different segments of its business activities. There are multiple decisions and trade-offs to be made. For instance, an in-depth case study of a speciality car firm (see Sect. 5.5) revealed that there were decisions affecting innovations in all areas of its basic product, the car, as well as in the type of assembly operations, payment systems, methods of design, paint plants and many other areas. Recent researches, such as those by Gold, demonstrate that the corporate user faces very considerable problems of gaining accurate information about new equipment, raw materials and new practices (Gold 1979). Gold has documented the immense uncertainties which exist under normal conditions. Within the corporation those key decisions have to be carried through several iterations and different states, during any one of which the original intentions may be drastically altered. Inertia, for instance, is a powerful feature of organizational decision-making (Clark/Starkey 1987).

Existing studies of the user neglect to analyse the power structure. So the typical internal struggles within management are discounted. Likewise the

---

[1] See all references under entry for von Hippel

whole area of industrial relations and management-labour relations seems to have been forgotten despite the fact that studies of British and American firms from an institutional perspective show that such factors have a pervasive effect. Innovation has been wrongly and naively treated *as a politically neutral process*. In everyday occurrence it is the political interests of factions and their active agency which is highly consequential.

Users vary in their capacity to handle innovations, both those externally sought and those internally constituted. The internal regimes and paradigms of those centrally involved in the choice of directions are sometimes very restricted in scope, mostly so with respect to situations which require alterations in the paradigms and consequent changes in the political structure of occupations. It is very hard for users to develop a format for organizing innovation to achieve radical and robust designs rather than lean designs (see Sect. 5.4) and to obtain profitable design hierarchies rather than elegant, but unprofitable ones. Finally, by way of introduction it must be observed again that the process of imitating existing successful innovations is very difficult and requires a sophisticated infrastructure – a tissue or *filiere* – between firms as well as within the adopting firm. The transmission of innovations requires an understanding of their crucial embedding features. These limitations to the existing model have been identified. This chapter addresses some of the directions which analysis and practice ought to be taking.

5.2 Multiple, Simultaneous and Diverse Innovations

One of the limitations of existing thinking is the tendency to define the adoption of an innovation as a single, precise event. Attempts are required to breakaway from that restriction and to identify the variety of subprocesses (Eveland 1979; Eveland/Rogers 1980). The decision processes in noticing a problem or opportunity to which an innovation may be related have been elaborated in the decision sequence suggested by Mintzberg and colleagues in Fig. 5-1 (Mintzberg/Raisinghani/Thoret 1976). The main features of the model are that a myriad of small decisions are set in motion when a problem or opportunity is identified. From then on the pattern of decision-making may be immensely varied. The only certain aspects are that any decision whether to adopt or not will be a political one and that the overall process will be full of detours, iterations, abortions, recycling and seemingly bizarre discussions. The movement along the decision line in Fig. 5-1 from left to right will be complex. The figure illustrates how the decision may start by considering an internal solution to the problem and then return at a later stage to the designing of a unique solution based on the introduction of external ideas, equipment and practices. Dependent on the degree to which the decision process activates latent political conflicts and hostilities will be the

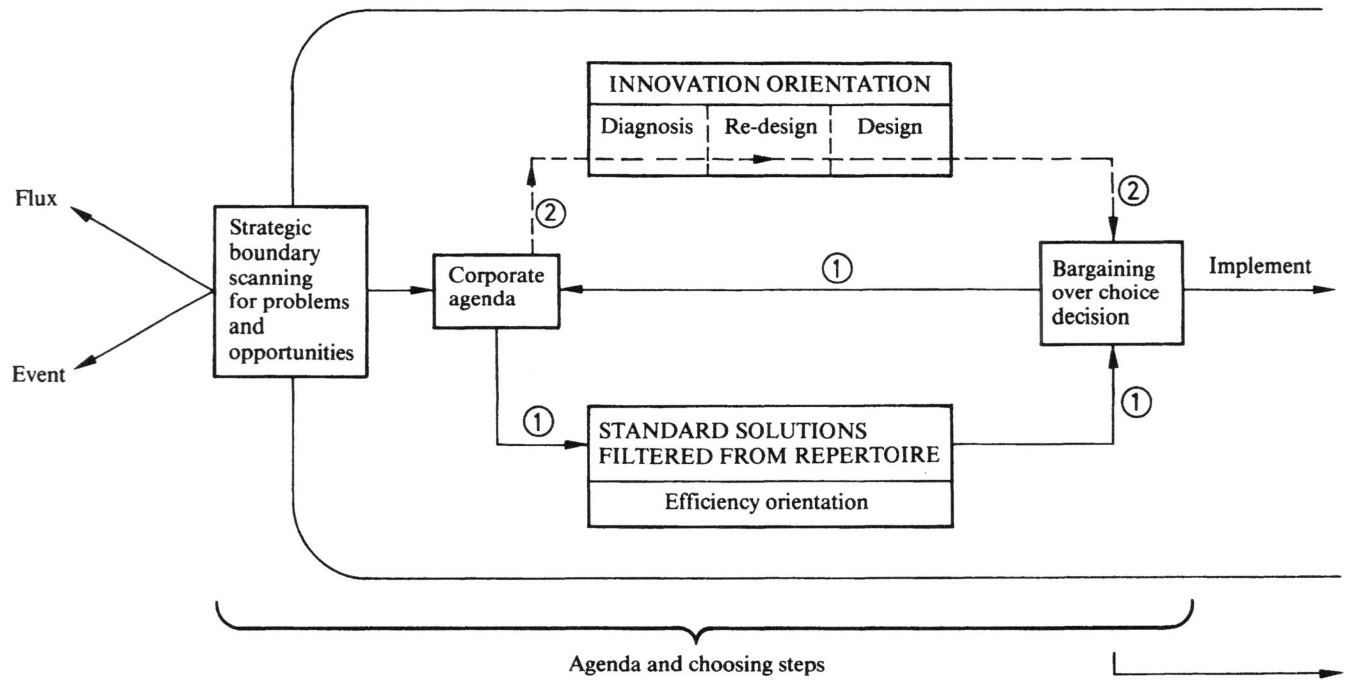

Figure 5-1. Top level decision processes: Choosing stages

extent to which the repertoire of structuring in the organization is run through in a relatively smooth fashion, or, becomes a vortex of processes (Hickson/Butler/Cray/Mallory/Wilson 1986).

For strategic decisions and major tactical decisions it is most likely that the decision process will be set in motion by senior executives within the organization. However, later, and sometimes sooner, that decision may be identified as significant by various factions not initially involved – who then attempt to affect the outcomes. For example, in a study of a hospital-design team working within the British Department of Health and Social Security, the designing of a new hospital was initially undertaken by the senior administrative staff and a small group of professional nurses[2]. They received a standard design package formulated in the DHSS which had been devised to constrain decisions along lines which were considered to be cost effective by the DHSS. At the regional level, within the DHSS, the specific design team responsible for taking the design to operating steps disagreed with the way in which the standard package placed all treatment activities in a central location, the central treatment area (CTA). Thus, at the regional level, the design committee introduced modifications to the standard package whereby the template for central treatment concept was disengaged from the total package of design templates. The activities of treatment were decentralized and *retained* in the various wards. Of equal relevance is the fact that the activities of the design committee were ignored by the medical professionals even though they occupied places on the committee. During the early stages of design, the medical staff on the design committee made no serious interventions or contributions even though the proposed designs had considerable implications for the medical hierarchy and for the allocation of space and resources to the senior consultants. Then at a very late stage, when it became apparent to them that the new hospital, if built as designed, implied considerable alterations in their working conditions and in their professional control of activities, they became much more active, thereby necessitating structural alterations during the commissioning of the new hospital.

The decision schema presented in Fig. 5-1 highlights these key points about the iterative and complex nature of decision-making for innovation. Yet even though that account probably understates the complexity which is involved, it does provide a useful means of depicting how different factions enter the decision process at different stages and with different opportunities. The influence of the medical staff, just discussed, to achieve alterations of a kind which they desired may be compared with the situation in the case illustration in Sec. 5.5.

[2] See reports from the program at ESRC, WORC, Aston University by Margaret Grieco

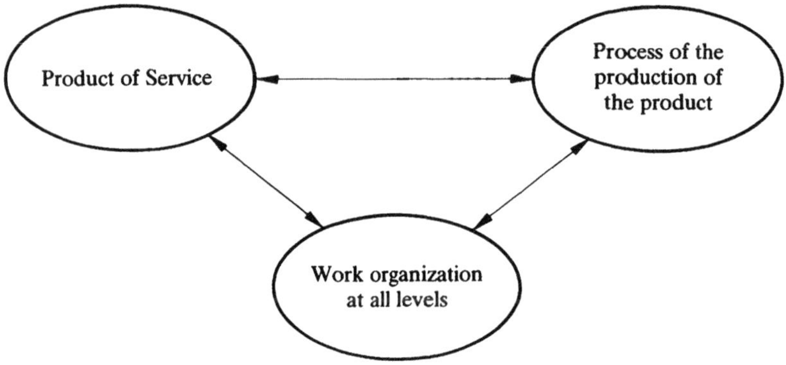

Figure 5-2. Decision areas

The diffusion of innovations approach tended to assume that only one innovation was being targeted at the organization. In practice, corporations are the objects of numerous suppliers of diverse innovations at the same time. Typically these innovations occur across the full range of activities and affect the product/service being offered, the technical processes through which it is produced/delivered and the overall organization of work (see Fig. 5-2). In addition, larger corporations will be generating and incrementally modifying some of their own internal innovations. For example, major car firms typically face innovations in a variety of aspects:

1. Raw materials: these range from the increasing weight of electronics to the substitution of plastics for wood as well as the requirement for tracking long-shot developments such as the ceramic engine.
2. In methods of designing the car: General Motors is investing billions in acquiring new computer-aided design systems and in integrated manufacturing.
3. In methods of planning and scheduling: these are tending to change from the "push systems" of materials resource planning (MRP) to the introduction of "pull systems" such as just-in-time (JIT) (Schonberger 1982). GM and external consultants have collaborated in the development of new systems which, like optimised production technology (OPT), have been experimented with in twenty-five plants[3]. The latter is designed to simulate

[3] Optimised Production Technology is a softwear system which aims to simulate a factory flow so accurately that bottlenecks can be detected in advance. Its creator, Dr. E. Goldratt, has even written a novel to sell the ideas. Applying the OPT approach undermines the conventional distinction between production and finance; see articles by R. E. Fox in the four issues of *Inventories and Production Magazine*, for the issues of August and December, 1982, and February/April, 1983

the workflow of a factory so closely that future bottlenecks can be detected.

4. New concepts of the operator and new forms of work organization are introduced at the plant level.

These are just some of the innovations which are being explored, adopted and implemented. The case of GM is common for large and medium-sized enterprises in all sectors. Health organizations, for example, have to take account of diverse new chemicals and drugs, the new ranges of which are being deliberately designed, as well as new technologies of health care and information processing. The examples are endless and they are not confined to large firms. Small firms and households are faced by a mushrooming of sources of innovations. Therefore, it is necessary to give much more consideration to the internal knowledge processes of the user than much of the diffusion of innovations literature has so far implied.

5.3 Organizing: Communities, Networks and Bodies of Knowledge

As the process perspective develops, there are clear signs of convergence around the importance of certain concepts which are closely intertwined: organizing, communities, networks and bodies of knowledge.

In the structuralist approach to organizations, Weber's concept of bureaucracy as an ideal type was dominant as a form of work organization which could be described and also prescribed[4]. Bureaucracy became designated as best practice even when certain of its features might seem to be unappealing because of their apparent restrictions on individual freedom. Weber analysed the growing separation in the West between the control of the family as a network of mutual obligations situated in a residential community and the growth of employment outside the family in, for example, public offices situated in urban centres. The latter provided alternative sources of control to the familial ones. Weber emphasised that the systems of rules in administrative settings which were independent of familial control placed a severe constraint upon the individual's ability to pursue personal individual goals. A similar common theme was also found in the early twentieth century Western theories of personality and identity construction (e.g., Freud) and in the insightful attempts at psychohistory by Elias (Elias 1978). Such approaches concluded that the future for mankind was becoming a matter of struggling against the externally superimposed rules of 'iron cages' (DiMaggio/Powell

[4] Albrow (1970); Weber (1947); Meyer, Stevenson, and Webster (1985)

1983). However, that conclusion, whilst important as a corrective to the highly individualistic thinking of the nineteenth century 'great man' theories, very much underestimated the extent to which human agency appropriates directions and shapes objectives (Giddens 1985). Weber's analysis tended to suggest that communities and networks would be replaced by relations based on impersonal association. More recent interpretations have emphasised the existence of informal networks and communities within all organizations. Their continued existence in new forms has considerable implications for the development of bodies of knowledge inside user organizations. It is those bodies of thought which drive the directions taken when choices are made about innovation.

The social network based on patterns of contact, interaction and feelings, can be used as a means of revealing the connections within an otherwise hidden community of persons (Mitchell 1970). For instance, Kuhn's theory of paradigms in normal science implied the existence of normative frameworks and controls by communities of scholars (Kuhn 1970). Later, Crane utilised the idea of invisible colleges when reconstructing specific instances of how academic communities were spread through different universities and institutes (Crane 1972). The implication is that the members of the network (or, invisible college) share similar bodies of thought. That is, there are stocks of knowledge within organizations which its members use routinely in their everyday activities.

Applying these concepts means analysing organizations as collections of communities and networks each with: *specific languages, forms of discourse and possessing specific definitions of problems and solutions* (Giddens 1985). Therefore it is necessary to extend to the whole organization the use of the concepts proposed by Dosi and others for the study of technical knowledge (Solow 1957). The treatment of whole organizations, and *all* their members, as constellations of bodies of thought provides a very obvious bridge to the claims that stocks of knowledge are highly specific to the firm, the sector, and possibly a national culture (see Part. III).

Within the organizations, networks of individuals construct bodies of thought and thinking practices by reassembling and organizing past experiences to build up a version of the past which provides a loosely organized, pretheoretical, recipe knowledge and heuristics for handling the future (Weick 1977). This local knowledge, which is quite unlike the prescriptive theories proclaimed in the universities, contains the following: key constructs, exemplars of how to map the external environment, standard operating procedures, blueprints for action, practices laid down in the past, vivid narratives and sagas. These linguistically formulated expectations are almost never submitted to critical discussion, yet the expectations enclose bodies of thought which control perception, filter experiences, define their meaning and influence the exchange of ideas (Giddens 1985). The expectations are, as March and Cohen show, information-seeking structures that accept and direct attention in particular ways (March/Cohen 1976). The development of an internal language for description and for argument promotes the selective elimination of counter perspectives without destroying the basic language which has been constructed.

The bodies of thought can be conceptualised as a retention system of rules which can be activated to interpret the contexts of the firm and to indicate which parts of the repertoire of behaviours, cognitions and emotions should be activated (Weick 1977). In Sec. 1.2, the example of the Cherokee was introduced and it can be imagined that these people devised bodies of thought for linking events into meaningful sequences which were used to decode the ambiguity and equivocality in events facing the clan. The bodies of organized thought would be used to formulate certain key questions like: How could the Cherokee be sure that another tribe was preparing for war? The rules in the retention system provide the basis for activating specific structural poses which would, in due course, encounter sets of conditions and external events which may confirm or not confirm the original interpretation.

The concept and approach of bodies of thought has been applied to the sector and to the firm. There is widespread implicit agreement that the languages of firms in the same sector will share common constructs and that these constructs will be woven into a series of recipes for action which all top executives would know (Whipp/Clark 1986: 25-34). Two studies may be cited. First the study by Spender[5]. His research, undertaken in the three different industries of metal forging, milk delivery and fork-lift trucks, relied on taped interviews with a small sample of key executives. Each respondent was defined as a representative of one firm and their constructs were extracted from the transcripts, then submitted to cluster analysis to reveal patterns. The results do provide strong evidence of shared patterns within the sectors examined. Second, Miles and Snow conclude that corporate postures to their markets and their technologies by firms in the same sector tend to share more features in common than across sectors (Miles/Snow 1978). Their fourfold taxonomy is illustrated in Fig. 5-3. However, the claims for sector-wide bodies of thought may require a degree of qualification. Even the original studies by Spender indicated some variation in the types and usage of certain constructs. Also many sectors contain several firms which have been the subject of excellent business histories which show that there are important dif-

---

5 See the doctoral thesis by J. C. Spender (1980): Strategy Making in Business. Manchester Business School

ADAPTIVE SEQUENCE	DEFENDERS	PROSPECTORS	ANALYSERS	REACTORS
1. Selection of areas for future innovation	Slight search Low priority	Regular searching for opportunities	Combine stable and innovative domains Close watch on competitors	Unfocussed awareness of uncertainty
2. Choice of product – market domain	Narrow search process	Broad	Two domains strategy	Narrow and unstable
3. Choice of technologies for production and distribution	Incremental evolvement of existing systems	Varied	Two technologies	Inconsistent
4. Reconfiguring existing process and structures	Focus on efficiency	Focus on innovation	Co-executives of efficient and innovative	Unfocussed reactions

Figure 5-3. Strategy and innovation: illustrative taxonomy

ferences between firms in the same sector in their conception of the market opportunities and in their approach to innovation[6].

The existence of sector-based bodies of thought receives anecdotal confirmation from many stories about how innovations occurred, particularly the many reported instances of how the shared bodies of thought in a sector had precluded certain possibilities. The closeness and fragility of those bodies of thought is often revealed by interlopers who are unaware of the implicit blinkers on everyone within the sector. Thus, traditional American brewers "knew" that diet beer such as the Lite beer introduced by Miller Brewing would never be bought by the consumers, yet an unknowing interloper introduced one of the most significant product innovations in that sector. Likewise, in Britain, with the development of the special Wilkinson Sword razor blade, the interloper, who was from the aerospace industry, introduced an entirely different body of thought about materials.

At the level of the firm there will also be many distinct types of knowledge and language developed in different departments[7]. One of the subsets of the stocks of knowledge will be developed to analyse the three major segments of the external context: the input systems (e.g., raw materials, purchasing), the clients and competitors, and the various regulatory agencies. Interpreting the external environment means entering an essentially socially created symbolic world.

The marketing department of a firm has been described as a centre of strategic time reckoning because its activities in the firm include the decoding of diverse external events in order to decide whether the immediate future is a reproduction of the recent past or a period of innovation and transition (Clark, P.A. 1985: 35-80). Firms in the consumer end of the textile sector experience complex and varied seasonal patterns which have to be decoded as illustrated in Fig. 5-4. The figure shows that the marketing executives attempted to discover sequences of events – their ordering, pace, intervals and durations – which could be matched against their collective and highly specific, unwritten library of templates (Clark, P.A. 1985). In the figure, their enactment of the situation has been simplified to three alternative definitions of the situation and the pattern of trading for the year has also been simplified into four seasonal shifts in production.

Those members of the marketing department at Acorn Textiles who were available met daily at coffee time when they chatted about the recent past

6 There are, for example, differences between the leading British chemical firms of Courtauld's and ICI

7 The concept of the logics of action is especially useful; see Karpik (1978: 46ff.).

(e.g., last season), the current season and the upcoming seasons. They were simultaneously editing their past, present and future judgements. For most of the time the main sources of variation were between:

1. Expecting a continuity of seasons, except for some variations in the degree of high/low production
2. Searching for indications that a major shift in the market was emerging

A correct decision meant that levels of profit could be retained. If the marketing men made a correct assessment and were then able to persuade the rest of management of their assessment, then an appropriate part of the structural repertoire would be activated. If a season of high sales had been correctly judged, then overtime would be introduced to build up stocks in the warehouses. Conversely, if a poor season were anticipated then all overtime would be suspended and various cost-cutting programs were activated. The judgements were difficult to constitute because it was necessary to have an estimate of the total situation for the sector.

At Acorn Textiles there were two groups of marketing departments. The one situated at the headquarters correctly anticipated a major shift in the consumers' preferences towards even lighter garments whose mixtures of colours and stitch geometry were becoming increasingly complicated. That led,

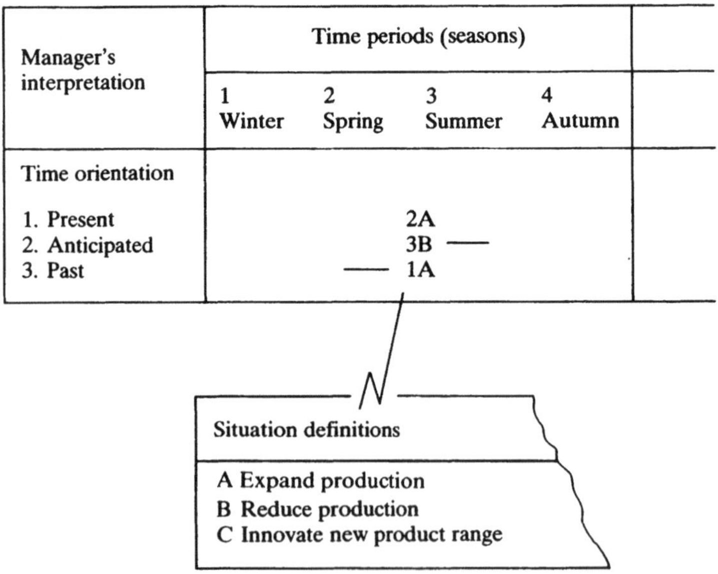

Figure 5-4. Seasonal aspects and design decisions (As Clark 1985)

after a degree of politicking, to the activation of the structural poses (refer to Fig. 1-2) for handling product innovation. Three seasons later it was very clear that the correct assessment had been made. The second department of marketing men at a different geographical location, yet facing the same market contexts, enacted the environment in a quite opposite manner. They anticipated a poor season with low sales from the existing styles and that assessment led to low production and no wastage. However, they failed to prepare for the innovations which came and their warehouses were full of unsold garments.

At Acorn Textiles, the headquarters group of marketing men were skilful at activating the repertoire of the firm, particularly in selecting structural poses suited to the processes of innovation and design. The process of adaptation lasted more than two seasons, but was successfully accomplished. Yet, the incidents and data on which the assessment was made seemed, at first sight, to be bizarre. Their enactment of the environment consisted of a complex series of fragmentary cause-effect linkages spread across the weather, guestimates of expenditure on household heating, interpretations of how the media were affecting tastes in clothes, the capacity of the industry to produce new colours (e.g., dyeing) and styles (e.g., new technology), the main retailers and the fashion expeditions. The second group interpreted low sales as a bad season: a common and unpredictable occurrence in the textile industry. Hence they failed to activate the structure for design and innovation in the product.

The example illustrates the firm-specific nature of the body of thought which was present. However, the marketing men were mainly one network amongst many others: production, engineering, transport and so on. The performance of the firm as a whole rested on a highly differentiated anticipation and handling of events and a complex adjustment of structuration, which was normal for anyone who had been at the firm for a number of years, and about one-third or more of the workforce, especially management, were familiar with these adjustments.

A prime requirement in process models is to build up a description of the phenomena of policy-making because it is events in that area which shape the driving forces of organizational activities, particularly of the degree and type of innovation being undertaken. Key actors with access to the strategic choice agendas normally enter arenas which are already heavily scripted[8].

[8] In the cognitive perspective on leadership the notion of script is used to analyse situations. Scripts are heuristic devices which define issues and which facilitate the understanding of new situations in terms of similar, yet different situations. Scripts have a paradigmatic role and their existence enables leaders to provide meaning to situations whilst simultaneously giving implied directions; a leader can use the script to generate principles

The scripts typically embody the sedimented and silently founded choices from the past. Recent studies of Cadbury-Schweppes and Imperial Chemical Industries (ICI) show that top board members were often divided into alternative camps, but that an established script was difficult to overthrow and replace by a viable alternative[9]. In the former company a slowly formed coalition shifted corporate strategy from product diversification to product concentration and accompanied the shift by a massive change in technology and work organization for both the managers in design and development and for shopfloor workers in every category. Likewise, in ICI, a period of very poor profits and the first-ever quarterly loss, led to the shift of strategy into speciality chemicals and to the development of an American market rather than domestic or European markets. In each company, the shifts involved particular individuals combining to alter the pre-existing script.

These driving forces have been neglected in earlier studies which have tended to be preoccupied with those simple prescriptive models which specify the so-called stages of processes rather than their contents. The specific driving forces will vary, but will include the ambitions by individuals and groups to take themselves and others in particular directions. Their capacity to achieve those ends will be influenced either by their creation of communities within the organization (e.g., British shop stewards creating a constituency in the interestices between the formal labour union and the local management) and/or the occupancy of designated positions in the authority structure, which are bases from which rules can be promulgated and resources can be allocated. These features can be illustrated by the striking examples of the influence of chief executives within the British system of 45 universities following the rationalization program introduced by the Thatcher government between 1981-1985. That program introduced cuts of about 10% overall, but were heavily concentrated into about five universities. Two of the universities (in the technological sector) had their block grant of financial funding reduced by around 40% because they were considered to be recruiting low-quality students and not to be undertaking sufficient basic research for which one-third of the block grant had been allocated. The two universities, Aston and Salford, had a strong image for applied work and the former was at the top of the league for having its students accepted by employers[10]. Just prior to the cuts each university obtained a new chief execu-

---

9 Typically three features are involved: an environmental jolt, the existence of alternative encompassing ideas for handling the jolt, and an authoritative motivational commitment from the members of the organization; these facets are being examined in the ongoing research of Dian Hoskings (Aston University) and at Alberta University by Bob Hinings and Royston Greenwood

10 The case of the situation at Aston University fits the three points of the previous note; see also, Walford (1987)

tive, both of whom turned the normally consensually oriented office into a base for massive alterations in the existing scripts leading them to change their own university, yet each started off in a different direction. At Aston University the new chief executive returning from a score of years' experience at Stanford sought to simultaneously handle the departure of half the staff whilst also shifting the university towards its former areas of weakness: basic research, advanced scholarship, quality intake. The quality intake of students was achieved in four years by a series of annual decrees which prohibited the recruitment of students with weak scores and the result was a massive 50% improvement in entry score, to well above the national average. The introduction of better students meant less remedial teaching and less need for resource-intensive teaching, thereby reducing the average teaching load for faculty and freeing time for research and scholarship. However, achieving an alteration in the research activity was much more difficult to achieve in Britain than in America because of structural rigidities: the slow-changing prestige system, the regulated control of funds from central agencies, and the low investment by British firms in the universities.

In contrast to Aston University, the chief executive of Salford University decided to refine and build on the existing image for applied work by increasing the funding from industry and by creating new posts, often at the highest levels, which attempted to co-opt key scientists and technologists into the university.

The two universities have been at the frontier of other means for attaining a more cost-effective arrangement of activities through the introduction of organizational innovations already familiar in the North American context: redesign of courses to obtain common elements and extensive modularization with the rationalization of unpopular courses. In each case, a dramatic alteration in the initiatory behaviour of the chief executive coincided with externally imposed reductions in resources, coupled to severe and critical assessments of excellence. Moreover, the two new chief executives were very adept at mobilising sufficient internal support, whilst inhibiting countermoves from the pre-existing political factions, as the basis for introducing near-radical innovation and considerable incremental innovations.

The examples just given demonstrate the existence and importance of the bodies of knowledge held by top executives. Moreover, it is evident from the earlier examination of technological trajectories in Chap. 4 that members of the white collar strata associated with design and development also possess distinct bodies of thought which are probably alike and unlike those held by top executives. Yet these two large groupings may account for less than 20% of the total number of employees in an enterprise. So, it must be recognised that the remaining 80% contains further subcultures. In short the whole enterprise contains a plurality of subcultures (Karpik 1978). The problem is: in what ways can the entire configuration of subcultures be conceptualised so that a useful theory of innovation can be constructed?

Two major strategies for conceptualising culture have been attempted: each very different and neither fully satisfactory. First, the individualistic and aggregative approach attempts to reconstruct the overall culture as a summation of scores obtained from questionnaires (Hofstede 1980). That approach provides a great deal of interesting material, yet as Brunsson observes, it provides little insight into the shared basic assumptions and recipes which guide action (Brunsson 1985). Second, considerable attention has been excited by the concept of organizing introduced by Weick (1969). Although very little systematic research has been informed by this perspective its conceptual apparatus has been utilised to interpret other studies. Also, the key concepts are regularly invoked in the plethora of articles attempting to embrace the analysis of organization cultures. It is therefore necessary to briefly identify the contribution of the Weickian solution.

Weick attempted to create a process perspective by two main strategies. First, he adopted a socio-evolutionary model which theorized processes by reference to a population of organizations each and all of whom are undertaking actions in their various environments. Those environments contain niches, some of which provide bountiful pleasurable and desirable resources; others contain few resources for renewing those already in the organization, whilst simultaneously leading the members of the organization to expend the existing reservoir of resources. Some organizations will prosper and others will disappear. Before developing this theme further it is necessary to introduce the second theoretical strategy. This commences with the theory that individuals:

1. Construct cause and effect maps linking together sets of events. These cause maps are used to *selectively* scan events and to select the stimuli to which the individuals will respond. The idea of cause maps can be used to describe the thinking of the marketing men at Acorn Textiles. The whole process is called enactment, that is, the situation is defined symbolically.
2. The cause maps contain concepts and rules for their application to enacted events and are located in the retention system from where they can be activated.
3. To varying degrees, members of the organization monitor the relationship between the stock of cause maps and their symbolic reconstruction of what is happening, what will happen and what did happen. That process of editing may lead some members of the organization to suggest that the cause maps are revised and replaced.

Hence we can imagine that within every organization, which is part of a larger population of organizations, there will be individuals attempting to determine the strategic directions to be taken by arguing the case for the use of their particular interpretation.

To return to the socio-evolutionary dimension, Weick argues that there may be some patterning to the processes by which some organizations receive rewards and others do not. When the patterns of reward, the *selection criteria,* remain stable, then certain individuals may consider that their own cause maps have provided the basis for success. Conversely, failures may also be attributed to possessing the wrong cause maps. However, the main thrust, the deep analytical cut, from Weick is to observe that the relationship between perceived success and the cause maps tends to be retrospective rather than concurrent with events and, that the selection criteria in the environment are in a state of flux for much of the time. The logical conclusion has been drawn by Peters and Waterman, namely, that there is a high element of luck and bad fortune![11].

Weick uses a small number of basic concepts, namely, retention systems, variations and selection criteria. The retention system contains the cognitive frameworks used to enact the environment as well as the cause maps utilised to devise action plans. The variations are the equivalents of the structural poses in the schema suggested by Gearing (1958) (see Fig. 1-2). When activated, these structural poses embody the intentions of key decision-makers and their unfolding reveals a part of the structural repertoire. The selection criteria are the unstable, hidden rules which operate in the external environment to reward or punish the use of specific variations. With these concepts several possibilities can be considered. For example, the retention system may or may not be finely tuned to monitor the performance of variations. Moreover, fine tuning may not lead to extensive learning because there will be a tendency to interpret the perceived failure of a cause map as the need for modification of the cause map rather than the severe editing of its position in the retention system. Radical shifts of the cause maps in the retention system are infrequent and unlikely (Argyris/Schon 1978).

The most striking usage of Weick's perspective has been by Roger Hall in a prize-winning piece of reconstruction of the collapse of the American weekly, *the Saturday Evening Post* (Hall 1984). Hall used the techniques of systems dynamics to examine the internal flows of cash, inventory, orders and advertising and of decisions about the format of the periodical. The reconstruction demonstrated that the various subunits in the firm each operated with a facet of the retention system and that facet was grounded in certain very basic assumptions which were entirely taken for granted. Hall uncovers the following basic assumptions which tied the firm together: the advertising

[11] Peters, and Waterman (1982); see Part I for the use made of the perspective of Weick

department was rewarded for getting advertisers, the editorial department always produced a set ratio of copy per full page of advertising, and the circulation department attempted to increase sales by finding new outlets. Each department possessed its own cause maps and the system worked reasonably well when the periodical was small and in the successful early period. However, within the complex of cause maps, there was a hidden recipe for disaster because at a certain stage the advertising department was able to increase the copy at the same time as circulation was becoming less bouyant. The increase in advertising required an extension of the editorial insertions, thereby increasing the physical size of the periodical and necessitating a switch from a foldover format to the more expensive squared-back format. Put simply, marginal costs accelerated past marginal revenue. Such was the rigidity with which the members of the company held to the established assumptions and cause maps in the retention system that they were uable to reverse its momentum and the periodical entered a period of serious financial losses. Hall's reconstruction does much to assist our understanding of the relevance of Weick's perspective, whilst also displaying its severe limitations. The strength of the perspective is its recognition of the symbolic nature of managerial work and in the provision of elementary concepts for starting a process analysis of phenomena in organizations. However, that strength is drained away until the perspective is more exactingly grounded in an extensive number of theoretical descriptions because Weick fails to adequately address the *contested nature* of organizational symbolism and his concepts lack content (see Sect. 5.5).

5.4 Innovation-Design Capacities

From the base established in the preceeding sections, it is now necessary to re-introduce and highlight the innovation dimension to organizing. This section starts by setting the emergence of innovation as a managerial process in its generalised historical contexts. Later it will be necessary to show how the content of the processes differed in the USA from those in Britain (see Part. III). Innovation originally arose by differentiation from other managerial processes, yet in the modern competitive period that differentiation has been reversed so that innovation has become integrated with marketing and design (Kanter 1984). I refer to that feature as innovation-design. The capacity to innovate depends on an understanding of the difference between two broad types of designing: robust and lean.

It is useful to start by simply considering how the nature of managerial work has developed over the past century. Three main stages of development can be identified (Kaplinsky 1984). First, the early management activities were mainly concerned with the flows of materials, equipment and men within a single production entity, such as the textile mills found in late eighteenth century Lancashire or the plantation systems found during the same period in the southern USA and in the Caribbean area. Managing operating systems was the initial area of development. Second, from the second quarter of the nineteenth century onwards and especially after the Civil War of the 1860s in the USA there were an increasing number of situations where several operating systems were brought inside the same firm so that the stockowners could achieve economies of scale. The second area of development was the coordination of the enterprise as an entity. The development of the cigarette industry in the USA is an excellent example (see Chap. 10). Duke gambled that cigarette smoking would become the normal way of taking tobacco when cigarettes accounted for less than 1% of all tobacco sales and he acquired the license for Bonsack's cigarette-making machine. It was clear to Duke that a small number of machines could supply all the needs of Americans and so he set out to discover how to dominate world sales. Within the USA his cost-cutting actions drove out competitors who were taken over and rationalized. Thus, large numbers of different kinds of operating units were drawn into the same firm: sales units, advertising, leaf buying, distribution and manufacturing. A new set of managerial skills were required to handle the flows between the units and to coordinate their activities around common cost profiles. Third, in the early twentieth century an additional area emerged, that of innovation-design. The two previous forms were largely concerned with managing the near future rather than the long-term future. The new kind of expertise was in designing long-term future situations. In automobiles, Sloan oriented General Motors so that a whole set of new occupations emerged. An example is that of Harley Earl who was recruited from Hollywood, where he customised cars for the early filmstars, to the new design studios of GM to provide a styling which could be centrally envisaged, yet applied with some discretion in each part of the market by the divisions within GM (Bayley 1983).

Each of these three areas of managerial work introduced requirements for new competences and new skills. The initial development of three facets of managerial work was a process of growth by differentiation and addition. The result has been a segmentation of roles and layers of management, particularly in the North American and British managements (cf. Germany and Japan). However, today those internal divisions are being substantially altered by two parallel sources:

1. The application of new technologies to white collar work has progressively transferred many skills of reliability and discretion into equipment. At first the equipment consisted of isolated machines (e.g., comptometers) in the operating facet, then similar equipment for the coordination facet (e.g., computer payrolls, inventory control) and more recently into the heartland

of development and design (e.g., computer-aided design). Further, the new technologies are linking those developments together, as in computer-integrated manufacturing (CIM). These developments have led to considerable reconfiguring of managerial manpower.

2. Comparisons between American and Japanese managements and between British and German managements tend to suggest that the former in each pair is less good at obtaining an integrated managerial team than the latter in the pair. That theme has been tackled in two high-selling texts: *In search of excellence* and *The Change Masters* (Peters/Waterman 1982; Kanter 1984).

Therefore, there has been increased concern with achieving integrated managerial teams. The development of innovation design as a facet of managerial work is in a transitional period where its importance is increasing, just as much as its shape and uses are changing.

In any sector there are certain designs which come to be recognised as state-of-the-art examples[12]. In the aeroindustry the Douglas DC-3 in the mid-1930s and DC-7C in the mid-1950s followed by the Boeing 707/727/747 series are clear examples. Also amongst European carmakers there has been the Renault 4CV after 1946, the Ford Zephyr, the Renault R5 and the Volkswagon Golf. On the whole, these examples are of designs which possessed the capacity to be stretched and refined over longish periods. In an analysis of designing as an activity, Gardiner distinguishes between *robust* designs and *lean* designs (Gardiner 1983). The model assumes that the different stages in a robust design must be spread across more than one firm, though usually they are concentrated in one firm. Gardiner contends that robust designs are accompanied by a good deal of basic redesigning. An interesting example of the lean design is the Ford T which is one of the few lean designs to become what Abernathy labels a dominant design (Whipp/Clark 1986, Chap. 3).

Robust designs possess certain distinctive features as shown in Fig. 5-5. A robust design is one which draws together several divergent lines of development which are relatively novel and then forms these into a composite design. Later, the composite design is internally elaborated and adjusted to formulate a new consolidated design. In due course that design is further developed by stretching, possibly into a design family. The composite designs tend to *filter in* many potentially relevant new ideas and lines of future development in a mixing that often combines good and bad elements. The composite and consolidated stages are based on convergent design processes

[12] The notion of state-of-the-art examples is heavily used by Abernathy (1978) and P. Gardiner (1983)

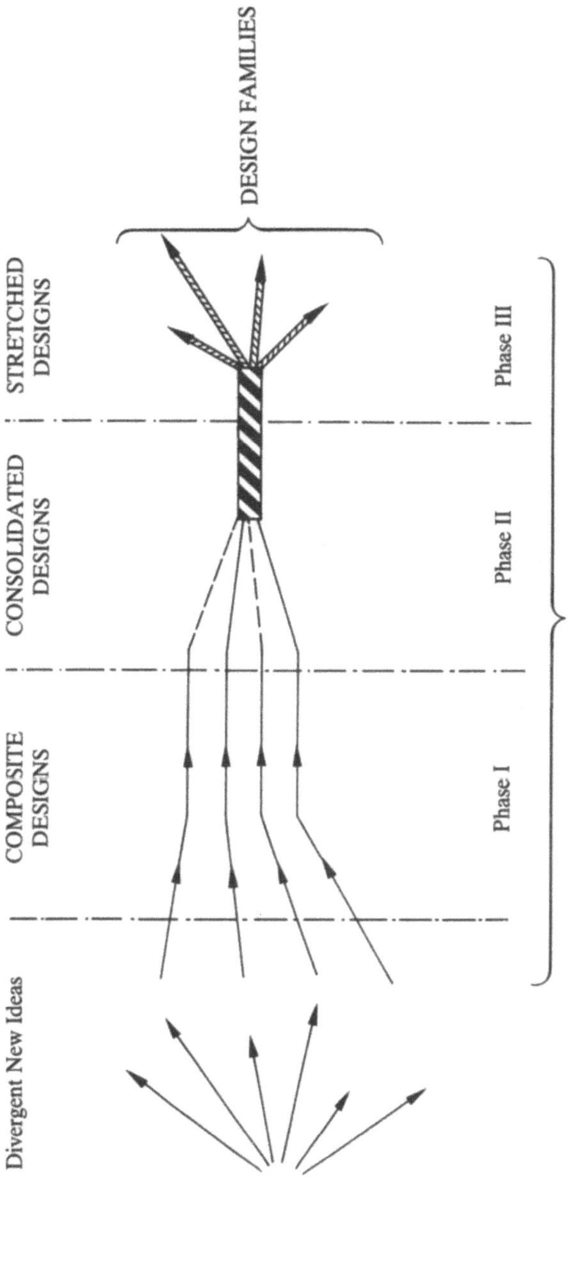

Figure 5-5. **Robust designing: characteristics (From Gardiner, 1983)**

from diverse sources, whilst the stretching is the creation of divergent designs. The consolidated designs of the second phase further rationalized the selected lines of development, giving an emphasis to some elements and downgrading other elements. The difference between a robust design and a lean design is that the latter is a process which lacks the diversity of inputs of the robust design and therefore rarely reaches into the later stages. In a study of the British car firm Rover during its crisis period of the 1920s, it is clear that that firm unwisely attempted to undertake the creation of designing across a wide band of the market and that their various designs were lean rather than robust (Abernathy 1978). Then in 1932, at the very moment when exit from the sector seemed most likely, the actions of its bankers and a major supplier led to the appointment of S. B. Wilks as co-chief executive. Wilks achieved a double shift by moving the market strategy of the firm into the high value added speciality sector, whilst beginning the process of robust designs. The early efforts were sufficient to ensure that the firm survived and obtained significant wartime contracts which deepened its design capacities. In the immediate postwar years, robust designs were introduced for the Land Rover, the Range Rover and the saloon cars.

5.5 Case Illustration: Contested Innovation

The main theme of Part II is that the orthodox approach to the diffusion of innovation requires revision along two main dimensions:

1. To accommodate the actual position of the user enterprise in its own right as a counterbalance and complement to the established view of treating the user as a location for adoption and as a testing ground for the supplier.
2. To gain a perspective on the longitudinal unfolding of innovation

This section illustrates the points already made about the user, whilst Chap. 6 illustrates the principles of the longitudinal unfolding of innovations in a transatlantic context. I shall examine the case of a European speciality car firm with an excellent reputation for robust designing, the former Rover Company, now merged into Austin Rover.

The Rover Company entered the mid 1960s after an immensely successful period of robust designing for its saloon cars and its four-wheel drives just at the point when the European market for speciality cars was developing, especially in Germany and France, whose standards of living were surpassing those of Britain. The directors believed that future growth depended on heavy financial investment and on the extension of its main markets from the domestic and commonwealth territories into Europe and North America. In 1967, the firm was voluntarily merged with the cash-rich Leyland Company, but then found itself brought into the newly created British Leyland

(BL) which the then Labour government had pushed through as a defensive move to protect the future of the British-owned automobile industry. For three years the Rover design team operated as an autonomous centre with BL and then, having produced a concept for the new range of cars, extensions to the existing production facilities were commenced in late 1970. However, within months these plans were altered by a decision from the corporate headquarters. Figure 5-6 provides a highly compacted process description covering the three areas of the car, its production facilities and the work organization through the four stages of design and innovation.

Before merger in BL, the Rover Company was a small-volume producter with robust designs so that a high-volume potential was obtained over the span of several years. The workforce consisted of semiskilled and highly skilled operatives, who identified themselves with craftwork and with craft unions. They had largely accepted the managerial paternalism and there was a fierce, family based, attachment to the image of the firm. As was the case throughout the British-owned car industry, managerial authority had pivoted around a complex and inconsistent array of payment systems whose legitimacy was taken for granted. In the early 1960s, there had been initial signs that the firm was shifting towards higher volume production and to semiskilled assembly work. Also there had been a series of spectacular strikes arising from *intra-union* conflicts over identity and the manifest anomalies of the payment systems. After a highly critical report from an external consultant, the directors had appointed the first specialist executive in industrial relations whose early actions centred on introducing a new protected earnings plan.

In February 1971, the directors of Rover were informed that the company was to be the site for the new speciality division and for its first car, the SD1. The car was to be aimed at the growing European executive market, and so had to be produced in higher volumes than the company had ever experienced. It was to be assembled in an entirely new production facility and paint plant which was heralded as "the most advanced in Europe". In other words an entirely new concept was to be introduced: one which brought with it a whole train of innovations in the car, plant and work organization as well as in the processes of innovation-design itself, because the new concept was to be produced on a company-wide basis. Figure 5-6 shows that the four stages of design and innovation occupied more than five years and that it was seven years before the new facility was operating adequately. Even then its projected sales were only two-fifths of the original estimate of the market.

Over the period from 1971 onward, a plurality of small-design groups were allocated some part of the overall design and translation processes. Those groups concerned with the new production facility had to engage in two

	Concept 1968/69 1970 1971	Translation 1972 1973 1974	Comissioning 1974 1975	Operation 1976 1982
Product	– Nov '68 Specialist Cars Division formed – Corporation rejects P8-10 programme Feb '71 – European Car concept emerges '69-70 including 50000 p.a. output targets and volume production; simple vehicle design – Absence of integrated product planning Heavy influence of Leyland and ex-Ford marketing – Immense changes and risks implied – Target date for production Aug '75	– Jaguar and Rover-Triumph separate Oct '72, rejoin June '73, separate Feb '74 – Feb '72 'programme approval' given to SD1 vehicle design by BL board – Difficulties in developing vehicle engineering; tension between Rover and other BL engineers. Dilution of Rover design capacity – Negligible personnel or production input – Use of suppliers and skilled assemblers	– Ryder Inquiry appointed Dec '74, Report Mar '75 confirms SD1 project status and corporate expansion plans Creation of Leyland Cars Division – 1-year delay 1975. Engineering audit and Liaison Engineer appointed. 1200 Job 1 changes – Time-cost pressure mounts for whole project	– Day 1, 30 Jun '76 – Early high demand. 6-month waiting list by Jul '76 – Pressure to meet output targets, yet difficulties in proving lead to 6% fit for sale Nov '76 and 20% below target. Rectification backlog. Worker protests at vehicle quality Nov '76 – Car: mixture of radical elements within a general profile of incremental innovations
Production Facility	– Feb '71 Rover plan for P8 North Works rejected. Aug '71-Nov '72 15 alternative schemes prepared, leading to: – Nov '72 decision to build Specialist Car Centre costing £95m with flexible capacity – Design brief in-	– Underdeveloped capacity of facility engineers for radical innovation. Dominance of traditional recipes. Little systems use. – Rejection of Volvo methods after May '73 Tension with design brief QWL features – Jan-Jun '73 debate on plant form after ID certificate granted Jan '73	Method build → Pre-production → 90-day run-in – loose cycle times and proving – Unravelling of uncertainties by architects, facility engineers and consulting engineers during commissioning leads to re-designs – Process sheets (late) embody traditional pro-	– Loose task times and proving problems compounded by breakdown of component supplies – Output fails to reach 50% of plant capacity 1976-78. Major overhead penalties mount – Standard progressive assembly line layout 1400 ft tracks (3) raised on mezzanine. 1 million sq. ft floor space

	cles, casino line)	community, contractors – Major interpretive role of architects. Diplomatic role of consultant engineers. – Special Projects Cttee Oct '73 – Jul '73 construction starts	– Special Projects Committee collapses Apr '74 – Plant ready Oct '75	
Work Organisation	– Corporate IR problem set: poor communication, fragmented authority, obscure corporate strategy – Decision to move to MDW 1970 and procedure agreement signed May '72. 'Pragmatic' view of QWL – Rover. New IR Director prime mover. 'Quiet revolution' to formal bargaining and MDW – Consultant's reports Jun '70 – PEP agreement signed Mar '71 Phase I	→ 1971–73 creation of corporate IR Reforms based on MDW, mutuality, participation. QWL – established practices tension unresolved Phase II Phase III Jan '75 Procedure Agreement →	– High turnover of Rover management and introduction of Austin-Morris appointees. – OD consultant attempts rescue of managerial morale – Corporate IR advisor appointed. Minimal impact – Disputes: Internal IEs Oct–Nov '75; BL toolmakers Apr '76. Interunion conflict TGWU-AUEW over recruitment	→ – 'Disoriented' production management and high turnover 1976–78. No senior managers from 1968–75 survive – Union fragmentation and work force sectionalism re-appear. 'Green' labour 40% → – Conventional work patterns emerge. Seventy work stations. Ad hoc training of new assemblers by workers. Work group re-shape tasks. High dispute density '76–78 – New corporate strategy of retrenchment leads to decision to close unprofitable plant. Closes March '82

Figure 5-6. SD1 Rover project: the whole designing process

massive decisions each of which involved many ideas which were unfamiliar to them: with the new paint plant and with the new assembly hall. There was a requirement for a new paint plant which would be able to cope with the new materials to be used in the SD1. So there was a requirement for bodies of knowledge which combined metal working with the chemistry of paint, but it seems that the specific problems were not recognised until the commissioning and operating stages. The new assembly hall was the biggest to be built in Britian since 1945 and there were sharp disagreements on its basic format. The production engineers from within Rover were most familiar with the design of slow-moving assembly lines whose major task was the carrying of heavy weights rather than the pacing of the operatives. They thought in terms of six-minute cycles, of loosely formed gangs of workers who self-deployed themselves, often under the direction of a shop steward who 'managed' a particular zone of the plant. Another set of production engineers, recruited into the corporate headquarters from Ford, thought in terms of one- to two-minute cycles on highly balanced assembly lines where each operative received a measured amount of work. The latter concept relied for its success on the imaginative productionising of the car, that is, the concurrent designing of the allocation of work with the designing of the car itself. The significance of that principle was recognised in Rover, but it had never been applied with such specificity and so analytically. In practice the chosen length of the work-cycle also determines the length of the line and therefore has very considerable implications. A further arena of conflict amongst the various design teams was over the relevance of the Volvo factory at Kalmar and there were quite basic confusions of the actual shape and uses of the assembly line at Kalmar. By the translation stage (see Fig. 5-6) the local union officials were informally aware of what was intended through their network of contacts in the various parts of the design occupations. Their planning for the future, much of it highly cooperative, shows that although their access to decision-making was confined to certain specifically designated arenas (e.g., Special Projects Committee), in practice their existence was constantly recognised in the designing activities. They were far from inert elements.

These points show that the user may be involved in a plurality of quite different kinds of innovation and that the activity of design in envisaged as the core decision process through which that diversity is ordered and assembled, i.e., *innovation design.*

5.6 Recurrence, Momentum and Inertia

In contrast to the optimism about transformational change which typified the early 1960s and became embedded in the quick-change models of Organizational Development, the models of the late 1980s seem to highlight momentum and inertia (Clark/Starkey 1987). The momentum theme emphasises the 'chronic recursiveness' of individuals and structured collectivities like organizations. Momentum as a theme suggests that organizations continue to unfold their actions recursively in certain directions because there are a small number of templates for organizing which become internalised in the structural repertoires. There is ample evidence of the pervasive influence of inertia and momentum. Consequently there is the issue of the extent to which training can influence inertia and its opposite, transformation (Pettigrew 1985).

There are several factors which affect momentum and inertia in enterprises. The specific factors can be illustrated from the state-run British health sector (Loveridge/Child/Harvey 1985). First, methods of funding can have considerable effects. The British state attempts to achieve certain kinds of cost-efficient transformation by hospital design and in the processes of decision-making at the topmost levels of the hospital through the insertion of new roles (e.g., replacing collegial decision-making by general managers). However, it is the private sectors of medicine where many of the innovations in new technology and work organization are occurring. For example, in the USA the high usage of new technology – arguably a cultural disposition (see Part. III) is underpinned through the private insurance schemes. Also, in Germany, a hospital receives payment from the sickness insurance funds for each test which is undertaken. Arguably such funding systems encourage certain practices and the use of the new technologies in testing provides a fast turnover for earning income. Second, certain occupational groupings have a definite influence. In the health sector the licensing of some occupations by the state may introduce structural rigidities into the labour market where the medical profession enjoys the highest degrees of market closure on entry. Such groups are often able to ensure that their definitions of common sense – hegemonic controls – have a pervasive influence. Third, especially in the public hospital system in Britain, decision-making is decentralized to the level of the theatre and ward. So, although the medical profession is the major taker of key decisions, their decision environment and everyday decision-making is strongly shaped by the nursing subculture. The theatre-ward as a subsystem is essentially an operating system rather than a reflective and design-oriented subsystem. Fourth, the health service lacks an infrastructure of design occupations. Until after 1972, the social services in Britain consisted of a fissiparous collection of small departments. More generally, there are various ways in which the *design-to-operating* cycle may simply be a mechanism for the reproduction of existing practices (see Fig. 1-3, Chap. 1). First, at the design stage, one might have expected that a carefully analytical and synoptic thinking was applied to the concept stage (Gold 1979). However, suppliers have vested interests in their products and may be uncommitted to serious modifications affecting the implementation of the new innovations.

Although most design textbooks emphasise the importance of the imaginative production of alternatives, in practice those in the design process usually tend to operate with a variety of decision heuristics which are slow to alter. A study by Lupton and Tanner provided a fascinating insight into how a designer with a choice of many couplings amongst a set of dimensions tended to approach design with a partial usage of the total set and many dimensions 'wired together' in a manner which precluded serious innovation (Lupton/Tanner 1980). A recent study of engineers in Renault suggests that their approach to the redesign of the production system as a site for flexible deployment and variety production was implicitly shaped by considerations inherited from an earlier period when mass production was the pattern (Clark/Windebank 1985). Further, often those groupings involved in the concept stage are in competition with one another. It is recognised by numerous commentators on the concept stage that those most centrally involved may experience very great difficulty in breaking from the bounds of existing thinking in the firm to giving consideration to the full range of possibilities which are apparent. Nadler cautions against two extreme positions: one is the simple reproduction of the past with only slight modifications and the other is the wild optimism about the future in which it is imagined that the enterprise can leap past its rivals by adopting the latest thinking in every sphere of activity (Nadler 1981). In the former case many competitive opportunities might be neglected. That would be very serious in major pieces of design as in any systemic innovation cutting across the product, the production process and work organization. The reason is that the concept stage actually dedicates the future of the enterprise down very specific trajectories which may take years, even decades, to amortise. For example, the choice of orthodox assembly plant at the Longbridge plant of BL for the Metro line would require a decade of car output to recoup the costs[13]. On the other hand if those in concept stage are over-optimistic about what new materials, new practices and new equipment might achieve, then serious difficulties can arise later. The RB211 concept for Rolls Royce involved a very serious underestimate of the problems of new raw materials. Second, the shift from the concept to the translation stage of the process may unfold with a high degree of routineness and with considerable continuity. That smoothness has been found in studies of innovation and designing in the British food and tobacco sector which provides clear illustration (Clark 1972; Smith et al. 1988). However, these firms were typified by integer-power centres[14] and by sophis-

[13] Based on interviews with the suppliers

[14] The concept of integer power centre is defined and used in Clark (1972)

ticated handling of the industrial relations issues. More usually the translation stage leads to a drift from the original concepts (Miller/Rice 1967). The translation stage may lead to further problems when the initial ideas of the concept stage encounter the resistance of middle management, coordinators, and other nonmanagerial groupings. Third, it is during translation that some of the problems in the original concept will be encountered and the internal Design and Development departments may lack the expertise necessary to reshape the original concept and to make better selections of raw materials and equipment.

Usually the shift from concept stage to translation stage involves a different set of individuals and therefore there are strong possibilities of a drift from the original concepts. This problem is more clearly recognised now than some years ago. One of the ways in which some firms handle the transition is to have extensive internal documents and briefing of those involved in the latter stages of translation, commissioning and operating. In certain construction firms this hand-over can involve several days of briefing, seminars and exercises devised to ensure that the new groupings understand the initial intentions (Clark, P.A. 1986b: 130-143).

The commissioning stage is always a source of anxiety, whatever the situation. Various studies, ranging from new hospitals to new petroleum plants, reveal high excitement amongst those involved and considerable tension. It is at this point that those who will be most centrally involved in the running of the new situation will become aware of the detailed features of the new organizational structure.

5.7 Cultures, Leadership and Adaptation

Corporate culture was a very minor theme in organization studies during the past two decades and therefore it is useful to examine a recent restatement by Schein (1985, Chap. 9) of the importance of culture. It should be noted that recent definitions of culture in organization studies tend to reflect a consensus model of the firm and they understate both the degree to which there are intramanagement conflicts and the extent to which there is widespread potential for schisms.

Schein defines corporate culture as the deep level of basic assumptions and beliefs which are learned as modes of handling the adaptation to external problems of survival and to the internal problems of collective integration, i.e., a learned outcome of experiences. He notes that organizations contain many layers, subgroups and occupational associations in addition to which there is usually some degree of geographical dispersal. Therefore, there could be many subcultures, yet it is possible to identify underlying patterns which arise from the cooperation and competition between subcultures. Schein rejects the established definitions given to culture in terms of climate, learning 'the ropes', the espoused values (e.g., product quality), corporate philosophy and atmosphere.

Organization culture has an impact on many facets of the organization. First, any corporate strategy is affected by deeply held assumptions and long-standing patterns of behaviour. In fact, there is a strong argument which states that culture constrains strategy and that the form of pre-existing cultures has adversely affected many recent attempts at mergers, acquisitions and diversifications. More attention should be given to the cultural risks of strategy, most certainly in the area of radical innovations. Second, culture affects the integration of new technologies because some of these new technologies have direct effects on the existing structure of power, or, conversely the existing political systems shape how the new technology is utilised. Third, previous studies of culture have attended to the surface manifestations rather than to the dynamics. Thus, there has been too much optimism about the possibilities of cultural change. Fourth, earlier studies of culture have given too much attention to the general processes of socialization and too little attention to the *content* of what is actually learned. Most existing categories for studying culture are ad hoc lists of dimensions which fail to address the whole organization and to tap deeper dimensions. Fifth, the topic of leadership can be most satisfactorily examined in terms of culture. The role of top leadership and the founders of an organization is central in the creation of a 'consensually' grounded grammer. It is important to emphasise the consequences of the founding culture and to note the great care which Japanese firms in overseas settings take with 'soft' variables when they are establishing branch plants.

Similarly there are basic internal problems of which the development and editing of common languages and conceptual categories is quite central, but is not the only problem. There are other problems, such as group boundaries and individual identities, power and status, intimacy and friendship, rewards and sanctions, and ideology.

Culture shapes technology, concepts of time and leadership. First, technology is culturally defined, yet it is difficult for observers to decode this aspect. For example, few visitors to one of Britain's most successful technological universities, Loughborough, would know that its preoccupation with technology was heavily grounded in its founding events, when the head of a local technical college led his students (in 1920) in the disassembly of a German submarine at Portsmouth and its re-assembly 200 miles away at Loughborough. That very practical project in general engineering was followed by others in civil engineering (e.g., the swimming pool). It was not therefore surprising that all students were required to have periods of practical experience, including their induction in the large, specially constructed workshops on campus. The workshops, apparently very ordinary to the outsider, and a source of qualified dismay to a director appointed for their running in the mid 1960s, were a taken for granted feature by the professoriat. Their closure would have been unthinkable. Yet that was proposed (and defeated) when Loughborough was upgraded to full university status. Curiously, the general engineering degree, although well thought of by industry, was easier to replace by more advanced courses than were the workshops.

Second, there are assumptions about time (and space). According to Hall, American management has institutionalised a view of time as an infinitely divisible linear arrangement that can be subdivided into bits, each of which contains a single activity and decision framework (Hall 1959, Chap. 9). The American culture on time has been labelled the monochronic culture and contrasted with Asian cultures in which bosses deal simultaneously with many issues. Third, culture forms through two core processes, leadership and learning. Leadership is about the creation, management and transmission of culture and requires three basic kinds of competence: an analytic ability to recognise and formulate problems, an interpersonal competence to build and maintain various relationships and groups, and emotional competence to cope with the emotional demands of the managerial role itself. Leadership will play an important role in *what is learned,* both positively in the choice of winning solutions and negatively in the actions to avoid certain kinds of situation and experience. Leaders play a central role in the embedding and transmission of culture through time and through primary or secondary mechanisms, each of which has considerable consequences for the potential of transformational change. The articulation of various visions by leaders will be embedded through several primary mechanisms, such as the criteria they adopt, when allocating rewards or sanctions and in key areas like recruitment, promotion and 'excommunication', Likewise what they pay attention to will be influential. Equally important will be their reactions to critical incidents or crises and their activities as mentors and in role-modelling. Their reaction to critical incidents will have powerful consequences. There is a great deal of evidence to suggest that the handling of corporate crises has direct effects on the commitment and motivation of employees. In addition there are secondary mechanisms which both articulate the culture whilst playing a major role in its reinforcement. These will include the whole area of myths, legends and stories about incidents and people. Recent studies have strongly emphasised their role in shaping everyday events and reactions. Other secondary mechanisms are the standard operating procedures, formal statements and the actual design of the physical space and its allocation.

The dynamics of culture require attention to the earlier distinction between recurrence and transformation[15]. The concept of momentum refers to the dynamics of momentum, whilst inertia may be construed as the overall tendency to reproduce established definitions of problems and to leave the repertoire of structural poses unchanged over long periods. The dynamics of culture need to address the problem of the adaptive coping cycle by which changes occur in the organizational paradigms and thinking practices as well as in the structuring of relations. Schein suggests that the typology of four functional problems known as AGIL can be used with caution (Parsons/Bales/Shils 1953, Chap. 5). The typology presumes that all organizations have to resolve four basic problems in sequence and in a balanced relationship to one another. The four basic problems may be described as:

1. Adaptation to the external environment: Adaptation includes the degree of agreement and acceptance of the core task, the extent to which there are the means to implement the core tasks, the scope of the mechanisms for editing cause maps and the strategies for developing new policies; changes in the organizational language and concept of its own history will be the targets for transformational change
2. Attainment of corporate goals
3. Integration: the handling of internal strife
4. Creating quiet passages for the reassertion of latent core values and latent identities

Each functional area becomes the specialised province of specific managerial skills. For example, the problems of the management of the production unit and of the coordination of a number of locations requires sequential attention to attainment of goals, to integration and to erase handling latency. However, it is quite possible for an organization to survive under certain conditions with only slightly developed capacities in adaptation. For example, in the early phases of a new sector the functional area of adaptation might reside entirely with the founder(s) and might not become developed as an autonomous subsystem until much later; such was the case with the car firm of Rover in Britain until 1932.

The functional problems of adaptation are those which relate very closely with the building of a corporate capacity in innovation-design. Yet, even when a separate subsystem devoted to adaptation has been developed the subsystem may fail to notice danger signals because its cause-maps are insensitive to the implications. For instance, in the post-1945 period the French car firm of Renault engaged in the massive ingestion of American concepts of car design, car production and marketing around a state of the art model

[15] See discussion in Chap. 1

known as the 4CV (Clark/Windebank 1985). Once established the adaptation subsystem began to articulate and to ossify its practices around the recipes of mass production and mass distribution. Thus, in the 1970s, when the consumer market began to change, its existing adaptive system had acquired such momentum and inertia that the transformation of its subculture became an enormous and lengthy task. The problem of Renault illustrates a limitation in the approach to adaptation of Lawrence and Dyer (see Sect. 1.3.3) because they concentrated upon the population of organizations and failed to uncover the means through which organizations achieve transitions in their cultures[16]. The extent of the problem was illustrated in the earlier examination of the discontinuities and the sector life-course model (see Sect. 4.4). From the perspective of specific user organizations, adaptation requires a corporate design capacity in which there is a high degree of appropriation of ideas, practices and innovations which relate to the core activities of the organization.

5.8 Appropriation: The Basis for Survival

It is not possible to cover all the significant facets of the position of user, thus the objective has been to emphasise how the internal processes and strategic decisions about innovation-design of user organizations might be more effectively analysed. The main threads may be drawn together by considering the degree to which the user needs to appropriate particular innovations (Wilkins 1974).

There is a simple scale which can be used to compare the extent to which users are able to ingest and develop innovations. Figure 5-7 shows a vertical continuum ranging from total failure to ingest a desired innovation through a process of ingestion by simple imitation, to the situation where the user has so appropriated the innovation that the original innovation has been modified and robustly redesigned as an integrated feature of the user. There are three broad areas: failure to adopt, partial imitation, and appropriation. Each state can be illustrated by taking the case of computer-aided production management (CAPM)[17].

Failure occurs when the user attempts to adopt a relevant innovation, yet is unable to make the innovation operable. In its simplest, minimum format, CAPM requires the user to establish certain basic databases for auditing the existing processes of production and these are then monitored against re-

---

[16] See the discussion in Chap. 1 of the perspective of Lawrence and Dyer (1983)

[17] SERC/ACME investigation of the extent of adoption of computer aids to production management in Britain, by Clark, P. A., Tann, Bennett, and Scrimshire at Aston University

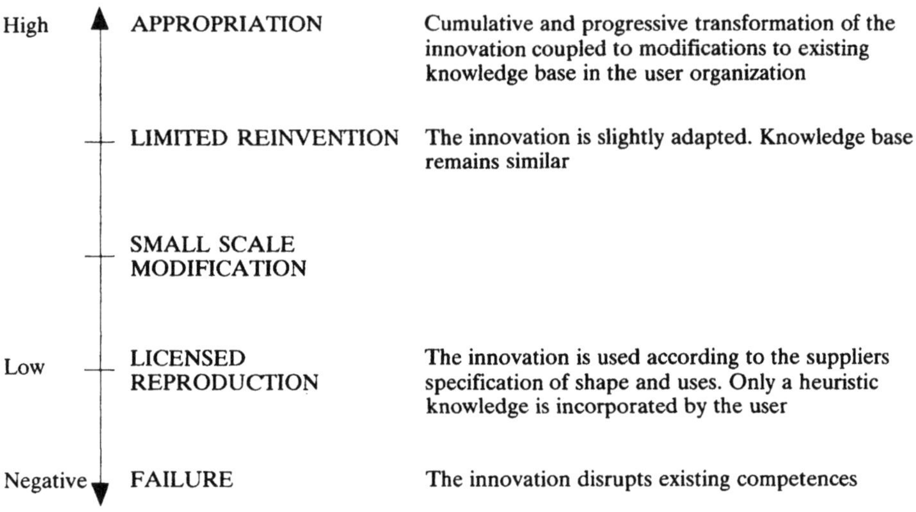

Figure 5-7. Degrees of appropriation of innovations by users

quirements. Failure would occur when the competition and conflict between marketing and production restricted the accuracy and the recency of the data which was available.

Partial imitation is a situation where an innovation is adopted in its simplest, minimal format. Thus, CAPM could be utilised by applying heuristics to the standard data sets on the basis of a turnkey system devised by outside consultants. In that instance there would be some degree of appropriation of the innovation, but that degree would be very slight and would remain static until the external consultants returned and devised updated systems.

Appropriation is a situation where the user starts by recognising the potential value of CAPM and manages to:

1. Narrow the absorption gap between the requirements of CAPM and its own limited capacities, then ...
2. To begin to creatively modify and refine the innovation so that the user develops the specific innovation on a continuous incremental basis

Success in appropriation is rare.

The problematic nature of appropriation is indicated by many examples of firms which license a process from its originator, yet fail to develop the process further. For example, the American firm Continental Can developed a method of mass producing the open-top can and licensed that process to the British owned firm, Metal Box, which in turn applied the process throughout their factories in the domestic market and throughout the former Commonwealth markets (Reader 1975). In the 1960s it became feasible to make cans by a new process known as extrusion and the new process was developed by the R and D experts in Continental Can and subsequently licensed to Metal Box. In that case the British firm, operating in its domestic markets, was able to secure very satisfactory commercial returns from licensing, though its own slight R and D investment was unable to appropriate either the old or the new technology to the point where its patents controlled the benefits from the investment. In contrast, Japanese firms (see Chap. 13) have been more successful in appropriating organizational innovations, like quality control and technological innovations.

The objective of the user should be to appropriate those innovations which are essential to its core skills and from which it can derive benefits. However, appropriation implies the continual, cumulative and incremental modification of an original innovation in all its details. The modifications may be so extensive that they cumulate into a radically different entity. That point is illustrated in the case study which introduces Part III in the next chapter.

Part III.
Anglo-American Patterns of Organizing

6 Transatlantic Evolvement I: Americans and the Absorption Gap

6.1 Introduction: Absorption Gap Illustrated

The preceeding chapters have established that innovations can take different forms and that they evolve and change their shapes and uses. It has also been indicated that the contexts of an innovation will alter when an innovation is moved from one space to another, especially when the change of location involves a different culture. That process of alteration in the innovation arising from changing contexts will be greater when the innovation is loosely formed and when there is an only partially developed agency for its transmission. Therefore, one can imagine that if any innovation flows from one context to another, then the changing shape and uses of the innovation might be revealing of the context. That possibility is illustrated in this chapter in two steps.

The chapter commences with a brief examination of the excellent study by Jeremy of the transfer of British textile technology and practices into the American context (Jeremy 1981; Kranzberg 1970: 255-276). Although that study refers to the period 1790-1830, rather than today, its basic conclusions apply equally well to contemporary innovations. The Jeremy study is used to introduce the second illustration which is of an organizational innovation, a form of "teamwork". Many years ago Riesman and Denney examined the game of American football and concluded that it had evolved from the British game of Rugby union[1]. Today it is possible to extend and deepen their study and to show how the evolvement of both games over a period of much more than a century is highly revealing of certain features of middle-class cultures in each society. Further, as North Americans know well, games in general and American football in particular (and its Canadian variant) provide a rich imagery of metaphors for everyday life, particularly in the male executive culture[2]. It is not for nothing that a leading feminist writing about the problems faced by the female executive noted that the games her mother never taught her were the field games of American football and baseball! Ideally the analysis of the two games should be accompanied by selected

[1] Riesman and Denney (1954); Bale (1984) 15 (1): 38-63; see particularly the essays in Hart (1972)

[2] See the discussion by Keidel (1984); Manning (1979)

and edited videos so that the reader can compare the differences and can better anchor the points being made. Fortunately, American football is now available on British television on a regular basis and is watched by several millions. Also, rugby union is shown on the public networks of American television, and there are numerous excellent accounts of the two games.

My analysis will show that – treated with care – the different trajectories of evolvement taken by the two games provides a valid and insightful view of the shaping subcultures for the executive stratas in the two countries.

The transfer of innovations across national boundaries may be either a passive process for the recipient society (e.g., South America in the nineteenth century) or an active process in which the recipient society gradually takes over the original innovation and develops it in a distinctive manner which is at least equal to the inputs made by the originators of the innovation. The former case is one of transfer and imitation, whilst the latter is a case of societal appropriation which Wilkins labels the absorption gap (Wilkins 1974).

There have been remarkably few systematic and longitudinal studies of the diffusion of innovations from one country to another. However, there are a handful of excellent examples which can be followed and refined.

One such case study which illustrates the principle of the absorption gap is Jeremy's reconstruction which examines the transfer of British textile technology, technique and knowledge into the USA between 1790 and 1830 (Jeremy 1981), and demonstrates how and why Americans in the east coast corridor between New York and Philadelphia succeeded in closing the absorption gap with Britain. As is well known the rise to economic and political eminence of Britain after the eighteenth century rested heavily on its earnings from shipping and from the commercial activities of the City of London. The rise of the cotton industry provided an important addition which acted as economic multiplier stimulating the growth of related industries, such as transport, energy and construction. The principles behind the case are general and it has been selected because the original study, undertaken in the two countries, by David Jeremy, is carefully documented and beautifully illustrated. Moreover, the study is suggestive of how other examples might be reconstructed and examined. For example, it is now evident that the Japanese absorbed American business practices in at least two waves, Taylorism after 1914 and Quality Control in the 1950s. It would be highly relevant to examine those cases following the framework used by Jeremy (Jeremy 1981).

There were considerable barriers to the diffusion from Britain to America, some deliberate barriers and others unintended. The British government placed restrictions on the transfer of machinery and of certain kinds of knowledge in an endeavour to protect expertise. Faced by that situation some manufacturers of equipment sought to evade the restrictions and their actions were complemented by American enterpreneurs who sought to observe and record British methods as well as to recruit key managers and millwrights. Moreover, there was a large number of textile workers who emigrated to the USA and might have taken their skills with them. These deliberate attempts at diffusion were hampered by the absence of standardized categories for describing the weights of yarns, which in turn were reinforced by the great diversity of unrecorded practices of manufacturing. Britain, in contrast to France and the USA, lacked the predisposition towards the encyclopaedic documentation of the 'how to' type of knowledge of the Americans or the more abstract knowledge found in the French encyclopaedias of the period. These barriers were heightened by the diversity of market segments served by the complex horizontal and vertical grids of the Lancashire-based industry (Lazonick/Mass 1983). Consequently, the early attempts to transfer the industry largely failed. Jeremy observes that in Philadelphia during the early 1790s the boxes containing an Arkwright mill remained unpacked and unassembled for lack of adequate skills and knowledge.

In fact the key knowledge and skills were brought into the USA by less than a score of key men who came into the development corridor between Philadelphia and New York, especially in the areas close to Philadelphia. There, an indigenous industry began to surface in a series of uncertain steps commencing with the build-up of a small nucleus of key carriers of technical knowledge who possessed plans and made models of equipment, with verbal descriptions. They used American cotton and made wooden machines to establish pilot production in highly localized plants connected through friendship and kin networks. Their development of an indigenous industry spread through a neighbourhood effect and became increasingly autonomous from British developments.

There were several quite crucial features of the emerging American industry. The raw material cotton was of a very high quality long staple which was strong enough to be usable with the very shaky and jerky wooden machinery. America had excellent supplies of hard woods well fitted to the purpose. In due course the wooden machinery broke and was readily replaced, each replacement affording the opportunity for the ingenious local craftsmen to further refine the basic equipment. Another change factor operating in that context was that the market for the output was quite unlike Western Europe. In the USA the market was for a coarse, very basic cloth which was used by the women in the homestead to make a variety of garments. Thus the market demand was relatively homogeneous.

Figure 6-1 interprets the configuration of factors which were operating to illustrate the interaction between the combination of contextual factors like cotton staple, wooden equipment and the type of market, with the specific

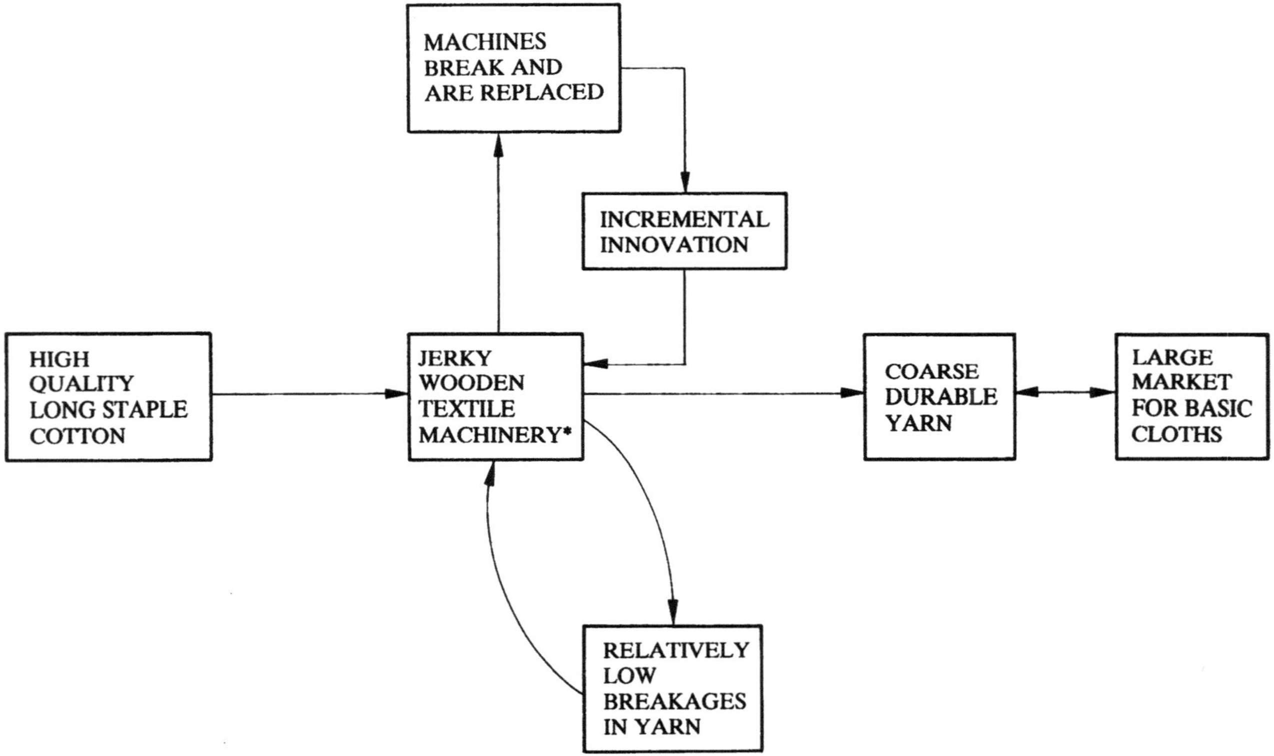

Figure 6-1. Contextual factors and agency in the American appropriation of textile production techniques and the creation of production knowledge (American equipment was specialised to a task and therefore non-versatile)

cultural skills of certain groups who were able to create a nonversatile technology and break from European, especially British practices.

During the four decades covered by the study these Americans managed to appropriate the requisite knowledge and to close the gap with Britain. Later, the American industry continued in its own distinctive trajectory of evolvement.

Jeremy's very extensive study illustrates a number of features which the later analyses will develop further. The transfer of cotton technique to the USA is clearly an example of the active closing of the gap between the two countries. By the 1830s the American industry had acquired a large part of the means of production for producing goods which were functional substitutes for those produced in Britain. In one way or another the development of the American industry illustrated three of the four lines of development indicated by Wilkins: importing products which are copied, adapted and then produced in the USA; observing best practice techniques and then modifying them for the new context; taking over British plants (Wilkins 1974). The fourth route, purchasing and developing patents, was also used through a process of smuggling and substitution. In the American case, the narrowing of the absorption gap was eventually facilitated by the growing size and structure of the American market which permitted economies of scale (see Chap. 10). Also the internal infrastructures of transportation and marketing skills were well developed.

6.2 Case Study: "Teamwork"

6.2.1 The Questions

Teamwork in the enterprise was encouraged and facilitated in America and in Britain throughout the nineteenth and twentieth centuries. It is not a late twentieth century discovery as some ahistorical accounts of work organization have implied. What is problematic, however, is the meaning of teamwork. Equally problematic is: do the British and the Americans have the same understanding of teamwork[3]?

An early indication of divergence between Britain and America is provided by the reaction of British workers in the late nineteenth and early twentieth centuries to the introduction of American teamwork into the British boot

[3] There is widespread assumption by students of work organization that teamwork is a modern development from the sociotechnical tradition; in practice forms of teamworking were well established in nineteenth century Europe, but these differed considerably from the forms which later emerged and were scripted by industrial engineers

and shoe industry. Church reports that British operatives, supervisors and employers were sceptical about the viability of American practices of team-working for the British context (Church 1968). The reaction of the British employers is especially interesting because many of them made explicit and implicit use of the ideology of being in a 'team'. Indeed a number of larger employers specifically supported company team sports to foster the team spirit.

6.2.2 Managerial Templates of Organizing

A template may be defined as a shape which provides the structure for the production of a new product. For example, the German monthly magazine for women, *Burda,* contains many templates for do-it-yourself dressmaking, each one of which can be modified to suit the individual circumstances of the user. Also the concept of the template can be applied to social activities like the wedding (Harre 1979). What kinds of templates for organizing do managers have in mind when they use the analogies of the team and of teamwork?

Analogies with teams and with teamwork are used extensively by managers (see note 2). In this chapter it is demonstrated that when managements in Britain and America refer to teamwork they have in mind quite different templates for organizing. The chapter examines two winter sports each of which can be shown to be influential as both a symbol of organizing and a source of concepts (e.g., "dropping the quarter back") for organizing:

1. Rugby union
2. American football

The two games are examples of how to organize for promoting responsible autonomy amongst a set of competent persons. In Britain, the nineteenth-century private boarding schools were a crucial arena for the development of British templates of organization and it was in that setting that two novel winter games were slowly articulated[4]. They became known as association football and rugby union. The former is a kicking game, whilst the latter is more of a carrying game. The games also differ in their opportunities for structured violence (Elias 1981). These two games were diffused, as templates, into Canada and the USA after 1850. Association football was the more defined template, possessing a specific set of rules. Rugby union was still somewhat in an embryonic state and its rules were heavily overlaid with hidden social meanings (Riesman/Denney 1954).

[4] Riesman and Denney (1954); Dunning and Sheard (1979); Elias and Dunning (1971)

The mid- and later-nineteenth century was a period when there was considerable interest by Americans in European, especially British, sports (Rader 1983; Baker 1984). In the American universities the two British ball games were taken up, taken apart, reformulated, and then constituted into an entirely new game: American football. That game then evolved from a mass formation game into a finely tuned exhibition of space-time synchronization and choreographed maneouvres. Moreover, in North America, association football and rugby union failed to attract players and watchers. Attempts in the past decade to adapt association football to the American context have so far failed[5].

This chapter explores the close interlinkages between the development of a strata of middle-class male executives and the sports they came to play[6]. It is demonstrated that the template of American football differs very sharply from that of rugby union. The evolvement of American football reflects the rationalizing spirit of American middle-class society and reveals its adaptive potentials. The slower transformation of rugby union is also indicative. These two games provide starkly different templates of teamness and teamwork.

Rugby union became, and largely remains today, a distinctly and deliberately amateur game. That feature is evident in many respects. For instance, the moral principles sustained by the authorities governing the game state that the reflective discussion of strategy and tactics during the match must exclude the active involvement of nonplayers. Thus, the taken for granted role of the North American coach is intentionally excluded. In rugby union it is taken for granted that the players should control the directions of play. That principle means that the appropriation of skills by nonplayers, so evident in North America, was inhibited until the 1960s. The moral principle is still active, yet is being challenged and gradually subverted. For example, in 1984 a correspondent of the *Sunday Times* reported on a game between a composite Franco/British team, known as the Barbarians – chosen to symbolize that moral principle – and the visiting Australian national team. At the interval, as the players stayed on the field, an Australian reserve came to the resting players with light refreshments and *also* a piece of paper, which looked remarkably like instructions from the travelling coach. The moral principle was being challenged and possibly restated.

American football gained a hegemonic hold over winter games at American universities and colleges from where its intercollegiate features were progres-

5 These have included altering the flow of the game and examining the possibilities for increasing excitement by widening the size of the goals

6 The broad principles behind this attempt are set out in Dunning and Sheard (1979); Eric Dunning and Rick Gruneau made helpful suggestions which I have attempted to incorporate

sively inserted into the high-school curricula and then down to the pee-wee leagues. From its earliest days in the 1870s, American football was significantly shaped by the actions of coaches who operated within contexts which led to the institutionalization of an orientation to play that transformed play into sport, thereby linking work and sport[7]. In America, sport became the social manifestation of nineteenth-century social Darwinism and its progressive choreographing, now undertaken for mass television audiences, transformed coaching and the ownership of activities in subtle and significant ways.

In this chapter it is shown that the encounter of rugby union with American university students in the 1870s reveals hidden cultural barriers to the transatlantic diffusion of teamwork. The pattern of evolvement of the two games is most significant for their separate trajectories of formation and institutionalization. The differences in internal spread and segmentation of the games in the two countries reflects crucial contrasts in the cultural hegemony of the social templates of **teamwork** in America and Britain.

The following sections present a comparative reconstruction of the evolvement of rugby union and of American football, showing how the general cultural features interacted with the distinctive milieu in which each game was intentionally and unintentionally shaped. As a preliminary, however, it is necessary to anticipate some later points by noting the basic assumptions which underpin the case study.

6.2.3 The Evolvement of Innovations: Recapitulation

As demonstrated in Part II, the study of innovation as a process is at a turning point. There is widespread dissatisfaction with existing paradigms and their influence on empirical research[8]. A new counter-paradigm is being formulated (see Chap. 3). The orthodox concept of an innovation tends to treat the process of innovation as if there was an object which was fixed in its attributes and uses, easily recognizable, could be rationally evaluated before use, and remained essentially stable in its attributes over time. Existing approaches are too strongly oriented to the interests of the seller of the innovation (see Chap. 5). Recent developments in research and theorizing suggest the following basic assumptions:

1. From the point of view of the user the shape and uses of an innovation of any kind are normally problematic. A recent study demonstrated that British experts in the building of automobile assembly lines found diffi-

[7] Broeg (1983); see also the essays in Gruneau and Albinson (1976); Howell and Howell (1969)
[8] See the discussion throughout Chap. 3

culty in describing the Volvo car plant which they had visited (Whipp/Clark 1986).

2. Innovations are sets of ideas with both technological and organizational dimensions which are embedded in distinct sociocultural settings (Riesman/Denney 1954).
3. Innovations are typically combinations and consolidations of many diverse elements[9].
4. Innovations are best conceptualized as bundles of elements with varying degrees of tightness and looseness. The generators of an innovation may consider the bundle to be tight, whilst the users may well prise it open and make a quite distinctive series of uses (see Chap. 3).
5. The users make significant and extensive contributions to the eventual shape and uses of an innovation[10].
6. Innovations are shaped by the unintended configuration of contending groups who surround their construction and adoption.
7. Innovations vary in the degree to which their initial, founding conditions affect their subsequent stretching and evolvement[11].
8. Innovations are often temporary solutions to an array of interconnected pressures[12].
9. The organizational context of innovations is often highly differentiated so that the struggles between contending groups and strata operate on the innovation long after its apparent introduction into any setting.
10. Innovations are influenced by many contextual factors. For example, the growth of the railway and the telegraph contributed to the rapid diffusion of American baseball after 1869 when the first club tour was undertaken by Cincinatti and after newspapers carried syndicated league tables.

There are two quite different sets of assumptions which have underpinned the examination of the intersocietal spread of innovations. The orthodox model assumed that there is a process of imitation (see Chap. 3). Now, it seems likely that imitation can explain an important part of the variance within a society, but not so much with regard to the flow of innovations between societies (see Chap. 12). The model based on the ten assumptions outlined above presumes that innovations are re-invented[13] and maybe appropriated.

---

[9] J. D. Eveland has made an important and insightful examination of this area and I was greatly assisted by discussions with him

[10] See Chap. 5 and the case which follows in this chapter

[11] This point requires much more extensive research and enquiry; see also the notion of robust design in Chap. 4

[12] This is implied in the approach of Mohr (1982)

[13] The concepts of re-invention, appropriation and innovation-modification require further specification

6.3 British Predispositions

6.3.1 Nineteenth-Century Societal Context

It is widely accepted that the private boarding schools established in the eighteenth and nineteenth centuries were major arenas of social change and that events within those settings were significant in shaping the templates of organization which the elites carried with them, both within Britain and to the former Commonwealth. Wiener, in his influential analysis of the development of English cultural predispositions (Wiener 1981), contends that the schools had a formative imprint on future leaders because of their values of gentlemanly behaviour. It is concluded that within the schools there was a struggle between two rival visions. One vision favoured the development of trade, commerce and, to a degree, industry. The other vision favoured the development of the humanities, gentlemanly behaviour and political control. Wiener contends that the aristocracy and minor gentry gained control over the value systems of the schools and were able to impose their template of gentlemanly behaviour. Thus, the boys became 'gentrified' (Wiener 1981). That template of behaviour was taken from the schools into the London clubs, into the major universities of Oxford and Cambridge and into corporations[14]. This section explains how the evolvement of two winter sports – rugby union and association football – both reflected the processes of gentrification and provided an unintended template of teamworking.

The significance of gentlemanly behaviour in sport from the midnineteenth century until the 1960s is clearly revealed in the distinction between 'gentlemen' and 'players'. The former play sport in such a way that rehearsals and practice should be hidden, casual and elegant. Winning is a matter of fate, good behaviour, furtive adjustment and brave physical responses. 'Players' want to win, are prepared to be seen practising, routinely examine their strengths and weaknesses, rely on coaches and are prepared to construct game plans to facilitate winning. In English cricket, a summer game established in the eighteenth century, the differences were symbolized in a regular three-day match (note the length!) between gentlemen and players, a practice which persisted until the 1960s. Coleman drew a parallel between that match and the struggles within the board room of Courtaulds, one of Britain's major corporations (Coleman 1959). The template of the gentleman and Wiener's suggestion that the British elites became gentrified are important points of entry into the understanding of British predispositions.

This section examines the development of gentrification in the private boarding schools attended by the sons of the established elites and by the sons of

---

14 Wiener (1981); Barnett (1976); Fox (1984)

the newly rich. The evolvement of the winter games within the schools, especially the contest between rugby union (at Rugby School) and association football (from Eton School) provides an important probe into elite cultures.

Examining the evolvement of rugby union presents certain challenges because Hobsbawm contends that the British are experts at inventing traditions (Hobsbawm 1983). In British life the cultural value of tradition is often used to legitimate the status quo. In the case of rugby union the history of the game was of little interest until the 1890s when its official governing body commissioned a report. That report attributed the origins of the game to a pupil at Rugby School, Webb Ellis. The report claimed that in 1823 Ellis broke one of the basic rules of the embryonic game when he picked up the ball and, instead of kicking it downfield, he ran with it and claimed a touchdown. More recent reconstructions, now accepted by the governing body of rugby union, show that the origins of the game cannot be understood by focusing only on Rugby School (Titley/McWhirter 1970; Dunning/Sheard 1979). Rather, it is necessary to examine the evolvement of rugby union and association football in the context of social struggles within the middle and upper classes over different visions of being English (Dunning/Sheard 1979; Elias/Dunning 1971). The outcomes of that struggle had unintended consequences[15].

6.3.2 Origins

In the later eighteenth century the aristocracy and gentry began to send their sons (and daughters) to board in establishments which had been initially established for the sons of impoverished gentlemen. By 1780 approximately 80% of the aristocracy attended four key schools, including Eton. This new practice of schooling created many unexpected outcomes.

There were often confrontations between the boys and the masters. Rugby School experienced two rebellions in the late eighteenth century and the militia had to be called to Winchester School in 1818. The boys sought to obtain a high degree of control and autonomy over a situation in which they were faced by their parents' employees. In the area of leisure pursuits the boys engaged in active, sometimes violent forays which they described as hunting expeditions and the boys obtained a high degree of control for themselves in the areas of the boarding houses. Here they instituted the practice later known as fagging by which younger boys had to act as unpaid ser-

[15] The unintended outcomes of struggles between competing groupings was discussed in Part II; the notion of *unintended consequences* is essential to the study of innovations and their evolvement

vants for senior boys. The older boys claimed the right to discipline the fags. To varying degrees the masters and other authorities sought to regulate and control fagging, but the institution remained relatively strong throughout the nineteenth century and survived well into the present century. In a further attempt to regulate the boys, some headmasters introduced the prefect system by which designated senior boys were incorporated into the authority structure, yet were in principle bound to maintain a less punishment-oriented form of governance. It was in that context that boys sought forms of winter activity which were satisfying to themselves and could be undertaken at the 'frontier of control' with the masters[16]. One of the forms of winter activity was a form of ball game.

Ball games had been played long before the opening of the public schools, yet those earlier forms were played very infrequently, often on a particular day in the year. There were a great many variants all of which were entirely local and very violent. The typical game was a rare event in which an uneven number of men from two adjacent villages either kicked a ball or attempted to carry the ball over a designated line. The two lines were very far apart and the area of play would cover acres. The encounter could last all day. In the kicking games the participants would play over a vast area of land and the ball would disappear for great periods within a mass of men who were shoving, biting, hacking and punching. There were few rules and a great deal of excitement. Many were injured, so when the game appeared in urban settings it was often banned. The carrying game was less violent and may have been more regulated. The evolvement from the folk games into modern sports commenced in the last quarter of the eighteenth century in the public schools and then, from 1840 onward, at Oxford and Cambridge, and later in a small number of specially founded clubs. From the 1850s onward the kicking variant (association football) and the running variant (rugby union) were increasingly established in local clubs. Association football soon became a popular working class activity and a spectator sport (Mason 1980; Bale 1980). It became a game very much more associated with players than with gentlemen.

In the public schools it seems likely that many of the initial games involved uneven numbers of boys playing across large, crudely demarcated areas. The methods of winning were to carry or kick the ball over a designated line. The two sides might involve a hundred or more boys and the ball would typically disappear from sight into a mass of players. The few onlookers could infer the position of the ball from the mass of shoving players and the game might last several days before one side scored. It was crude and violent so many boys were injured.

---

[16] The notion of the frontier of control is adapted from Goodrich (1975)

The original template of the folk game was highly unspecific. Consequently, the boys had to formulate local rules. An account by Strutt shows that by 1801 the folk game was adapted to the local topography of buildings and fields. At Eton, for example, the wide corridor and buttresses between certain buildings were utilized for a rough game of carrying the ball whilst an area of wide fields was used for the kicking game. At Rugby School an area known as Bigside was about 100 yards by 200 yards[17].

The control of the games was in the hands of the boys. At Rugby School they recorded and edited the rules and it is therefore interesting to note in an account of a game in 1801 by Strutt that he reports that the two sides were evenly matched and played between two lines one hundred yards apart. It is evident that certain developments had taken place most notably in the regulating of the areas of play, in the defining of a score and in the reduction of the numbers playing at any one time. Sides still varied in numbers, yet the reduction in numbers meant that the ball was now visible from time to time and that it sometimes came out of the maul of players.

It is important to stress that the rules were devised to assist the captains of the two sides in resolving differences. There were no impartial referees because gentlemen do not require referees. Infractions of the rules could be decided between the captains or by relying on fate, e.g., by the tossing of a coin.

6.3.3 The Embryonic Games

In the first quarter of the nineteenth century there were indications that the games at Rugby School and at Eton were diverging. Both schools played games based on kicking. However, at Eton the objective was to kick the ball between two posts placed 14 feet apart and with a bar across the top. At Rugby the two posts were much taller and the bar was placed to create the shape of an 'H'. The method of scoring was to kick the ball over the bar between the posts. The winning side was judged on the goals scored. In the Rugby game a goal could be scored either by a direct, drop and kick movement, or, by working the ball over the line defended by the opposition and claiming a touchdown. After the touchdown, which did not score, there was an opportunity to take a placed kick at the goal posts.

At Rugby School scoring a goal required that a player drop-kicked the ball. Consequently, the game permitted handling, but did not normally permit running with the ball. The general form of play was for the one side to come close to the other's goal and then attempt a drop-kick. The rules at Rugby

[17] Calculated from Titley and McQuirter (1970: 29)

were variable, and were subject to revision. After 1800 when the ball came out of the maul more frequently the captains would designate small boys (the fags) known as dodgers to wait for the ball and then to punt it towards the opposing goal where the new maul would form until a score was achieved. The rules prohibited players taking the ball out of the maul. *It had to exit unintentionally* and the two captains were expected to settle any disputes[18].

The rules were slowly refined in two directions, with the purpose, first, to gradually delimit the areas of the opponent's body which might be the target for hacking, and, second, to reward the occasions when one side pushed the ball over the opposing line and grounded it firmly. Then, the offensive side received a direct kick at the goal. It was this rule which provided the key link to the scoring of a touchdown. At various times it seems that the rules at Rugby School became so fluid that occasionally boys picked up the ball as it exited from the maul and carried it downfield. That manoeuvre was greeted with the shout "hack him down".

In this period the ball games were almost entirely played within a school, between boys from different boarding houses. It would have been difficult to arrange matches between schools because of the variations in the rules.

6.3.4 Framework of Analysis

It is useful to examine the subsequent evolvements within a common framework which can then be applied to the American case[19]. Figure 6-2 portrays four interconnected features.

In the public schools the *ethos* of gentlemanly behaviour was central, yet its meaning was contested. There was strong agreement that gentlemen should accept the authority of captains and of prefects. Being a fag was part of that template of social order, namely, the largely unquestionning acceptance of decisions from those holding positions of legitimized authority. It became more important to play the game than to win. An important difference in ethos concerns the degree of violence which was permitted. Here there were evident differences between the kicking game and the running game. The latter retained opportunities for violence and hence for a display of bravery. However, exponents of the kicking game (e.g., from Eton) came to regard rugby union as a game for barbarians. Hence the symbolic significance of the composite team known as the Barbarians.

---

[18] The point is dealt with very effectively in Riesman and Denney (1954)

[19] Denzig (1956); Gent (1974); Smith (1972); Weyand (1955); Hoch (1972)

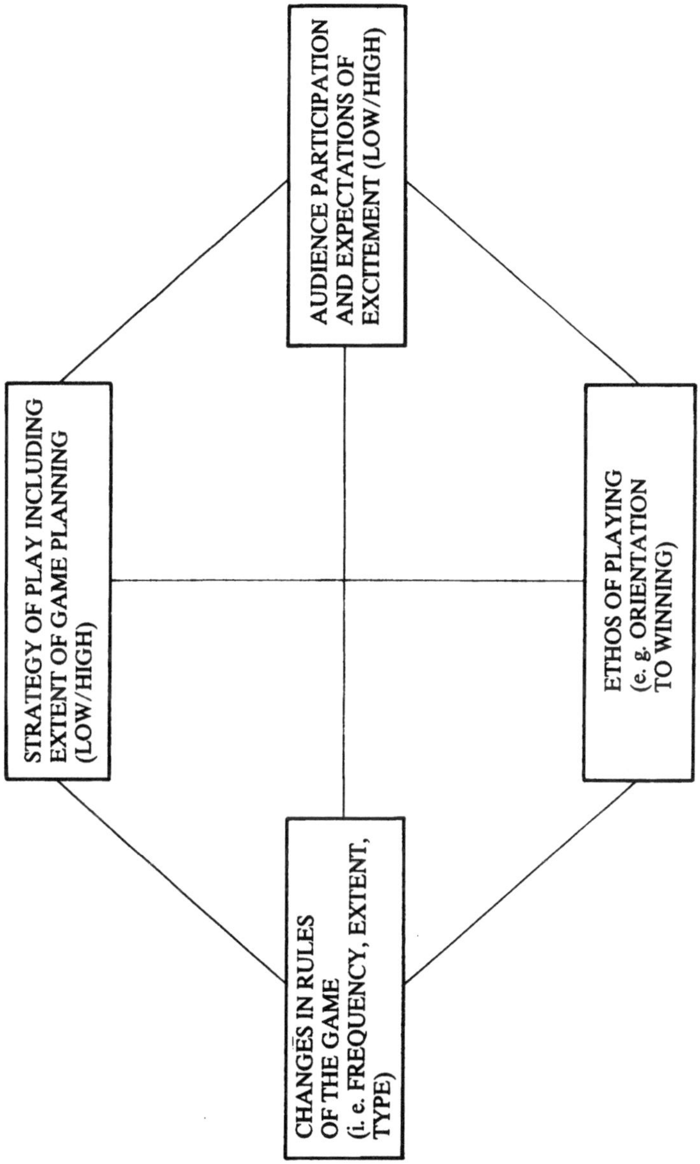

Figure 6-2. Framework for comparison of Anglo-American teamwork

In the public school setting the evolvement of rugby union was closely associated with a minimum amount of *game-planning*. Some planning of tactics did occur, yet its extent was slight and rehearsals were unknown. This feature persisted into the late nineteenth and twentieth centuries in the middle-class areas and clubs. However, in South Wales (e.g., Cardiff Club) and in the North of England the game was played by large numbers of men who did not share the gentrified notion of playing. In those areas strategy was developed. Yet, the moral principle expounded from the governing bodies of the game minimized the opportunity for strategy.

The possibilities for strategy were also affected by the *rules,* especially by their specificity and the great difficulties of achieving changes likely to increase the possibilities for standardized plays. The rules were so unspecific that matches between schools were difficult to maintain before 1850. Also, the rules were often known only to the players and the small number of onlookers may well not have understood when the rules were being broken. The rule governing the unintentional exiting of the ball from the maul required a high degree of social acceptance of the behaviour of the opposition and the roles of the two captains.

Rugby union was a game for those on the field. There were *few spectators* and their significance to the game in Britain did not become important until the last two decades. The small number of onlookers certainly cheered one side rather than another. However, a key dimension of their relationship to the game centred on the gentlemanly behaviour of those on the field.

It may be anticipated that American football is played with the ethos of winning, through game planning and rehearsals. The spectators and their excitement levels were and are an important influence on changes in the rules.

6.3.5 Crystalization and Schism: 1830s-1850s

In this period the centre of attention shifts from the boarding schools onto Oxford and Cambridge, where the boys from different schools sought to establish a wider basis for the winter ball games. Within those settings there was a struggle between those who supported the kicking game (led by former Etonians) and those who favoured the carrying-pushing game. It seems certain that the crystalization of rugby union owes more to the events in these two universities than to the influence of the game at Rugby School.

At Rugby School, their variant continued to be played in its distinctive form, yet it was still very much a mauling and kicking game rather than a running game. Indeed, and quite contrary to the impression created by the Ellis legend, the practice of picking up the ball when it exited from the maul was against the rules during the 1830s. Certainly, in the late 1830s the boys wavered in their statement of the rules, so that in 1841 and 1842 a strong, big boy named Jem Meckie, was able to run with the ball. By that stage the number of players on each side, still a matter of variability, was getting smaller, so that there was more opportunity for open play. After 1842 running with the ball was made lawful under certain restricted conditions; after 1846 running was permitted.

The changes at Rugby School were probably stimulated by developments at Oxford and Cambridge. For example, when Ellis left Rugby School he went up to Oxford University where the game of rugby did gain adherents, though several attempts to found a rugby club failed, even as late as 1839.

Meanwhile, at Cambridge University a running-and-carrying game became established in the 1830s, as did the kicking game played according to the rules of Eton School. There was an acrimonious attempt to establish a single common set of rules in the 1840s, but in 1846 the Old Etonians objected to the crude, violence of the 'Rugbian game'.

Two years later in 1848 the 'Cambridge rules' for the *kicking game* were formulated and agreed upon. The kicking game became labelled association football, which later became known as soccer. That game was more fluid and less violent than rugby union. In 1863 the basic rules for the kicking game were formalized and association football was fully established with agreements on the number of players – eleven – and the mode of scoring, which was to kick the ball through and under the goalposts. Association football was rapidly diffused to the urban areas where its basic features were much better adapted to playing in the restricted topography of street corners. In these areas players could play a form of the full game without much personal danger from the surface conditions.

At this period the game of rugby union was still only partially crystalized. There was still a lack of standardization over the number of players which could range from 15 to 20. Even so, by the 1850s the game had been diffused to the Scottish public schools and numerous public and club sides had been founded with a heavy concentration of new clubs in the south of England in this period.

6.3.6 The Rugby Football Union (1871)

In the 1860s rugby union became gradually established outside the South of England in South Wales, where it was played by miners and in the North of England. Yet, compared to association football the game of rugby was less widely spread. In 1871 the Rugby Football Union (RFU) was formed and 59 basic rules were agreed upon. These still permitted the number of players to vary between 15 and 20. Also in 1871 further innovation was introduced at

Cambridge when the loose, amorphous, dull formation of the maul was replaced by a scrummage in which an even number of players from each side were permitted to engage in orderly, nonviolent pushing. In the scrummage situation, one side is permitted to enter the ball by rolling it between the two sides of opposing players. That formation reduced the violence. However, by placing an emphasis upon heavy forwards the new formation reinforced the relatively static features of the game, so Cambridge students continued to experiment with the game in attempts to develop more fluid moves. In 1875 they reduced the number of players to 15 to create a more open potential. At Oxford the students experimented with attempts to make the static, hidden play more fluid and open and they developed passing movements between the forwards (1878-1882).

It is evident from the foregoing examination that rugby union was, in the 1870s, a game which was still in the process of refinement:

1. The rules were still diffuse, tending to retain the spirit of their origin as guidance to the captains. Consequently when the first international occurred between England and Scotland in 1884 there was considerable disagreement. Immediately after the match, a new board was formed to govern the rules of the game. The number of players was consolidated at 15 and this remains the number through to the present time.
2. The method of scoring remained that of kicking the ball over the cross-bar from drop-kicks and place-kicks. It was not until 1886 that a point-scoring system was adopted and only in 1905/1906 did this scoring system connect the scoring of goals to the scoring of tries. The preference for goals remained for four decades until 1948/1949 when the points for a drop goal were reduced from four to three. Even later, in the 1970s the points for a touchdown/try were increased from three to four.
3. From the earliest period the kicking of goals from a placement was taken in a line from the place of touchdown/try. That feature places a premium on scoring a touchdown near to the posts in the centre of the scoring line.

6.3.7 Diffusion and Further Schism: 1880s and 1890s

In the late 1870s and the 1880s the game spread from its concentration in the south to the north and to Wales. In Wales the game was taken up by coalminers who introduced a new dimension – a concern for winning and an interest in tactics. Hence, they gave much attention to learning the basic skills, to experimenting with moves, to having practices and to the possibility of coaching. In Wales the Cardiff Club occupied a special place through its development of new moves for the backs rather than the forwards.

In the North of England the game was taken up in the towns and particularly in the various closed communities based on coalmining and on textile mills. The roughness of the game seems to have appealed to the masculine imagery. Membership of these clubs was open and the matches between them were often very competitive. Also, crowds were attracted to watching the games (cf. Fig. 6-2). Thus, as early as 1876 a Northern cabal attempted to introduce a structure for formalizing the competition between clubs into leagues. Also there was an interest in promoting the crowd-pulling aspects of the game. Then quite quickly, by the 1880s, the Northern clubs had gained an ascendancy in their skill at playing and could defeat Southern clubs. The Northern and possibly the Welsh clubs may have acquiesced in the practice of 'boot money' by which a player gained some recompense for the costs of playing and possibly a small element of payment. In the 1890s the Northern clubs proposed that players should be compensated for 'broken time', that is for the potential loss of earnings arising from the time of travelling and practising. This proposal fuelled an underlying tension between North and South (see Wiener 1981; Dunning/Sheard 1979) and the acrimony between the bourgoisie led to the separation and segmentation of a collection of Northern clubs in 1895. The Northern clubs soon found that the size of the gates was insufficient to sustain their costs and so they began to examine how the game could be made more attractive to paying crowds. In a period of a dozen years they developed a new game, rugby league, which was the focus of a number of alterations (Macklin 1974).

6.4 American Predispositions

6.4.1 Players and Winners

Riesman and Denney suggest that rugby union was the major, possibly the only influence on the emergence of an American game which later became known as American football (see note 19). However, it will soon become apparent that rugby union and association football were influences which intertwined with indigenous American developments. It seems likely that rugby union was a key template for the influential and prestigious East Coast universities, yet the shape and uses of the template created so much confusion and dissatisfaction that the original template was taken apart, re-invented and appropriated (Riesman/Denney 1954; Denzig 1956).

There are many indications that the American upper classes were guided by English and European templates of social behaviour towards games. Rader claims that in the midnineteenth century the top East Coast elites sought to imitate the British profusion of amateur sports (Rader 1983). English clubs served as a template in games such as tennis. The American upper strata tended to use these clubs to build subcommunities differentiating themselves from surrounding strata. For a time the English game of cricket was played and in the 1880s the City of Philadelphia acquired four cricket grounds and was thereby better served than London in the same period. It seems that the American social elites did espouse the British principles of amateur sportsmen and therefore endeavoured to distinguish play from work. Yet, relative to the British, they tended to be more oriented to winning. This theme is most evident in the preparations of the prestigious New York Yacht Club for winning the key event at Cowes, England in 1851 and which gave birth to the highly competitive Americas' Cup race. The middle-class Americans were even more competitive and they gradually penetrated the subcommunities espousing amateurism, until by the 1920s only the area of Olympic athletics retained any dimension of amateurism (Rader 1983). The American universities and colleges were an important location for the contest between strata who contended that games should be playful, therefore unlike work, and those strata who contended that games should be approached systematically in the pursuit of winning.

Baseball was established as a game in the 1850s and its results were recorded in the local newspapers. The availability of the telegraph facilitated Chadwick's creation of the score box summarizing national results. The Civil War brought large groups together and there was much playing of baseball. In 1869 the first known baseball tour became a focus for commercial exploitation. There was ruthless competition for spectators after the infant season of 1876. By the end of the nineteenth century there was a crisis in the organization of that game; there was rowdyism on the field and the players struck for better conditions.

In order to examine the position of American football it is important to focus on the colleges and universities, particularly on the pivotal role played by leading East Coast colleges (e.g., Yale, Harvard and Columbia). In those colleges the theme of amateur sport and playing the game was important and this orientation probably explains why intercollegiate football became so important in the twentieth century. College football could not be exploited by capital as could baseball (which was not a college sport) because there were certain barriers, including the roles played by presidents and administrators. Even so, one of the striking features of the evolvement of American football is the role of the coach and the absolutely crucial role he plays in the hegemonising of skill and knowledge and its learning. American football also emerged from the configuration of competing interests and pressures in the colleges, but once having surfaced, the game became a major social template for teamwork.

6.4.2 Re-Invention: 1876-1886

Up to the postbellum period, a variety of winter ball games were played in the colleges and there were some intercollegiate games in late 1860, for example, between Notre Dame and Michigan State. In the 1870s the number of intercollegiate games increased, but there was a problem because the colleges played different variants of games.

The period from the 1870s may therefore be seen as a contest between parties seeking to impose specific sets of rules. Harvard, Yale, Princeton, Columbia and Rutgers were all experimenting with winter games. Princeton played a variant of the kicking game with 25 on each side. In a match arranged with Rutgers, the temporary rules permitted some handling of the ball. The Eton game was influential at Yale who invited Princeton to a convention in 1873 which had been established to consolidate one set of rules. At Harvard a variant of the local 'Boston Game' – banned in the city – permitted the handling and the carrying of the ball, but the Harvard version made it impossible to have matches with Princeton and Yale, so in 1874 a team went to Montreal to play McGill using the rules of rugby union. Harvard then attempted to apply these rules in a match with Yale and had to agree to there being 11 players on each side. That was an important accommodation which anticipated the eventual number of players in American football. Then in 1876 representatives of Harvard, Yale and Columbia held a convention and agreed to adopt rugby union as a template.

The three years which followed contained a high degree of 'learning-by-using' (Rosenberg 1982: 16ff., 122ff.), which led to the transformation of the original template. The rules of the English game caused considerable difficulty. That was partly because the rules were diffuse and did not cover every eventuality and so there was tension between the American and British understanding of the rules and their intended usages. That problem was most obvious with respect to the requirement that the ball should exit unintentionally from the maul. The rule covering unintentionality was a legal fiction which could be made operable within the social contexts of the English boarding school and club (Dunning/Sheard 1979), but in the American context, where legalism and winning were central predispositions, the rule of unintentionality was continually breached by behaviours disguised as unintentional. After three dissatisfying seasons with the amorphous British variant of rugby union came a period of modification and of wholesale transformation (1879-1882). One of the key agents in the process of re-invention came from Camp (Denzig 1956). His influence was considerable for a period of two decades. Camp began to experiment with a series of modifications:

1. It was agreed that a designated person on each side could intentionally remove the ball from the maul.

2. Next the maul was replaced entirely; the new formation was a static scrimmage in which one side was given possession of the ball.
3. The side with possession started a movement by passing the ball backward to commence a **carrying** movement.

These radical alterations transformed rugby union and also introduced a quite separate pattern of evolvement. Now the entire morphology of the original game of rugby was changed. The orderly scrimmage could be used for each side to redeploy itself. Also, the orderliness of the new approach introduced the potential for greater standardization of strategy and tactics. However, the new rules meant that the side with possession retained the ball for long periods, usually until they had scored. So teams developed offensive plays with the objective of retaining control over the ball. These developments became very unpopular with the players and with the small number of spectators.

In 1882 another radical change was introduced, the down. A down is defined as a number of episodes or moves in which the side in possession must advance a given distance up the field or concede the ball to the opposition. That permitted the offensive side with the ball to retain possession provided that they had made a specified advance down the field of play in a defined number of downs. If the offense failed to make the specified yardage, then the ball was conceded to the other side who then became the offensive side.

Within six years, by 1882, these changes had created the basics of a carrying game: American football had been invented. For the next 25 years it was a game of mass formations such as the 'flying wedge'. At that stage the game was still significantly controlled by the players, though there were influences from the college administration. The new rules were used creatively by the players to introduce increasing standardization in strategy and tactics. American students gave much more attention to the strategy and tactics than the students at Oxford and Cambridge.

Three further areas of evolvement may be noted:

1. The 11 players began to devise a **sequence of contingent** moves which became the repertoire of that side. The moves, or plays, were rehearsed and their ordering was designed to take account of the extent to which the offensive side had achieved its required yardage for retaining control of the ball. The moves were based on carrying and forcing the ball forward rather than on passing. One tactic which emerged was the use of the kick when the offensive side had not made sufficient ground forward.
2. A degree of role specialization amongst the 11 players began to crystalize. At that stage the same 11 players were on the field all the time. The role of the quarterback as pivot on the field had become established, though considerable time was to pass before the role became what it is today.

3. The college game began to attract large crowds and intercollege matches were becoming events of great rivalry.

By 1886 the shape of a new game was established. Through the energies and the tendency to formalization by Camp, the rules had been copied and agreed amongst the prestigious colleges. These rules had been rapidly adopted throughout the American universities. The game spread rapidly from the social elites – that is, those students probably closest to the British amateur orientation – into the middle class sectors of employed professionals and technocrats at state and private universities. Outside the founding colleges, the role of the coach and the influence of the college administrators became more powerful and consequential.

6.4.3 The Coaches

From 1886 to the early 1900s the game grew dramatically, but unlike baseball it was significantly confined to the colleges and universities. The commercialization which affected baseball was contested and its consequences delayed until the 'Golden Age' of the 1920s and the coming of television in the 1950s.

American football in the late nineteenth and early twentieth century was an exciting game of mass formations. The offensive side would adopt configurations designed to protect the player carrying the ball and would charge into the opposition. It was a series of violent, short episodes. The defensive side could link arms to counter the offensive moves. Winning was all important as an ethos.

The role of the coach became increasingly significant, especially at the prestigious major universities just outside the top elite. Whilst the President of Cornell could publicly mock the effort otherwise devoted to American football, that the game had been transformed into sport and teamwork was the basic principle. The coaches became the repository of 'best practice'. They sought to insert the same analytic orientation into sport which the management science movement sought to introduce into the organization of work. Arguably their task was easier because a sport like American football lent itself to standardization and could be transformed in that direction. Also, as Noble indicates, the ambitious, middle-class salariat and independent professionals believed in and actively constituted a particular vision of the design of organization (Noble 1977). In the 1890s the coaches took over and certain innovative coaches, such as Stagg from Pennsylvania State University, developed plays for which their teams became known.

The college administrators began to realize that as a spectator sport American football could generate the revenues which could carry the whole intramural program of sports and make their universities famous. There were secret subsidies and players were selected for the university on the basis of their prowess at the sport. By then sport was more evident than scholarship on the national scene. At that stage the rehearsed plays were violent and so players were often injured and deaths became a regular occurrence. By 1900 the game was in a state of crisis. In 1906 eighteen players died from injuries. President Theodore Roosevelt spoke out against the dangers of injuries and the deaths and the game became the focus of critical debate over its level of violence. In the ensuing struggle the influence of the founders of the game was overturned.

Meanwhile the professional and inventive coaches scrutinized the rules to find loopholes which could be exploited to provide a winning team and to increase the excitement of spectators. Pressure from the coaches influenced changes in the rules (see Fig. 6-2). In 1903/1904 the number of offensive players in the line was specified. A further major modification came with the alteration of the scoring system to give a higher number of points for 'run-ins' (i. e., touchdowns) than for goals (cf. rugby union). Further innovations were introduced to break up the dangerous mass formations. Then in 1906 the forward pass was introduced in a limited, yet crucial change in the rules. It was some time before the coaches discovered a play to exploit that new rule. In fact it was used sparingly until in 1913 when Notre Dame rehearsed the forward push with the quarterback and used it as a secret ploy to surprise and defeat a powerful Army side, by 35 to 13. The down was respecified in 1912 so that ten yards had to be made in four downs and the 'gridiron' marking was formally introduced. By that period the game had become a major spectator sport. Many stadia were built. The Rose Bowl had been started in 1902 and in 1905 forty-three thousand watched Harvard versus Yale.

From the early 1900s into the so-called Golden Age of the 1920s the sport continued to consolidate its position and to spread downward into the high-school system where it gained a central position as the winter sport and as a major feature of the school year (Burnett 1969). The spread of the game facilitated the spread of the role of the coach as a respected member of the community. Even Harvard appointed a professional coach and gained good results for a brief period. From then on the game became the focus for detailed analysis and much reflective planning. By now the notion of a 'game plan' was more obvious. The pace of the game permitted the spectators to discuss the previous moves and to try to guess what would be next. Consequently, the spectators were drawn into the analysis in a way which association football and rugby union could not offer. For the coaches, the analysis through films became commonplace, but it came later than Riesman and Denney suggested (Riesman/Denney 1954). Even prestigious, small universities in Canada such as Queen's University employed a coach (from 1924), adopted film analysis, rehearsed moves and experienced a short period of success as a leading Canadian side in their variant of the game.

Coaches became general to American sport, just as they were absent in British sports at the same time[20]. The significance of the coach in American football became symbolically and hegemonically crucial to the legitimation of the template of coaching throughout American society. Pulos contends that there is an 'ethic of design' in middle-class American society applied by experts (Pulos 1983). If so, then tastemakers diffused a template of teamwork which was immensely important in North America.

6.4.4 The Professional Game

In the 1920s the commercial pressures for a professional game equivalent to baseball became much stronger, but the college game was now so well established that it had a decisive influence in the next half-century. The early professional game consisted of the franchising of clubs. Investors often lost their capital, but there was just sufficient encouragement for them to seek accommodations with the governing institutions for the college game. In 1935 the professional clubs began to draft college players through a formal mechanism. Much later that mechanism came to embody an interesting and revealing principle, that the weakest professional club had first choice of the best players from the colleges. That principle creates the exciting possibility that teams can arise from a weak position to a powerful one in a relatively short period of time. By the 1950s the professional game was established, yet it was the college game which still carried the symbolic influence for legitimating the discourse of teamwork: game plans, "letting you go", morale, the team, dropping the quarterback, leadership and discipline. In the next quarter of a century the professional game refined the earlier developments and in association with commercial television, elaborated the distinctive choreography of today's visual spectacle. Today television produces about two-thirds of the revenue. The professional game introduced another major refinement when in the 1950s the practice of substitution was permitted on a minor scale. Then in 1964 teams were permitted to have a pool of players from which an offensive or defensive side could be drawn. The introduction of the pool consolidated the trajectory of evolvement by increasing the potential for functional specialization. Thus, the 'special team' could be created in the pool and placed on the field for specific tasks.

---

[20] Note the tentative use of coaches in the 1920s by British athletes as depicted in the film *Chariots of Fire*

Twenty years' experience of teams using a pool of players has altered the game. It is now much more open and varied than the game of six decades earlier. For example, the visual beauty of the forward pass can be enhanced through having specialized rehearsed plays. These spectacles symbolize and legitimate the achievements of coaching.

6.5 Comparing American Football and Rugby Union

Figure 6-3 compares the two games along 20 dimensions. These show sharp contrasts between the two games. The two games provide templates of how to organize, how much to plan, who has the 'taken for granted right' to do the planning, the nature of authority, the extent to which plans can be examined reflectively through various means (e.g., films, charts) and the ways in which corrective actions can be introduced through training.

Feature	American football	Rugby union
I Game		
1 Carrying or kicking	Carrying game, with *forward throws;* kicks from certain positions; few passing movements.	Running, kicking and occasional dribbling; forward throws penalized. Extensive passing movements.
2 Synchronization and co-ordination	Very tight synchronization. Co-ordination achieved by pre-planning and rules.	Looser synchronization. Co-ordination achieved through mutual adjustment, feedback.
3 Time-Space framework	Highly defined. Pitch is marked as a 'grid-iron'. Precise measurement of 'downs'. Time precisely measured (e.g., '30 second rule', 'time-out', 'two minutes'.)	Less delimited. Some allowance of time for injuries.
4 Fluidity	Episodic; short episodes of great action followed by re-forming. Static re-starts at scrimmage line.	Longer episodes, frequently involving changes of control over the ball.
5 Tackling	Robust, highly defined and closely observed. Some protection to the receiver of forward throws.	Highly robust (e.g.: rucks) and continuous tackling.
6 Players/spectator balance	Considerable attention to the spectators' excitement and enjoyment	Still a player-oriented game, but with increasing attention to the spectators.

Feature	American football	Rugby union
II Players		
7 Pool of players	Each side has eleven players, with a pool of up to forty. Substitution.	Fifteen per side plus two substitutes for injuries.
8 Roles	Highly specialized, generally and specifically. Specialized roles used for certain infrequent situations.	Roles differentiated, yet more fluid and situationally interpreted (e.g.: forwards playing scrum half to a ruck).
9 Leadership	The quarter-back calls the moves and is *the thinking hub*. Authoritative and centralized. *Coach* continuously adjusts according to the game plan.	Player-centred game with a plurality of decision centre and sources of initiative. Coaches precluded and proscribed from involvement during the game.
III Stocks of knowledge		
10 Stocks of knowledge	Highly formalized, initially into team 'bibles' and then more formally on to charts, films and use of computers. Repertoires of moves stored. Use of detailed manuals. Hierarchy of knowledge.	Oral tradition amongst players. Much less formalization and recording.
11 Strategy	Extensive consideration of team strategy and the strategy of the opposition. Game plans laid out. Strong future orientation.	Less formalized. Top clubs do develop broad game plans.
12 Tactics	Continuously monitored (e.g.: 'quarter-back in the sky' giving advice and information). Coach-centred.	Player-centred.
13 Repertoire	A central concept. Extensive; many moves, some used very infrequently. Each move has designated code which is learnt in practice and on paper.	Important, less specific. Becoming more important. Moves coded.
14 Analysis and critique	Fantastically developed. Films of games even at college level. Recording and review. Some public accountability (e.g.: discussing the game with supporters). Professional players may watch ten hours of film per week.	Much less developed in a formal way.

Feature	American football	Rugby union
15 Training	Very extensive, based on elements learnt and rehearsed with specialist coaches. Focus on discipline and synchronization. Constant rehearsal.	Increasingly important. Stress on general features (e.g.: rucking, wheeling) done with forwards. Less focus on synchronization.
IV Infrastructure		
16 Diffusion	Very extensive. Into high schools; 'pee-wee' leagues and vigorous professional game. Limited to within the USA and Canada.	More socially and geographically encapsulated. Secondary sport to soccer in schools. No professional game officially, though hints from 1973 onward. An international game (e.g.: 'Five Nations').
17 Linkages	Very tight (a) trajectory from eight years old, through college into pro-fession, or early retirement to watching; (b) coaches and manuals linking levels.	Linkages less tight, especially in England.
V Spectators		
18 Centrality	Absolutely central. They are 'organized' and choreographed through the shaping of the plays and through cheer-leading, band playing (see linkages above). Umpires explain plays through signals and microphones. Television supplies high proportion of revenue to the professional game.	Increasingly important to survival of the bigger clubs and infra-structure. Live attendances represent 90% of revenue. Television supplies very small proportion. Spectator left to interpret the game.
VI Ethos		
19 Orientation	Skill, spectacle, winning.	Amateurism, excitement, the 'gentleman/barbarian'. Playing.
VII Technology and raw materials		
20 Investment	Very high indeed: stadia, special surfaces; extensive, specialized practice equipment; classrooms; expensive gear; expensive tape; filming; data analysis.	Much less investment. Simple gear, less specialized training technology.

Figure 6-3. Comparing Contemporary American Football and Rugby Union

American football is a game of carefully designed episodes, many of which are preplanned and rehearsed and most of which are activated in accordance with an overall strategic game plan. The episodes exemplify tight synchronization between designated players, so that the moves are disguised, the opposition is disabled and specific configurations come into being at a required moment – yet there are opportunities for inspired, unintended alterations, but these are confined to rare moments. The game is planned by the coaches and interpreted very largely by the quarterback. There is a hierarchy of organized knowledge to inform decisions.

American football is prime television time and its matches at the professional and college levels are usually major events. American university sides have stadia which hold more spectators than could attend a major international in Britain and France. American football is a significant event in the schools where matches provide key pieces of punctuation in the school year. The game provides a form of discourse which is learned in school and used in adult life.

As a template the game of American football is highly visible. Few onlookers would question the symbolic significance of planned synchronization or the right of coaches to dictate the game plan. It provides a major template for teamwork.

The significance of the template lies in a high degree of fittedness to those features of work organization which have been central to the shaping of large-scale corporations in the twentieth century. The division of labour, concepts of strategy and tactics, controlled by a hierarchy, knowledge-based bureaucracies and notion of the team found in the corporations which underpinned American commercial and industrial success were paralleled in American football. The game provided a legitimate and authoritative presentation of taken for granted ways of organizing. Training and the required self-discipline are highly visible as parts of American football. Training for a role in the overall team can begin at a very young age, can be undertaken in very small groups on predesigned equipment, and can be monitored against written performance criteria.

Rugby union is a sharp contrast. Yet the game is, it will be shown, a powerful indicator of the templates of organizing found in many British corporations. There are many contrasts to American football.

The patterns of evolvement and the periods of innovation for these games can be reconstructed as in Fig. 6-4 which shows how British folk games were appropriated and re-invented in the public schools. In the British context two major variants emerged and became recognizable some three decades before American colleges were experimenting with winter ball games. Given the prestige of English sports it is not surprising that there were attempts to

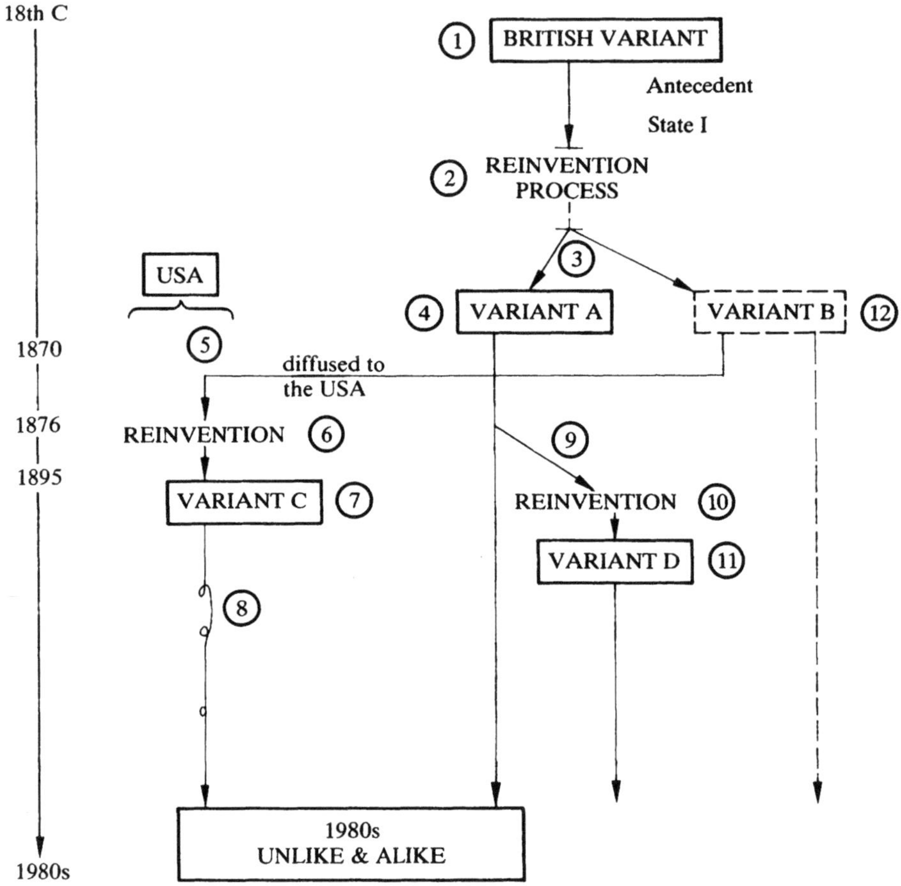

KEY

1. British folk games
2. Appropriation of folk games by public school boys in England
3. Separation of the running-carrying game from the kicking game
4. *Formalization* of Rugby Union Football
5. Diffusion of Rugby Union Football and Association Football to the USA
6. American appropriation of Rugby Union Football and its modification
7. *Formalization* of American Football
8. Further evolvement of American Football
9. Schism in Rugby Union Football in England
10. Separation of professional rugby including modifications to Rugby Union Football
11. Formalization of Rugby League
12. Formalization of Association Football

Varient A = Rugby Union Football
Varient C = American Football
Varient B = Association Football
Varient D = Rugby League

Figure 6-4. Carrying and running sports: Evolving innovations

transfer them into the American setting, yet as Fig. 6-4 shows, their introduction was soon followed by re-invention and by the creation of American football.

Figure 6-4 summarises the specifics of the evolvement of different templates of teamwork and also illustrates the varied and parallel trajectories which can co-exist and compete. These features are probably common to all innovations.

7 Economy, Structuration and Region: A Basic Framework

7.1 Introduction

The previous chapter provided strong illustrative material in support of the general thesis advanced in the book. 'Context' has been identified as the crucial embedding feature which previous studies of innovation have treated in an emasculated fashion and which, even when identified in a model (e.g., Rogers antecedent factors), is actually neglected in the research work (Clark, P.A./Ford 1970). The two case illustrations of the textile industry and of sport provided clear confirmation of the role played by certain contextual dimensions when innovations travel from one region and society to another. The context can include key nonsocial factors such as the type of cotton, the use of wood in the making of machinery, the topography of buildings and fields at English public schools. Yet, those nonsocial factors are themselves the objective of human action (Giddens 1985). They are not neutral. Nor are the actual outcomes of the interactions between social factors necessarily intended. The more typical situation might be that outcomes are unintended, yet also are often close to the reproduction of existing patterns (Giddens 1985).

The extent to which cultural predispositions are a significant input to the specific contexts is a matter of debate (Lammers/Hickson 1979). It is noticeable that the debate takes a different shape when comparisons are being made between Britain and America[1] and between America and Japan[2]. There are several positions in the debate. First, there are those who contend that culture is relatively unimportant and that its features neither facilitate specific innovations, nor do cultural features inhibit the transfer of innovations between societies. Schonberger, for example, in his energetic account of the Japanese JIT systems (Schonberger 1982) considers that JIT could be introduced into American plants provided that American managements have the will to apply them. Second, there are those who suggest that culture is in-

[1] Anglo-American comparisons have been largely undertaken by historians, mainly economic historians, and have covered the period 1760 to 1940 with few comparisons stretching beyond 1945

[2] Japanese-American comparisons have largely focused on the period after 1945 and have a heavy focus on two areas: the economic infrastructure and the corporate organization of production

teresting rather than important and that its role in model building is mostly as a residual factor whose saliency should not be exaggerated. Perhaps the most sophisticated version of that case is the excellent book by Hickson and Lammers (Lammers/Hickson 1979) examining alikeness and unlikeness between organizations. Culture, in their interpretation is simply an additional residual variable to be introduced at the end of the analysis to cover whatever variance is left to be explained. Their approach is designed to uncover the common features which organizations seem to acquire irrespective of region and time. Even so, their conclusions do suggest certain regionally located differences such as those between the Anglo-American model and the continental Europe model. Third, there is a growing school which emphasises that cultural differences are an under-researched and under-theorized dimension which lies behind many of the specific ways in which variables are ordered (Sorge 1983). The position of this growing school requires some amplification before attending to the general framework which is presented in this chapter.

In the immediate postwar years American social science had a pervasive influence on the whole social science community. In sociology it was the period when general very abstract theories were being developed[3]. Those theories reflected the role of social science in the Second World War where sociologists had contributed to the area studies[4] and had sat on committees planning the social and political reshaping of Germany and Japan. Underlying those theories was a belief in the development of societies along a single, broad trajectory at the forefront of which was the USA. The case of the USA was taken as the exemplar of what would be appropriate developments in the economy and social structure of all industrialising countries. The USA played the same role as Britain had in late nineteenth and early twentieth century political theories. Eisenstadt's theories of modernisation (e.g. Eisenstadt 1966) represent one of the most subtle and revealing examples of the best achievements of that style of theorizing about the present and future. Eisenstadt was able to subsume immense amounts of empirical material within the themes of protest and change to show that all societies were faced by external and internal agents who sought to overthrow traditional practices and replace them with modern forms. In fact Eisenstadt's interpretation did indicate that many societies were taking distinctive routes (e.g., Japan), but that cautionary point tended to be lost in the general push towards convergence theorizing (Eisenstadt 1966: 75-82). As stated in Chap. 2, organization studies became the vehicle for a community of scholars who pursued thinking which emphasised the convergence of organization structures. Those

[3] For example, in the work of Parsons, Eisenstadt and Smelser

[4] The program of research for a decade or so after 1945 sponsored by the American military was known as Area Studies; some of the analysis was at a very high level

tendencies, particularly strong in international sociology, also influenced most of the European scholars who had spent part of the wartime period in the USA. Gurvitch, the French sociologist of process, was a key example[5]. An early exception to that pattern was Crozier and his career is an interesting cameo for the whole treatment of cultural differences (Crozier 1964, Part IV).

In postwar France there was a strong and largely sympathetic interest in American institutions, ideology, forms of organization and technology. In Renault, for example, the whole program of postwar renewal centred around appropriating American technology and practices (Clark, P.A./Windebank 1985). In the 1950s Crozier began to study French state-owned enterprises (e.g., cigarette making) and public bureaucracies through the application of survey methodologies adapted from the approach of American psychologists, especially from ISR, Michigan[6]. Later, whilst still mulling over the findings, Crozier returned to the USA to spend a period at the think-tank in Palo Alto. He had previously visited the USA just after the war and had returned as an enthusiatic admirer. At Palo Alto he was struck by the distinctiveness of his French data.

Crozier accepted the general claims of the Carnegie School[7] that individual decision-making was subjectively rational (Crozier 1964, Part IV), but inserted the additional claim that rationality was shaped by cultural features. That is, French rationality differed from American rationality because, reasoned Crozier, French institutions predisposed action in different directions. The study of French workers, at all levels, suggested that they were less willing than their American counterparts to engage in face-to-face consultation and negotiation. There were many other differences, all of which Crozier attributed to basic features of French institutions.

In an extension of the previous points Crozier argued that the process of transformational change in France is rarely incremental. Instead, rigid structures are maintained in position by factions, even though all members might realise that that led to the accumulation of small problems into a major cri-

---

[5] Although the work of the French sociologist Gurvitch was quite central during the decade before his death in the mid-1960s, his extensive and insightful thinking did not penetrate Anglo-Saxon sociology; he was particularly influential in the development of temporally anchored processual analysis

[6] The Institute for Social Research at the Ann Arbor campus of the University of Michigan has been a major centre for survey research, methodology and theory for nearly four decades; its key members have formulated a distinctive recipe for survey research and for attaining the utilisation of survey data by client systems.

[7] The Carnegie School is the label placed on a collection of theorists and researchers who were members of Carnegie Mellon University, Pittsburgh during the 1950s and 1960s, e.g., J.G. March, R. Cyert, H.A. Simon, H. Leavitt, W. Dill and R. Cooper

sis. He stated that transformational change occurred infrequently and radically in sharp outbursts: small-scale short periods of revolution followed by long periods of high stability. That pattern of transformational change, Crozier stated, was typically French. Obviously, the events of 1968 provide a degree of confirmation of the thesis about the French predispositions. Moreover, Ladurie, the distinguished French historian has drawn a similar conclusion and he refers to the the pattern of structure-event-structure (Ladurie E. LeRoy 1979).

Crozier subsequently argued that France was a society characterised by structural blockages and that French institutions should acquire certain American features (Crozier 1973). More recently, and very significantly, he revisited the USA and concluded that American society had lost many of the characteristics which he had earlier felt were so favourable and desirable (Crozier 1984).

Until the mid-1970s Crozier's analysis of culture was frequently cited and rarely followed. A focused comparison of British and French tobacco firms confirmed his claim that although both were alike in the routineness of decision-making, they were unlike in patterns of authority, group organization and conflict patterns (Clark, P.A. 1979, Chap. 14).

By the end of the 1970s whole new schools of cultural studies were surfacing, several based in France. In ways which were complementary to Crozier, and more directly comparative, two French groups[8] completed a series of paired comparisons between France and various other European countries: Italy, Hungary, Poland, Germany, Britain, and Sweden (e.g. Dubois/Monjardet 1979). These studies described important differences in the core institutions, particularly of the educational institutions. Their claims, about intra-European differences, were strongly supported by the Industrial Democracy in Europe (IDE) project which revealed patterns of alikeness in participation within the firms of any country, but unlikeness between European countries (I.D.E. 1980). These large-scale, survey-like studies were complemented by other studies examining particular countries in depth. Gallie undertook a series of influential studies examining French firms and society in comparison to matched British firms (Gallie 1978; 1983). All these studies provided clear evidence of patterns within countries and differences between countries. So, behind the apparent alikeness of organization models which orthodox organization studies had highlighted there were also variations which were important to the competitive performance of the sectors of specific countries. Moreover, it was theoretically and empirically possible to attribute those

[8] One group was based at Aix-en-Province under the leadership of Marc Maurice and another at Paris 7 with Pierre Dubois and Dominique Monjardet

variations to national predispositions and to specific configurations of *institutions*.

Parallel to the growth of interest in cultural differences amongst European countries during the 1970s came two further significant shifts in the study of organizations: the historical dimension and the political economy perspective. First, there was a widespread interest in longitudinal studies[9] and to a lesser degree in what Clark and Whipp term the analytically structured narrative (Whipp/Clark 1986, Chap. 2). The historical dimension was taken up in a very distinctive way. In the main it has taken the form of a backward glance at the antecedents of the contemporary situation[10]. However, there are clear signs of the development of process descriptions, some of which bear on the analyses of firms undertaking actions with varying degrees of innovativeness as part of a strategy for adaptation and survival[11]. There has been a growing emphasis upon the concept of structuration and the idea that the short run is shaped by the inherited options of the long run as in the central vertical dimension of Fig. 7-1. That will be examined further in this chapter.

Second, and equally important, there has been a revival of interest in the very different political economy perspectives emanating from Marx, and from Schumpeter. In each of these perspectives attention is focused on the dynamics of the relationships between the economy and a social structure which is located in a particular geographical region. These two areas – economy and region – are located on the horizontal dimension of Fig. 7-1. As a consequence, the role of the economy as an allocative mechanism is coming to occupy an increasingly central position. The revival of interest in Marx, on both sides of the Atlantic[12], has led to the conjoint analysis of pulsations in the economy and changes in structuration for society (e.g., restructuring capitalism (Massey/Meegan 1982) and for enterprises (Aglietta 1976; Friedman 1977)). Within the political economy perspective an important debate has centred on the extent to which economic pulsations follow the cyclical patterns postulated by Marx and by Kondratiev because that thesis is being utilised to provide a powerful analysis of both the rise of the USA and of the changing balance of power within the North Atlantic region (Aglietta 1976; Boyer/Coriat 1986). Clearly, a political economy perspective which

---

9 Most recent critiques of organization studies have promoted the benefits of an historical perspective

10 Many longitudinal studies adopted simply extended cross-sectional studies by a few years or added a brief, ahistorical perspective to the front of a case study; systematic analysis of the long term is still developing as a methodology

11 Contributions to the new perspective were identified in Part II

12 For example, with respect to the American Radical Economists (e.g., Richard Edwards and James O'Connor) and the French Regulation School (e.g., Robert Boyer and Benjamin Coriat)

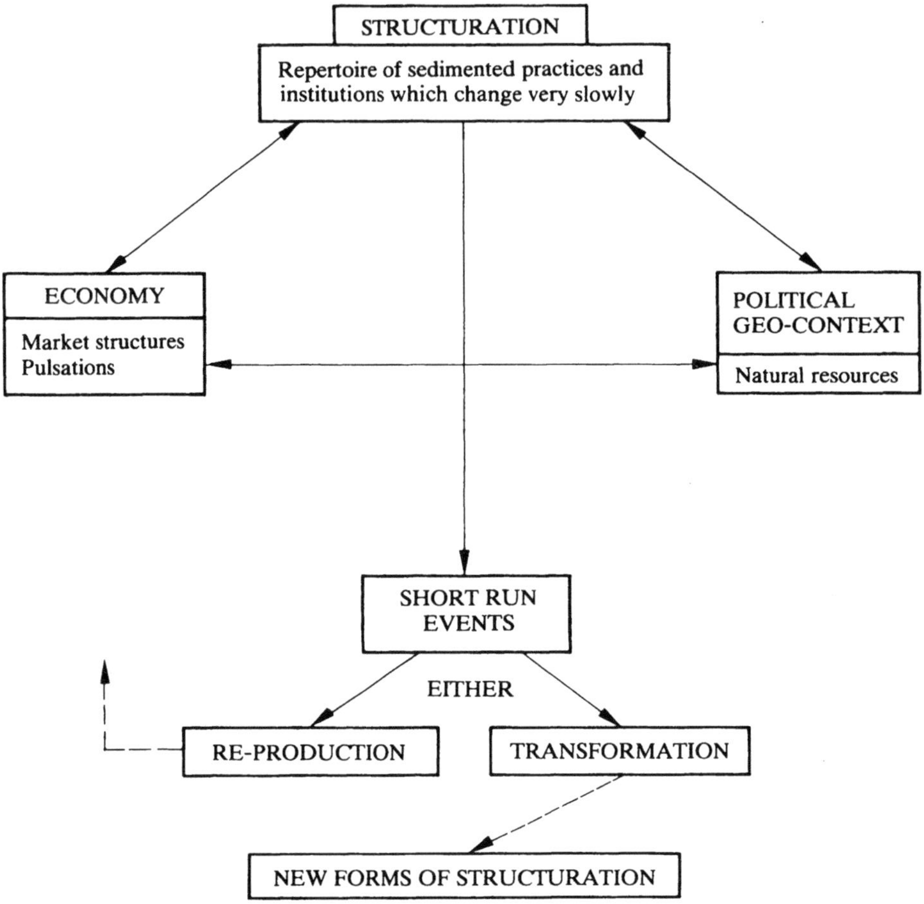

Figure 7-1. Structuration: The long-run framework

develops from Marx will tend to place the civil society and its structuration in a systemic relationship of dependence. The view taken here is that the interactions of economy and structuration are more complex: that the structuration of society is an authoritative mechanism whilst the economy is an allocative mechanism (Giddens 1985). That view is depicted schematically in Fig. 7-1 and the objective of this chapter is to elaborate that framework and then to demonstrate its relevance to Anglo-American patterns of innovation by combining the study of innovation with culture and with organization studies.

7.2 Economy: An Allocative Mechanism through Time and Space

7.2.1 Introduction

Economic analysis in the neoclassical school is pre-occupied with equilibrium models covering short-run events within specified and stable boundary conditions, in short, comparative statics (Clark, N. 1985). These models have been criticised for the manner in which they treat growth and neglect the unevenness of capital accumulation, especially the fluctuations. Approaches by economists to the analysis of the long-run variations in economic activity has centred on two major kinds of model:

1. Longwaves
2. Transaction costs

The significance of each of these will be briefly established and illustrated by reference to Anglo-American comparisons.

7.2.2 Longwave Theories

Four main schools of interpretation may be defined:

1. The capital accumulation perspectives deriving from the French Regulation School (Aglietta 1976; Boyer/Coriat 1986)
2. The longwaves perspective of Mandel (Mandel 1978)
3. The Schumpeterian perspective as interpreted by Freeman and the relevance of longwave analyses to understanding the changing position of Britain and America in international competition (Freeman 1983; 1982; Freeman/Soete 1985)
4. The Annales School of Braudel and colleagues[13]

These four approaches contain a great deal of variety. The position taken in this book is that the ideas of longwaves are a useful heuristic way of dealing with certain limitations in the comparative statics approach, particularly with respect to the analytic structuring of long-term adjustments, conjunctures and turning points in capitalism. Therefore, the most deterministic versions of longwave theory are rejected (e.g., Mensch) and the preferred interpretations are those of Freeman and of the Annales School[14].

---

[13] Fernand Braudel, who died in 1986, played a major role in presenting history as the subject which could integrate the social sciences around specific macrolevel problems; for a sympathetic review see Hexter (1972)

[14] Mensch (1979); the determinism of this work is modified in the later approach – see, for example, articles contributed to the *Journal of Product Innovation Management*

First, the contribution of Marx to the analysis of technical change and capital is extensive and significant[15]. Marx argued that the processes of capital accumulation were inherently uneven and he adopted the notion of the seven to ten year business cycle to express the idea of pulsations (Marx 1954). He set each cycle within a finite, linear, stream of cycles where each successive cycle 'determined' changes in the social superstructure of society until the nature of that superstructure was itself transformed from one whose principles sustained capital to one whose principles would sustain socialism. Cycles normally began with an increase in the rate of accumulation, but ended with falling rates which force part of the population of firms to exit from the market.

Marx argued that the economy pulsated according to readily detectable laws of motion. The owners of capital who wished to stay in the market could only do so by removing those areas of their activity in which the rate of accumulation was low, that is, the areas of porosity[16]. In an extensive discussion of wine-making he shows that the lengthy fermentation process was an example of the slowest rate of accumulation and that it was to that area which considerable applied scientific analysis had been applied. The removal of areas of porosity could be handled in any or all of several ways, including by tightening up on the working practices of employees, increasing the velocity of capital, introducing new technologies which gave some firms economies of scale whilst driving their rivals to exit, or by applying scientific knowledge.

The modern French regulation school of Aglietta, Boyer and Coriat have followed the perspective of Marx very closely applying and revising his thinking in a variety of studies. For example, Coriat has analysed the introduction of teamwork onto the assembly lines of the car firm, Renault, and interpreted their presence as an instance of the tightening-up process arising from competition between large car firms (Coriat 1976). Also, in studies of the construction industry he has argued that the introduction of variants of time study is a similar development (Campinos-Dubernet 1986). Aglietta has applied the Marxian perspective to the labour history of the USA to write an analytically structured narrative which examines the reaction of wage labour and its 'real subordination' through Taylorism and through the joint strategy of Fordism by which, according to Aglietta, workers receive high wages so that they work hard and consume faster, thereby creating capital accumulation (Aglietta 1976). Boyer and Coriat have recently proposed a theory of in-

---

15 Marx (1954) see in particular the use made by Mandel (1978) of that analysis

16 The notion of porosity has been used by Palloix to highlight the competitive nature of capitalism; Porosity as a concept is intimately linked with the commodification of time – see Clark, P.A. (1985)

novation and applied this to the contemporary situation (Boyer/Coriat 1986).

Second, in the first quarter of the twentieth century a diverse group of economists have attempted to analyse the long-term patterns of economic pulsations by the search for periodic patterns of cycles and of so-called longwaves[17]. Simiand detected broad movements in prices, both upward and downward, and over long periods of around half a century[18]. These became known as the 'A' and 'B' movements and were very influential on the French Annales School. These were movements rather than cycles. Gerlderen, a Dutch economist, claimed to have found evidence of cyclicity in price changes. In the USSR between 1919 and 1925 Kondratiev (at the Institute for Conjunctures) arranged several indicators of prices and made the definite claim to have discovered recognisable long cycles lasting between 46 and 50 years (Kondratiev 1976). He claimed to have discovered two complete cycles and that a third cycle was underway. Later, Kondratiev's interpretations, which have been hotly disputed were generally referred to as longwaves.

Kondratiev's longwaves consisted of almost equal lengths of upward price movements followed by downward price movements. The upward movement and the downward movement were reported to last about 23-25 years. It seemed that these longwaves incorporated the shorter cycles reported by Marx. Moreover, if there had been several longwaves, say two and half, then it might seem that capital was safe from an ultimate transformation into socialism. The projections of Marx might be replaced by the observation that capital was subject to periodic bursts of prosperity and decline.

The idea of longwaves gained some interest in the 1930s, yet economic thought was much more dominated by equilibrium than by disequilibrium and so, apart from the work of Schumpeter, interest in Kondratiev remained largely dormant until the publication of Mandel's analysis in 1972 (Mandel 1978).

Mandel argued that each longwave was set in motion by the uneven, periodic commercial exploitation of a whole range of generic raw materials and technologies. That is, that there were definite periods when *high rates of innovation in generic types were bunched.* The upward wave was carried on a

[17] The work of Kondratiev has become the most widely cited, so much so that it was re-issued, Kondratiev (1976); a useful review of the various contributions may be found in Mandel (1978, Chap. 4).

[18] Simiand, F. D. (1935) attempted to apply the methods of natural sciences to movements in prices and wages and produced a major review in 1932 which identified two major phases in price movements; these became known as the 'A' and 'B' movements and were influential on Braudel

	UPSWING	PEAK	DOWNSWING	ORIGINS
1.	1793–1825	1825		COTTON IRON Expansion of world markets
			1825–1847	Competition between England and Western Europe reduces profits. Rate of expansion of world market decreases
2.	1848–1873	1873		RAILWAYS STEEL Growing industrialisation and massive expansion of world market. Railways constructed in Europe and North America. Introduction of machine-made machines
			1873–1893	Profits fall. Diffusion of transformation machines reduces profit levels
3.	1894–1913	1913		ELECTRICALS CHEMICALS AUTOMOBILE OIL REFINING Extension of world market. Slow rise in price of raw materials
			1914–1939(45)	War disrupts world trade
4.	1940(45)–1966	1966		ELECTRONICS AEROINDUSTRY Intensification of the international division of labour
			1967–?	Falling levels of profit because of intensive international competition

Figure 7-2. Longwaves

radical innovation based on new items. It occurred every 50 years or so and contained an upward swing in capital accumulation during which the radical innovation gave way to more incremental innovation. Then after 23 years (or so) of upward movement came a similar period of downward movement when the competition between capitals drove down the levels of profits based on the generic innovations. Mandel reconstructed the history of longwaves as shown in Fig. 7-2. One of the striking features of Mandel's work was that it correctly anticipated the economic disruptions of the 1970s and in fact categorised these as a decline starting in 1969, in other words, some years before the 'oil shock' which most Western commentators regard as the key event. Mandel also argued that the 1970s and 1980s would be a period of massive restructuring of capital, but that the capacity of the political superstructure to accommodate transformation could not be predicted from the study of the economic base.

Mandel follows Schumpeter's claim that capitalism is inherently unstable and that the analysis of innovation requires an alternative to neoclassical economics, but concluded that the political economy perspective of Marx provided the most satisfactory basis. He contends that longwaves are obvious and cannot be disputed (Mandel 1980, Figs. 1-1 to 1-4). Four major long cycles can be identified.

Mandel re-examines Marx to derive his interpretation of the relationship between the economy and society. By referring to the second edition of *Capital* (Marx 1954, vol. 1, Chap. 17) Mandel distinguished between the dynamics of the economy and society:

1. The economy is characterised by objective trends whose dynamics hinge around the variations in the level of capital accumulation. There is still the constant tendency of the rate of accumulation to fall, yet because of the uneven development of sectors there are always capitals which can create short-run opportunities for renewal by investing in new equipment which can be amortised through the growth in size of the firm and the economies of scale.
2. The social structure is the subjective dimension whose dialectics *cannot be predicted* from the economic system because it is autonomous and also dialectically intertwined with the economy.

The relationship between the objective trends of the economy and the subjective trends of the social structure are depicted in Fig. 7-3.

Mandel's own work in Germany predated the revival of interest in Schumpeter and in Kondratiev by defining 1969–1970 as the turning point of the fourth longwave. Consequently, he maintains that the recession of the 1970s possesses its own deeper logic which is more significant than the movement of oil prices. He also suggests that followers of Marx have underestimated

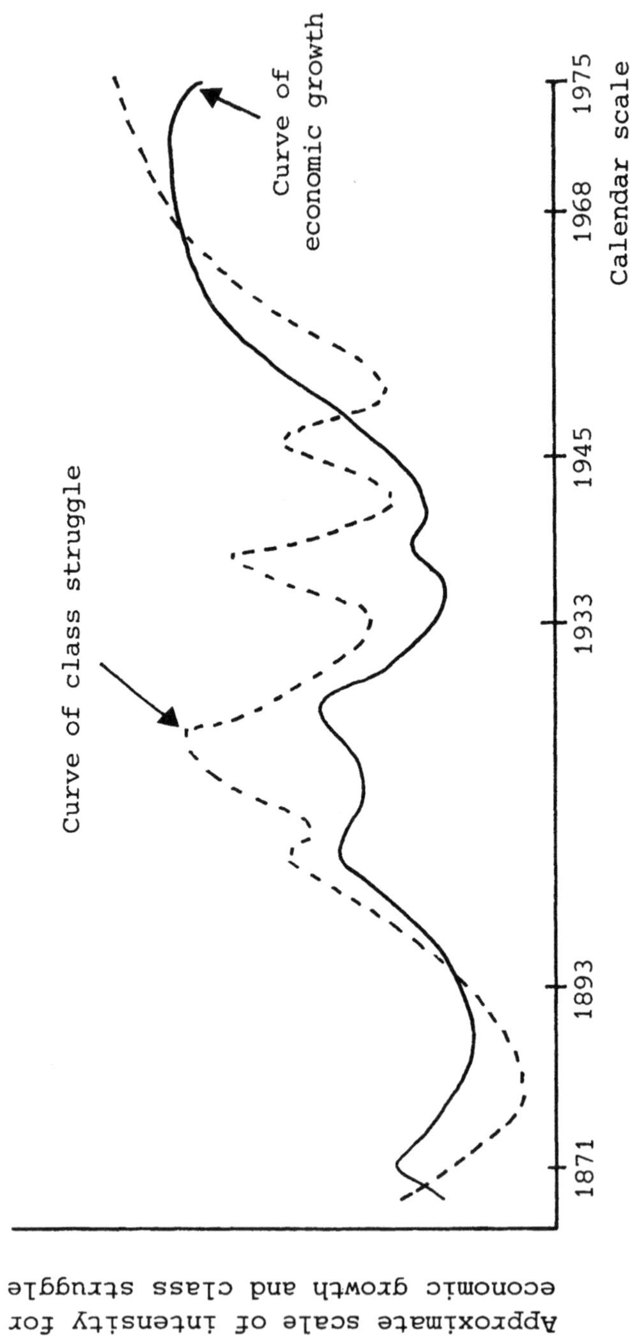

Figure 7-3. Economy and social structure: Autonomy of the social structure, 1870–1975

the flexibility of capital. That leads into an analysis of the issue of whether capital can achieve the 'required' creative destruction in the 1970s and 1980s to provide sufficent basis for the next longwave to occur. He notes that there are two major mechanisms through which the downward tendency of profits is countered: first, through massive restructuring including chronic unemployment, erosion of the value of real wages, reduction of the power of organized labour, reduction in the costs of raw materials, and wholesale devaluation of the investments of the least efficient firms – a major source of creative destruction will be the introduction of new technologies and working practices; and second, by the entraining of scientific knowledge to capital through the deskilling of all the workforces and the growth of a capitalist-owned science, especially in further growth of corporate laboratories.

Mandel is sceptical of the capacity of capitalism to achieve the changes which he theorizes are currently required. He suggests, probably to the surprise of many, that the strength of the American working class is likely to mitigate against major transformations except through catastrophes such as wars and the collapse of market exchanges.

Third, the idea of longwaves derived from Kondratiev deeply influenced one of the few Western economists to focus on the analysis of capital from the perspective of disequilibrium, the American economist Schumpeter[19]. From his earliest writing (circa 1911) he was preoccupied with economic pulsations as processes and with the attempt to refine their analysis with economic concepts and theories (Schumpeter 1911). In that early work Schumpeter attributed the pulsations in the level of economic activity to the behaviour of entrepreneurs and he identified them as heroes in the economy. However, his later work reflects a greater awareness of how in the twentieth century the large corporation increasingly becomes the engine for economic activity and he sought to refine the early theory by taking account of the longwaves discovered by Kondratiev.

Schumpeter argued that longwaves existed and that their origins could be explained by the existence of short, intense periods when whole arrays of unexploited inventions from diverse fields were compounded together in the creation of new generic sectors: railways, chemicals, electrification and automobiles, plastics and electronics. Those intense periods provided the transformation of inventions into innovations through the commercial acumen of entrepreneurship. In other words a small number of entrepreneurs would perceive the possibility of a new generic item: the car. In order to build the car they would draw together previously separated inventions and patterns of working. For instance, the development of automobiles had a profound impact on the development of textiles. Some of these early entrepreneurs

[19] Schumpeter (1939, 2 vols.) Kuznets (1953: 104–124); Schumpeter (1934)

would make a great deal of money and their activity would be noticed by others who would swarm into the new sector thereby reinforcing its tendency to draw many inventions into commercial applications. Schumpeter linked the theory of longwaves to technological innovation and argued that each longwave was set in motion by new generic technologies.

The theory of longwaves was used by Schumpeter (in the 1930s) to explain the long-run impacts of economic pulsations on the fabric of society. His concept of creative destruction should be used with care, but its broad meaning is that during the downward movement of a longwave its generic technology will be decreasingly able to act as a carrier for economic growth because its core technologies were being standardized (Schumpeter 1939). Therefore there would be intense competition between firms leading to closures, bankruptcies and debt. Although that process might seem to be destructive Schumpeter indicated that it was in fact clearing the ground for the new wave which would follow. Periodic structural crisis of economic adjustment would, in Schumpeter's perspective, be part of the normal long-run existence of capitalism. Freeman and colleagues (Freeman 1983) have followed the broad claims of Schumpeter whilst rejecting his claim that innovations are bunched. As economists, their approach tends to treat innovations as possessing a natural ordering rather than the approach advanced in Chap. 4, which emphasised cultural shaping and a high degree of unforeseen outcomes (Compare Freeman 1974). Freeman's major concern centred on questions of whether there are periodic deep structural crises of adjustment of the kind suggested by Schumpeter, and if so, what part do new technologies take: are they bunched, for example?

Freeman observes that a neo-Schumpeterian interpretation of the period of growth from 1945-1970 would emphasise that there had been a simultaneous and explosive growth of a number of new technologies, new industries, new materials and new products (Freeman 1982: 20). However, it is not necessary to adopt the idea of cycles, nor to assume that they have an exact periodocity which is fixed by calender time. Instead the notion of longwaves can be used in the context of the notion of a *major structural crisis of adjustment* where those crises vary in their form, intensity and timing (Freeman 1982: 26).

Freeman argues that the concept of capital accumulation is central. Short-run variations in investment can be assumed to be shaped by the expectations of potential investors and the overall level of interest rates:

> If investment demand is highly interest-elastic (i.e., responsive rapidly to small changes in the rate of interest given a certain level of expectations), then depending on wage and price flexibilities, a full employment general equilibrium could be easily obtained. (Freeman 1982: 27)

Expectations can be assumed to be highly volatile so that tendencies towards slumps and booms are exaggerated. If expectations fluctuate, then situations will arise when the hopes of rapid accumulation will attract a bandwagon effect, yet will be faced by various inflexibilities which create delays (Freeman 1982: 29). Those inflexibilities will produce disequilibria and might well lead to periodic crisis:

> ... technical change plays a crucial role in creating some of these inflexibilities, lags or discontinuities (Freeman 1982: 30).

Freeman contends that on the supply side there is likely to be dynamic competition amongst entrepreneurs who will play a central role in allocating resources between sectors. Their actions lead to structural change.

Freeman also gives emphasis to the role of scientific discoveries and their interrelatedness (Freeman 1982, Chap. 4), especially in the notion of families of innovations which incorporate radical steps and continuous modifications. He concludes that innovations occur in clusters because there is a bandwagon effect when the expectations of potential investors are raised, leading them to collectively invest at particular moments in given new technologies. Their actions reinforce existing tendencies by generating further waves of small modifications to existing new radical innovations. The key contextual macrolevel process is *when* the interaction amongst and within a cluster of innovations permits a new market to grow rapidly because large amounts of capital can be raised and invested.

Freeman contends that *if* there are longwave pulsations, then it is likely that the expansionary period of the upswing will contain two quite separate areas of the bandwagon effect:

1. The diffusion and exploitation of minor, yet important, innovations arising from the downswing of the immediate, previous longwave, that is, innovations awaiting diffusion
2. Innovations occurring even earlier

Thus, any clustering which might occur would arise from the *linking together of technological systems*. It is that phenomenon which creates the bandwagon effect rather than the swarms suggested by Mensch (Mensch 1979).

The bandwagon effect is a period of the rapid diffusion of new products, processes and services. It is believed that new basic innovations will generate enormous levels of profit. Further, there is likely to be the creative destruction of older products and processes. Thus, the longwave period of expansion from 1945-1970 can be interpreted as the simultaneous unfolding of multiple trajectories from computers, drugs, television, plastics and consumer durables.

	Previous Kondratiev	'Main carrier' Kondratiev		
		Recovery and boom	Stagflation	Depression
Research, invention	Basic invention and basic science coupled to technical exploitation. Key patents, many prototypes. Early basic innovations.	Intensive applied R & D for new products and applications, and for back-up to trouble shooting from production experience. Families of related basic innovations.	Continuing high levels of research and inventive activity with emphasis shifting to costsaving. Basic process as well as improvement inventions are sought.	R & D-investment becomes less attractive. Despite the fact that firms try to maintain their level of research it becomes increasingly difficult to do so with the slackening of their sales. At the same time the volume of sales required to amortize the cost of R & D is steadily increasing. Basic process innovations still attractive to management but may meet with social resistance.
Design	Imaginative leaps. Rapid changes. No standardization, competing design philosophies. Some disasters.	Still big new developments but increasing role of standardization and regulation.	Technical change still rapid but increasing emphasis on cost and standard components.	Routine 'model' type changes and minor improvements of cumulative importance.
Production	One-off experimental and moving to small batch. Close link with R & D and design. Negligible scale economies.	Move to larger batches and where applicable flow processes and mass production. Economies of scale begin to be important.	Major economies of scale affecting labour and capital but especially labour. Larger firms.	The slowdown in output and productivity growth leads to overproduction and excess capacity in some of the modern industries. These structural problems are 'cumulative and self-reinforcing', with repercussions for the economy at large, and lead to a further decline of economic activity.
Investment	High risk speculative. Small scale. Some in-	Bunching of heavy investment in build-up of	Initially continuing heavy investment but shifting	Relatively low levels of investment. Underutilisation of

	Previous Kondratiev	'Main carrier' Kondratiev		
		Recovery and boom	Stagflation	Depression
	ventor-entrepreneurs. Some large firms. Fairly labour-intensive. Problems of venture capital.	new capacity. Band-wagon effects. Large and small firms attracted by high profits and new opportunities.	to rationalization. Continuing rapid growth, but increasingly large sums required to finance R & D and rising capital costs. Rising capital intensity.	the capital stock in some of the most modern sectors of the economy; low profit margins and the general 'pessimistic mood' with regard to expectations lead entrepreneurs to be very (over) cautious in relation to new investment opportunities. Investment which will take place will be primarily directed towards rationalization. Search for new investment opportunities abroad.
Market structure and demand	Innovator monopolies. Strong consumer resistance and ignorance. Some new small firms to promote basic innovations.	Intense technological competition for better design and performance. Falling prices. Big fashion effects. Many new entrants in early build-up.	Growing concentration. Intense technological competition and some price competition. Strong pressure to export and exploit scale economies.	Even stronger trend to oligopoly or monopoly structure. Bankruptcies and mergers.
Labour	Small-scale employment generating effects. High proportion of skilled labour, engineers and technicians. Training and learning on the job and in R & D.	Major employment generating effects as production expands. New training and education facilities set up and expand rapidly. New skills in short supply. Rapid increase in pay.	Employment growth slows down, and as capital intensity rises, some jobs become increasingly routine.	Employment growth comes to a halt. Unemployment rising. In addition to the continuing labour displacement effects of rationalization investments, employment suffers (in the first instance) from the general recessional and depressional tendencies in the economy at large.

	Previous Kondratiev	'Main carrier' Kondratiev		
		Recovery and boom	Stagflation	Depression
Employment effects on other industries and services	Negligible, but imaginative engineers, managers and inventors are thinking about them and planning and investing accordingly.	Substantial secondary effects, mainly employment generating but gradually swinging to displacement.	Labour displacement effects, as new technology now firmly established and strongly cost-reducing.	Continuing labour displacement as new technology penetrates remaining industries and services.

Figure 7-4. New technology systems: Freeman

The unfolding of the new technological systems in conjunction with the periodic crisis of adjustment can be schematically presented as in Fig. 7-4. The four vertical columns express different states of the previous and the current longwave. The horizontal rows refer to (1) research and (2) design, (3) the scale of production, (4) investment, (5) the market structure, (6) the organization of labour, and (7) the effects on employment. In that schema the longwave pulsations have been interwoven with a variety of other processes[20].

The consequences of the schema for particular sectors and specific geographical regions will be very uneven. There is likely to be a *changing regional locus* for each new longwave (Freeman 1982: 167 ff.) and extensive effects on employment. These two aspects are examined more fully in Sect. 7-4.

Fourth, and rather different to the previous schools, is the French group known as the Annales School (Braudel 1972), which through its leading members represents perhaps the best-known contemporary historical perspective on economy and society. So much so, that the recently deceased Braudel has an Institute named after him at the Binghampton campus of the State University of New York[21]. Braudel's seminal study and an early exemplar was based on the A/B series of price movements applied to the analysis of the archives of Mediterranean ports. In a remarkable early use of the cine camera, the young Braudel recorded trading data covering the medieval and Renaissance period and the period when Philip II was the ruler of the Spanish-Hapsburg Empire. He concluded that price movements were irregular,

[20] The figure is based on Freeman (1982), Fig. 4-1 and on an earlier Working Paper

[21] Fernand Braudel Center for the Study of Economies, Historical Systems, and Civilizations, State University of New York at Binghampton; its specialist journal is known as *Review* and is edited by I. Wallerstein

jagged pulsations whose upward and downward movements were often highly consequential for the material life of cities and empires.

Braudel was immensely dissatisfied with the writing of history according to the scale of the calendar, because like Vico and Marx, he considered that historical processes contained many branches and alternative trajectories whose courses could only be situated temporally by reference to each[22]. In that thinking he followed Bergson and was profoundly influenced by early twentieth century physics and the debate over time (Bergson 1903). He was also deeply influenced by the concepts of social morphology advanced by the sociologists Durkheim and Mauss and so he sought to show how the economy and society interacted[23].

Braudel wrote his analytically structured narrative on three temporal registers, one of which was the 50-year A/B price movements and their turning points or conjunctures. The three registers were:

1. The plurality of times found in the social structure, which was given primacy. Braudel argued that social practices and social thought change very, very slowly indeed. He would not be in the least surprised by the claim that certain British patterns of individualism can be shown to have been evident in the twelfth century and to have underpinned conceptions of liberty for almost a millenium. According to Braudel there is immense stability in what he terms mentalities. Within that stability there are "brownian movements" where different areas of the mentalities are activated according to seasonal patterns and to other life courses.
2. The time of conjunctures. These are the economic movements of the price series and they tend to last about 50 years.
3. The short periods of a half-decade in which various political events occur.

There are important relations between these three registers.

As indicated in Fig. 7-1 the short term is a period in which there is a high degree of reproduction of the past repertoire and patterns. Moreover, the consequences of decisions taken in the short term and of events happening in the short term (e.g., nuclear accidents) take many years before they become evident. However, the pressures towards reproduction deriving from the mentalities and structuration are always subject to historical contingency and to interruption from, for example, the economic register.

As indicated earlier in Chap. 1 when reference was made to the decline of Spanish economic and political power in the seventeenth century, the men-

---

[22] See Hexter (1972); Braudel (1958); for a recent attempt to restate Braudel's approach to time in history, see Arieas (1986)

[23] Durkheim (1915) Mauss (1904); Braudel (1972: 679-681, 756-757)

talities interact with the economic pulsations (Braudel 1972). In the case of Philip II, his own thinking and that of his advisers, was strongly embedded in the material life of the Mediterranean world, including the ways in which its economic activity was shaped by the geographical features and the marked seasonal shifts. Consequently, it was not noticed that there were different patterns in the North Atlantic. Nor was it noticed that the consequence of Spanish plunder for the Mediterranean was to create inflation. In other words the policies of Philip II actually contained the seeds for the demise of Spanish power and the progressive bankruptcy of his exchequor. Meanwhile, the Dutch entrepot of Amsterdam was rising to influence as the beneficiary of an economic conjuncture of upward movements in prices.

7.2.3 Transaction Costs, Ideology and the State

There are two main lines of development of the transaction costs theory as a revision to neoclassical economics, one centring around O. D. Williamson and the other as formulated by the economic historian, D. C. North (1981; 1985).

The transaction costs approach represents a development from within neoclassical economics in respect of the close scrutiny of the mechanisms by which resources are allocated. Neoclassical economics had, until the seminal article by Coase in 1938, focused solely on the market as a costless and frictionless system of allocation based on price signals and differential accumulation of capital (Coase 1937). Coase, whose arguments were barely noticed for three decades, asked a simple question: if markets are so efficient, how is it that large firms have been able to emerge and sustain their position in the twentieth century? In due course the market mechanism was examined in comparison to two alternative mechanisms: the management hierarchy in the firm - the visible hand – and the mechanisms based on personal trust and/or subordination.

Neoclassical economics had concentrated upon the market mechanism which it was concluded would lead through the price system to provide the best mechanism through which real demand and supply would adjust in the long run. The neoclassical theorists constructed their models around the assumption that the fixing of prices was a relatively simple process, involving two sets of actors each of whom had a rational interest in searching out their own marginal benefits. The theory did not explore the costs associated with making economic transactions. Nor did the neoclassical theory sufficiently examine the societal features necessary to sustain a market mechanism. Instead the market mechanism was presumed to be obvious and unproblematic[24]. Likewise it was assumed that the only important costs worth examining were the costs of the supplier providing the goods and services.

As indicated the analytic breakthrough came from an apparently simple question. One so simple that all the early seminal classics promoting the study of organizations implicity inhibited its asking[25]. The question? Why are there organizations if the market is such an excellent allocative mechanism?

Coase observed that markets do involve buyers and sellers in costs over the *specification* and the *enforcement* of contracts which underpin all exchanges. The costs include the search for appropriate partners and, once found, the specification of what is to be exchanged. That is, a contract, explicit or implicit, specifies who has certain property rights and what are the attributes of those rights. The making of the contract is only the beginning and has to be followed by the performance of the contract, including what happens if there are contingent circumstances (e.g., the weather affects the contract). Finally, there is the problem of enforcing the contract and dealing with unsatisfactory performance. In the Western world the enforcement of contracts is largely carried out through the agencies of the state, though in practice contracting parties typically devise additional enforcement mechanisms. The role of the state is quite central. Its effects, positive and negative, include the role in creating the ideological context of market activity (North 1981, Part I).

If a theory assumes that the market mechanism involves transaction costs, then situations can occur when the costs of market transactions will be above those borne by alternative mechanisms. Under such conditions there will be market failure (Williamson 1975). An alternative mechanism is the firms within which transactions can be specified and enforced by the actions of the principals and by top management acting as proxies for those principals. To an extent transactions can be specified and enforced. However, the ease of achieving that objective may be even more complex than has been suggested in Chandler's seminal study of the role of employed American managers as the visible hand behind American success (Chandler 1977).

Chandler argued that two activities developed within the late nineteenth-century American corporations that facilitated the growing strength of the visible hand. First, managers devised languages which cognitised the activities of the business into components to which costs could be attributed. Second, top managers and their college-trained advisers devised internal costing methods which evolved from the orthodox historical accountancy into a future-oriented standardized costing which permitted them to develop a useful

[24] Interestingly the International Encyclopaedia of the Social Sciences did not have an entry for the market in the 1968 edition (E. A. Shils, ed., New York: Macmillan

[25] See the introductory paragraphs to the seminal March and Simon (1958)

understanding of which activities contributed least and which most to the accumulation of capital. Combining these two features permitted certain managements (and their employing firms) to develop internal systems which could act as alternatives to the market mechanism in their sensitiveness whilst also being relatively easy to operate in the American culture. However, Chandler's analysis does not satisfactorily deal with the enforcement of contracts within and outside the firm. Also, his analysis totally neglects the institutional relationships between capital and labour. Equally, Chandler's analysis seriously understates the role of the state apparatus and style in providing an appropriate context.

In a significant contribution, D.C. North reasons that the rulers will tend to specify a structure of property rights which is consistent within their interests (North 1981, Chap. 3+4). The state structure is itself an outcome of an interplay between configurations of political forces and these will strongly shape the political system within which economic activity occurs.

The enforcement of contracts within firms raises the issues of the relations between capital and labour, particularly of how specific institutional arrangements will impact on transaction costs and possibly affect the relationship between them and production costs. Institutional frameworks within the firm have been at the core of North's analysis of the problems of comparing existing practices with new, alternative and seemingly innovative practices. In a transaction costs perspective, the institutional framework may be defined in terms of four elements:

1. The terms of the labour contract at the moment of employment
2. The existing framework of moral and ethical norms governing behaviour and the ways in which these influence the specification of rules and the costs of compliance
3. The rules and regulations which are intended by the principals to act as constraints on behaviour
4. The procedures designed to detect deviations from the rules and regulations and to enforce compliance

The normative framework underlies the stability of any contract whilst also affecting the costs of the enforcement of contracts.

In the premodern world the distinctive feature of the costs of transactions centred on the costs of enforcing the contract. In the case of long-distance trade, specific forms of institutional organization developed, most of which also involved a heavy investment in *trust,* that is, the personal bonds between an investor and the trader. Often these personal bonds were anchored in kinship ties, in common ethical and moral values (e.g., Quaker businessmen[26])

[26] See the extensive references to Quaker employers in Child (1969)

and in ethnic solidarity. Thus, any organized exchange involved elaborate rituals whose purpose was to reinforce the contractual arrangements over and beyond the support which arose from the actions of the state.

Within the personal network of ties a new form of organization slowly emerged so that within the modern age impersonal exchange was increasingly undertaken through formal contracting. Moreover, the state undertook two activities. First, it established agencies through which a supportive structured ideology was created. Milton Friedman contends that the minimalist contribution of the state is to ensure that education is heavily oriented towards socialising populations into the acceptance of an economic chivalry (Friedman, M. 1962). Second, because the state acquired a monopoly over the use of coercive power the ruling strata also tended to foster certain aspects of impersonal exchange through legislation and the judiciary.

With respect to the role of the state in affecting transaction costs, it may be observed that in the post-Renaissance period of European economic growth the seventeenth century state institutions of the Netherlands and of the United Kingdom gave a definite impetus to the enforcement of contracts. These features were refined, speeded up and extended in the New World.

In medieval Europe, especially the earlier period, Braudel shows that the emerging forms of economic activity were only slightly built on impersonal exchange (Braudel 1983; 1984). A practice known as The Commenda dominated the sea trade of the Mediterranean, whilst fairs and towns slowly developed as marketplaces in which a degree of direct contact between buyer and seller took place (North 1985). The shift in the locus of power during the later medieval period – probably influenced by changes in military technology – facilitated the growth of the towns which were prepared to exchange protection from the king in return for revenue. Towns gradually took over the codes of exchange which had previously been operated by the merchants, replacing the personalised ties by the provision of the courts.

The regularity of exchange which occurred within and between towns provided four key contributions:

1. Means were devised which formalised the problem of measuring the attributes of the items which were to be exchanged, thereby providing a basis for measuring the performance of the agents involved in the exchange. Developments included the establishment of uniform weights and measures which provided an indirect and imperfect surrogate for the attributes of the items being exchanged.
2. With specialization between towns, especially between those of the Mediterranean and those states facing the North Sea, there was an increase in the linkages which made up an exchange (Elias 1978) so that exchanges

became increasingly impersonal[27]. There was an enormous increase in the volume of transactions.

3. There developed an elaborate body of statute and common law which underpinned the role of the state in the enforcement of contracts. Even so, enforcement was imperfect.
4. The normative framework governing exchange emphasised constraints on maximising profit and increasingly indicated areas of transactions where the offender could anticipate swift social and economic punishments. Yet that left other areas for opportunism.

Gradually, then, with increasing pace in the seventeenth century, the size of the area governed by impersonal transactions widened and deepened through specialization along the lengthening linkages of economic activity. That led to the development of those sets of new occupations involved in the wholesale and retail trade as well as in the two rapidly growing occupations of auditors and accountants (see tables in North 1985).

7.3 Structuration

7.3.1 Introduction

The concept of structuration will be utilised in two ways:

1. To characterise the duality of structuring and of agency
2. As a generic category embracing the repertoires of values, normative frameworks, rules of the game and institutions which characterise the various levels which simultaneously come into play in any society to examine the structuration-event-structuration thesis

The place of structuration in the basic framework has to be understood as part of the total configuration shown in Fig. 7-1.

First the duality of structure, which derives from Giddens (1985), means that structure is the medium through which individual agency unfolds and also the outcome. It is because of this duality that the concept of structuration is preferred to the use of structure (see Sect. 7.1.2). The stress of the duality of structure represents a definite attempt to counter earlier theorizing which portrayed human agency as almost entirely determined by preexisting struc-

---

[27] This feature is central to the analyses of Marx, Weber and Durkheim; for example, Marx observes that the initial form of trading involved the transformation of commodities into money and then back into commodities, the so-called CMC, but that evolved into the use of money to buy commodities which were turned into money, the so-called MCM. The evolvement of contractual relationships in the USA is dealt with in North (1981), and Chandler (1977)

tures. To counter that determination, attention is given to subjective agency. In order to examine agency it is necessary to examine the construction of individual lives in their social settings. We know from time-budget researches of individuals[28] as well as from a small number of organizational studies of structural repertoires[29] that the activities of individuals are so patterned in time and space that there is a high degree of recursiveness. Giddens refers to 'chronic recursiveness'[30] in the social practices of everyday life. The point was partially illustrated by Hagerstrand (see Sect. 7.3.2) in his reconstructions of individual trajectories, but that conception understated the potential of the individual to interpret the implied scripts within existing structures and to introduce alterations. Individuals are both constrained and enabled. They are enabled by their everyday knowledge of the conditions of reproduction of the social contexts in which they reside.

The knowledge that individuals possess has various different modes. There are unconscious sources of cognitions, the practical consciousness contained in the tacit stocks of knowledge which are regularly drawn on, and a discursive consciousness consisting of knowledge which individuals can articulate and express (Giddens 1979, Chap. 2). In any situation the individual brings these modes into operation. Also, individuals possess sets of intentions which are both long term in character and also open to alteration (c.f. Hagerstrand). Moreover, individuals may encounter new contingent circumstances so that the outcome is novelty rather than the reproduction of structure. Thus, the individual trajectory is situated in many ongoing flows of action in time and space. There are a multiplicity of individual trajectories in time and space. Unintended outcomes are very important.

The emphasis upon agency should not reduce the importance of tradition, ritual and routinisation in social life and one very significant, inescapable feature is the *sedimentation* of institutional forms in the processes of societal development. Social sedimentation means that the everyday practices of today were often laid down at some time in the past. The facticity of clocktime, for instance, is so taken for granted that the nineteenth century struggles between capital and labour over the application of clocktime are largely taken for granted. Indeed, a great deal of nonwork social practices in the area of entertainment rely on clocktime. The winners of past struggles are able, as Foucault and others observe (Foucault 1977), to have the outcomes silently entered into later everyday action so that the selection criteria underpinning

---

[28] On time budget research see the analysis of individual trajectories already examined in Chap. 3, and also Szalai (1971)

[29] See the discussion in Part I of chronic recursiveness

[30] Giddens (1985); the extensiveness of chronic recursiveness in organizations was examined for supermarkets, can-making factories and sugar-beet factories and reported in Clark, P.A. (1985 25–61)

Societal cognitising

I. STYLE OF THINKING	1. Theorizing 2. Noticing variations 3. Efficiency concept 4. Commodification 5. Logistical tightness Mastery of flows 6. Double loop learning
II. EDUCATION	7. Coupling to capitalism 8. Extensiveness of access 9. Specialists 10. Specialised generalists 11. Designing of
III. ENTERPRISES	12. The ownership of jobs 13. Professional work organizers 14. Script construction 15. Script learning 16. Boundary controls 17. Attachment/traditionalism 18. Structuring of activities
IV. DESIGN	19. Specialised function 20. Design: deliberateness 21. Design: productionising
V. DESIGN 'RECIPES'	22. Experimentation and evaluation 23. Market, competition 24. Choices focus/foci 25. Rationalization 26. Standardization 27. Orientation to technology 28. Manufacturing consent Incorporation of social sciences 29. Rewards and time horizons 30. Recognition of contingency concepts

Figure 7-5. Knowledge bases and thinking practices: Framework for Anglo-American comparisons

action are often unrecognised[31]. It is for that reason that many contemporary theories of society have used the imagery of archaeology to express the search for the silent origins of certain contemporary cultural features.

Sociologists are agreed that one key area of sedimentation in Western societies is property relations[32]. Moreover, property relations are both sustained by and taken for granted practices and are also embodied in the law as was indicated earlier in the examination by North of the societal contexts of transaction cost structures (see Sect. 7.2.2). Therefore, the analysis of the contemporary cultures of Britain and America has to involve a certain degree of penetration of those cultures through a form of historical analysis. The issue is what kind of historical analysis (see Sect. 7-5).

The sedimented social practices become embodied in institutions and the institutions of each society are potentially distinctive. Institutions are the most deeply sedimented facets of structuration. How can they be analysed?

Analysing the institutions requires the choice of a minimum array of concepts. What array of concepts will assist in the comparison of structuration between Britain and the USA? To gain focusing, the following sections will centre on three sets of concepts:

1. Asymmetrical power relations; domination, the commodification of space and time, orientations to the future, creating tradition, comprehensive control concepts and hegemony
2. Knowledge bases; educational systems and their interfaces
3. Institutions of capital, management and labour

7.3.2 Asymmetrical Power Relations

First, *domination*. It will be theorized that domination is achieved through control over resources and over rules, their making and application (Giddens 1985). Moreover, domination is inherent in the duality of structure and agency and is embodied in the reproduction (or, otherwise) of social practices. Relations of authority in society are essentially asymmetrical. Some interest groups and factions have more resources than others and more potential to activate and create the means to legitimate their actions.

The commodification of time and space is one of the most pervasive features of modern society, yet there are important differences between Britain and

---

[31] See the attempt to reconceptualise structure in Chap. 13 of Clegg and Dunkerley (1979); note the attempt to synthesise across the ideas of Offe, Karpik, Weick and Habermas

[32] North (1981); there is also revealing analysis of the influence of property relations amongst farmers and their employees in Newby (1978)

the USA (Clark, P.A. 1985). By commodification I refer to the attachment of monetary values to units of time and space such that the monetary values become integral parts of the social definition of time and space. The commodification of time and space has developed slowly over the last millenium and its articulation was greatly facilitated by relations within the earlier medieval cities[33]. In those settings, both employers and wage labour used the time derived from public, equally divisible time-reckoning systems to negotiate contracts. There came a very slow recognition that the rate of accumulation of capital and the rate of wage earning were being calculated in respect of these standardized units derived from the calendar (e.g., usary over 30 days) and the public clock. Not all factions in society welcomed these slow changes. It seems likely that the landed strata preferred the time reckoning systems which were least exacting[34]. Yet here is where the economy and society interpenetrate in the manner suggested by Marx in his analysis of the porosity of capital accumulation.

Likewise space became defined in standard units and to varying degrees those units could be purchased. That was especially the case in the USA. Any unit of space could be scrutinised in terms of its earning power. Two obvious examples today are in the routine calculations made by all supermarket chains of their earnings per unit of space and the calculations made of the costs of office space by reference to the costs of the area occupied by the wastepaper basket. Each of these are simply instances of widespread practices.

Orientations to the future (Bell 1972) and the treatment of *tradition* are intimately related to domination and to the commodification of time. Societies vary in the extent to which they are oriented towards the future or to tradition. A British visitor to the USA may be struck by the everyday American practice of asking "what are you going to be doing?" or "where are you taking your vacation?". For the British the questions are about "what did you do?". Also, the more prevelant question in Britain relates to where you took your holidays last year. Yet and paradoxically the British are great inventors of tradition. Many traditions owe a great deal to the shrewd marketing of Britain in the USA by the state tourist boards; history is business in Britain and tourism is the third largest sector!

Orientations to the future may incorporate *comprehensive control concepts*[35], that is, directions chosen to service the interests of a constellation of factions

---

33 See particularly Giddens (1981, Chap. 6) for a cogent introduction

34 The differences between strata and groupings in their commitment to specific time orientations is prised open in Gurvitch (1964)

35 The notion of comprehensive control concepts is central to Noble (1984) and to Van der Pijl (1985), as well as to the whole approach of Gramsci and to the perspective of the politics of production currently being developed by Burawoy; see, for example, Burawoy (1984)

and to portray those interests as synonymous with the societal good. In other words, these are future targets which come to acquire a high degree of public legitimacy in the authoritive mechanism of society, whilst also being very much angled to the allocative mechanism of the economy. The position of the agricultural sector in Britain has been presented to the public through highly favourable imagery of food, diet and health, all of which has understated modern methods of manufacturing foods. In the USA there is a striking mixing of future utopian projects with financial interests (Segal 1984). Noble reconstructs how the image of the workerless factory has been used by a variety of factions, many of them opposed to one another, yet all benefitting from the allocation of resources and from the making of favourable rules (Noble 1984). Thus, the whole machine tool industry and certain famous universities have been involved. A similar pattern, distinctly American, can be detected in the presentation of Star Wars.

It will be noted that although the notion of comprehensive control concepts is tied in with domination and the structuring of the future, the interpretation offered here does not accept the necessity of hegemonic controls as the one-sided imposition of a faction. That may occur in certain situations at specific occasions. However, the point being made here is that domination is simultaneously asymmetrical and also contested even within those factions who have relatively strong access to resource allocation and rule making.

7.3.3 Knowledge Bases and Thinking Practices

Second, reference has already been made to types of knowledge and to the role of the *educational system* and *its interfaces* with other institutions in society[36]. The significance of this institution in societal adaptation is increasingly emphasised and it will be impossible to do justice to the range of relevant debates and suggestions.

In the absence of extensive and systematic studies of these areas in the USA and Britain it is only possible to construct a strong hypothesis of the differences between the two societies from diverse sources. Figure 7-5 sets out five subareas which are of relevance, each of which contains several headings[37].

Styles of thinking have already been identified in earlier chapters as a key source of inputs to the processes of innovation. Several distinct aspects are important:

---

[36] See the treatment in Parsons (1951)

[37] Figure 7-5 is a preliminary framework for examining differences and similiarities; it requires further development and modification

1. To what extent are styles of thinking theoretical, rule of thumb, or pragmatic?
2. To what degree does thinking look for and detect variations from existing cause maps and what happens when variations are noticed? Can they be incorporated into existing thinking? Is thinking constructed around 'how to do it' issues?
3. Is there a sense of thinking as an efficient practice which can be improved by various social and technical devices?
4. To what extent is thinking evaluated by reference to monetary values? Is there a concept of value for money about thinking and knowledge?
5. To what extent is thinking characterised by a concern with logistical flows and their smoothness? Are disruptions to flows cognitised and rejected as unacceptable?
6. To what extent are existing paradigms open to the development of alternative interpretations previously rejected? Is there a potential for double-loop learning? Is that potential recognised?

These six facets of thinking styles can be used to sketch and highlight certain differences between the USA and Britain with reference to the managerial and executive stratas. Certain differences in thinking styles will be important in stratas which might be expected to play a leading role in modernising and adaptation.

First, it may be argued that the American style of thinking highlighted the role of pragmatic "how to do it" thinking and regarded bodies of thought as having an evolutionary role in the survival of the brightest. There was considerable stress on recording how activities were undertaken and on examining the extent to which practices conformed to the notation of best practice. Consequently there was some potential for noticing variations from the expected. That tendency became embedded (in the twentieth century) in the management sciences. The idea of the efficiency of thought and of knowledge has been very strong. Currently that interest is reflected in attempts to pool knowledge through conferences and to create expert systems. American thinking has been very capable of analysing processes and flows to define and to discover "bottlenecks" or areas of low benefit to capital accumulation. There is considerable debate over whether these styles of thinking are amenable to reflective revision through double-loop learning. Exponents of the importance of double-loop learning highlight failures in American managements' performance, yet it should be observed that a variety of methodologies and practices have been developed, particularly in the management sciences (e.g., dialectical thinking and the work of Churchman and Mitroff)[38].

---

[38] Churchman (1968); Mason, and Mitroff (1981); Mitroff (1981)

British executives have a quite different style of thinking. It tends to be antitheoretical. There is an accepted polarization between academic thinking, which is often regarded as unnecessary and impossible to digest, and the rule of thumb empiricism which seems to have a firm grip in many sectors. Evolutionary thinking and its associated concepts were never as strongly incorporated in Britain as in the USA. Management observation of existing practices, especially their own, was undeveloped until the entry of American consultancies in the 1970s. Sometimes observation was inhibited by trade union activity, but equally it was considered less often. Reflective thinking has been disrupted by the lack of a systematic base in observation. It is pertinent to note that in the 1950s and 1960s leading members of ICI, the major chemical firm, rejected production engineering thinking as inadequate and developed work study as a focused and abstract, systematic approach to the investigation of problems of productivity. More attention is given to explaining the alleged facts than to ascertaining their veracity. The efficiency of knowledge and of thinking is only taken seriously by a few isolated firms and even the business schools tend to adopt a pseudopractical recipe knowledge based on current fashions rather than trying to develop analysis and diagnosis of practical problems. So-called management research groups are more often political 'talkshops' than radical searches for better practices[39]. Knowledge is less commodified. As would be expected, the abstract and pragmatic concept of flows is less established. More often there has been a crude iconic thinking. The notion of the organization as system of flows is very hard to accept. As in the USA, double-loop thinking is practised in leading firms, but it is not well established.

Second, *education* may be compared by examining the following dimensions:

1. The closeness of the interface coupling education and capitalist institutions
2. The extensiveness of access to education at any one time and the return visits over the managerial career
3. The extent to which education aims to create specialists
4. The consciousness of using the design of courses as a means of retaining best practice and of introducing alternative practices

American education involves a close coupling between the educational system and corporate capital and the mutual interpenetration of each by the other. Moreover, there is extensive, highly developed education within many

[39] The foundation in London of the Management Research Group No. 1 in the 1920s brought together representatives from leading British firms who discussed their practices of labour control; the archives, known as the Ward Papers, are held in the Business History Unit, London School of Economics – see further comments in Chap. 12

corporations which is staffed by faculties of highly trained staff. The land-grant college was an early model from the nineteenth century and there is ample indication of the extension of that model. Various studies show the active role which the university sector plays in corporate research of many kinds. The notion is that education should serve the interests of its students and that those interests will be oriented towards future careers with employers. So there is widespread occupational education (e.g., for media in general and particular sectors of the media). Education is regarded as widely open, highly competitive and continuing through life. The American student takes it for granted that he will be compared against others on a bell distribution and assumes that failure is the fault of the individual. Education is conceived of as an investment in human capital which has to be maintained. The educational system produces both highly trained specialists from vertical knowledge systems as well as highly trained generalists whose knowledge bases lie across different areas each of which contribute to the common problem (e.g., the MBA). Courses are typically designed as totalities and compared with other courses. They are often monitored and subject to revision. Knowledge is packaged in modules and increasingly presented and assessed through the use of film (e.g., tutor video instruction) and modern technologies (e.g., personal computers).

In Britain the educational system has been in conflict with capital. There has been a long struggle over autonomy and control which is reflected in the rights of school teachers and in the autonomy of university faculty to exercise their discretion, even in the face of low demand for courses. These features were the targets for the right-wing radicals of the Thatcher administration during the 1980s, but their impact still leaves a very large gap between the British and American approaches. Access to higher education is limited to the top 15%, many of whom have considered the process complete after the Bachelor's degree. Fewer would expect to take Master's degrees or to follow regular updating courses. Within employing organizations internal education, whilst increasing in capacity, is still small by American standards. The small number of employers who are highly conscious of training are constantly mentioned: John Sainsbury (supermarkets), Tarmac (construction), and Marks and Spencer (retail). The emphasis has been on individual learning. There has been a low use of technology. Much more attention has been given to the unproven tutorial system. Systematic testing of students is on a much lower degree and the systematic monitoring of courses is virtually unknown outside the Open University format. British students make far less use of set books, particularly in the arts, humanities, management and the social sciences.

Third, societal cognitising is embedded in the operations within *enterprises*. The following dimensions are important:

1. The location of the ownership of job knowledge. Is occupational knowledge primarily owned and constructed by the employing organization? Does the organization have the power to alter occupational content?
2. Are there specialized, professional work organizers who design the content of jobs?
3. Do jobs have defined scripts which have to be learned?
4. If so, how easily are scripts redefined?
5. Are there boundary controls to occupations and who controls movements through ports of entry?
6. Is attachment to the organization primarily through traditionalism or through instrumentalism? To what extent is attachment choreographed by the employing organization?
7. To what extent are activities structured through standardization and formalization of processes?

The American situation for the medium- and large-scale employer is that the employing organization tends to own the job knowledge and the whole space-time framework within which jobs are located. Jobs are more likely, matching sector with sector, to be designed by internal experts who apply their skills of job analysis and design to the construction of scripts. These are more likely to be highly specific and recorded. The job holder will be willing to learn the scripts through a process very similar to the one used in university to learn the set textbooks. Further, new scripts are learned as readily as new texts supercede previous ones. The learning and discarding of occupational scripts is taken for granted and it is quite possible that the ego-attachment to any script is quite low. The internal training departments (or their functional equivalents) will play a central role in the learning of scripts by role occupants. Employers will regard themselves as largely owning the script-writing process and as being able to redefine the boundaries of the job. Overall activities are likely to be highly structured.

In the British context, the boundary of the job may be implicitly set by the stratification system of society rather than the firm. Or, the occupants view themselves as the definers of job knowledge. Managerial jobs are rarely the target of work organizers, their scripts are still less formally constructed than American equivalents and the learning of scripts would be less obvious. The design of roles in Britain is generally only specified with respect to shopfloor jobs and it is usually undertaken by nongraduate members of the technical institutes. The structuring of activities is likely to rest on precedent rather than formalization.

Fourth, *design* consists of:

1. Degree to which these are specialized functions
2. The deliberateness with which design is undertaken

3. The extent to which concepts of future services or products are deliberately productionised to achieve a tight fit with the modes of their creation
4. The extent of systematic experimentation and evaluation through reflective methods, such as computer-based simulation

Designing in America is a highly developed set of activities which is given a specialized and deliberate status in most contexts. Principles are extensively taught. Even in the recreation industry the wilderness concept tends to have a design format. There is a considerable level of achievement, especially with static projects (e.g., theme parks).

In Britain, design as a deliberate process of anticipating the future in a detailed way is less highly developed, though there has been a sharp increase in awareness of the problems. The style of thinking in design tends towards identifying individual design contributions rather than sustaining an overall corporate imagery.

Fifth *design recipes* are well developed in the American context where the analysis of the market is an early feature (e.g., for household items, TV programs provision of sports). Attention is given to the overall structure of the market, to the extent and size of niches leading to the choice of design focus. The focus affects all aspects and is pursued rigorously in the rationalizing of unnecessary features and in the standardizing of the designing sequence itself (see Fig. 7-1-2). There is a strong orientation to technology both in the product (or service) and in the processes through which it is produced. Handling the employee – the manufacturing of consent – and the provision of rewards are integral elements. In the whole process the social sciences are involved in most facets, from marketing concepts to the implementation process. Contingency planning is developed, but not necessarily in a flexible manner.

In Britain design recipes have become very much more established in the 1970s across a variety of areas. Yet marketing occupies a lower centrality in both the firm and in the educational system. There is less belief in the mastery of markets and more reliance on a network of information which may be hard to organize. For example, until recently, car firms relied heavily on the dealers as much as on clinics and market research. There seems to have been hesitancy about focusing and a preference for keeping several different items in the portfolio. The whole process is much less infused with social science elements and there is a more empirical approach to areas like implementation. Contingency planning may be less well developed.

7.3.4 Institutions of Work: Capital, Management and Labour

The economic analysis of the firm neglected just those areas in which other social science disciplines have specialized, namely power, authority, conflict, managerial organization and labour organizations, i.e., the institutions of work. However, the transaction-costs perspective (North 1981; Williamson 1975) has provided an important focus for the examination of the institutions of work by theorizing the firm (and similar) as an alternative mechanism for allocating resources to the market and to trust through personal relationships. In other words, the firm as an authoritative mechanism has to possess certain features of an internalised market otherwise there will be organizational failure and markets will take over. Chandler and others contend that American managements discovered how to handle the internalised market of the firm and that claim provides a useful theoretical addition to the framework for the comparison of Britain and the USA (Chandler 1977).

Managerial hierarchies, in order to survive in the face of alternative mechanisms, have to develop specific institutional processes of work. These centre on two of the facets which the economic study of innovation neglected, production knowledge and production institutions. The former was discussed in the previous section. Attention is now given to the production institutions. It has to be recognised that the form which a specific set of institutions takes has considerable consequences for any economy; here is where the authoritative mechanism of society interacts with the allocative mechanism of the economy.

In economic theory, an institution is an *arrangement which guides competition and conflict between economic principals* (North 1981) and it therefore has an important part to play in providing information which is relevant to the operation and unfolding of the transaction costs in that part of the contract which refers to the relations between employers and employees. The centrepiece of analysis becomes, Baldamus argued, the effort bargain and the various processes through which labour time bought in the market is transformed into labour output at the the point of production (Baldamus 1961). There are immense difficulties for management in monitoring labour contracts at the point of production. It is clear that the power of the market mechanism is not sufficient, even when there are large numbers of unemployed. Therefore economic principals within the firm have to create non-market structures of authority, compliance and consent. As demonstrated in Chap. 5, it cannot be assumed that labour is a passive, neutral agent in those processes of innovation. Labour is, in practice, an active agent in the processes through which labour time is converted into labour effort.

The effort bargain will be an internal focus of competition and conflict between the buyers and sellers of labour time. Each may invest in different ways of monitoring the effort bargain and in handling deviations from the assumed contract. Each party will, to varying extents, invest in institutions to shift the effort bargain in their preferred directions. Management, as the buyers of labour time, will be concerned with the amount of effort and its costs. The sellers will combine a concern for long-term job prospects with an attempt to maintain their inputs at what is deemed a desirable level. The form of this competition and struggle over labour time has considerable consequences for the use of new technologies, new raw materials and new forms of work organization. Each and all of these three will emerge as new forms of knowledge to be added into the existing pool of specific knowledge which characterises the situation of learning by doing inside the firm (see Chap. 5). Learning by doing is the crucial base for capital accumulation. Its enhancement requires that the production techniques (e.g., equipment), production knowledge (e.g., learning and paradigms) and production institutions favour adaptation rather than a state of competition in any given sector and region. Production institutions are therefore just as important a part of adaptation as is technology and knowledge.

The question which can be formulated is this: to what extent did British and American firms develop the production institutions which were compatible with both the nature of competition which they were entering and with the kinds of production techniques and knowledge which they required? In the next chapters it will be demonstrated that the British set off on a *learning path*[40] which retained existing, nineteenth century forms of production institutions and that these played a central role in affecting the extent of learning by doing and the extent to which new production techniques could be envisaged and absorbed. The nature of that learning path did not become obvious until well into the twentieth century, by which time, it was so heavily sedimented that its transformation became extremely difficult.

7.3.5 Structuration and Transitions

As indicated in Fig. 7-1 the framework permits the future to be either a reproduction of past forms of structuration or to be a case of nonreproduction, that is, transition. The presumption is that structuration is like a tissue of elements – a filiere – which is so pervasive that its short-run transformation is unlikely, yet transitions do occur. Two major points have to be made:

---

[40] Attention has been given to populations of organizations in the aggregate in Britain by Freeman (1980, Chap. 9). Also attention has been given to individual organizations, but little attention has been given to the trajectory or tracks through time and space taken by specific organizations. The possibilities for such work are indicated in the approach of Lawrence and Dyer as discussed earlier in the book and with the approach of the ESRC, WORC group; see Clark and Starkey (1987)

1. The framework is based on the claim that there are transitions from one form of structuration to another form
2. Different aspects of structuration change at different rates and in different directions[41]

Moreover, intended transitions are extremely difficult to achieve because they occur in a configuration of contests, struggles and competition. The unintended outcome is more likely.

One of the major criticisms of Braudel's model of structuration is that it seems to leave too little room for the nonreproduction of existing mentalities. That criticism has been specifically addressed by another member of the Annales School, Ladurie, who uses the concept of structure-event-structure to convey the possibilities for transition (Ladurie E. LeRoy 1979). The concept was illustrated in a study of the changing voting pattern within one French province over a short period of time in the late eighteenth century. Ladurie uses the concept of event to refer to the transition in voting patterns and examines how that was a manifestation of deeper changes in alliances. Thus, there was an existing form of structuration followed by a short period ('the event') of transition followed by a new form of structuration.

In the SDI case cited earlier that approach has been adapted in a study of a British car firm between 1896 and 1968 (Whipp/Clark 1986, Chap. 3). For example, it was quite clear that the firm entered the period of the 1920s with the same forms of structuration which it had established in the pre-1914 period when the market was in a high state of product innovation. However, in the 1920s there was the growth of the cheap small car assembled on large volume principles (e.g., Austin and Morris). Faced by exit from the sector, the leadership in 1932-1934, made two core changes in innovation-design by establishing a design competence at the top of the firm and by shifting the corporate strategy into the upper end of the price market. That was the period of transition. However, it must be noted that although there were key changes in corporate strategy many features of the previously existing structuration were retained and carried forward. Thus, the notion of transitions in structuration requires closer specification. Yet, it may be noted that relatively focused changes were highly consequential. In other words, the new forms of structuration will contain considerable areas of structural survivals side by side with major changes in certain areas.

The previous points emphasised that some areas of the structuration are particularly significant for transitions and that the transition of the entire struc-

---

[41] Considerable work needs to be done on disaggregating the different rates of change amongst varying dimensions of technology, strategy and work organization and in the external infrastructure. There are some interesting ideas in Gurvitch (1964); also, the Cambridge historian, Peter Laslett, is developing a relevant perspective

turation may be both impossible in the short run (see Braudel), yet achievable over much longer periods when the core areas have provided the leadership.

Structuration is an extensive issue which is much more complex and more embracing than the earlier studies of innovation have assumed. Transitions are likely to begin by being very localised. At the organizational level it may be easier to attempt to shift corporate strategy than to alter the main fabric of structuration.

7.4 Regions: Britain and the USA

Geographical location and its resource characteristics represent the third corner of the basic framework. As already indicated their inclusion does not indicate that geographical determinism is implied (Mauss 1904; Braudel 1972). Instead, it is presumed that position and natural resources interact with economy and society so as to effect the comparative advantage which might be enjoyed by specific regions. Wallerstein, in a controversial account of Western capitalism, argues that the regional division of labour of Europe since the midsixteenth century can be best understood by reference to the growing centres of commerce facing the North Sea, namely, Amsterdam, Antwerp, London (Wallerstein 1980). These became the core areas in which capital accumulation was concentrated in the Mediterranean ports (e.g., Venice). The North Sea became part of a wider network of international trade embracing both the Baltic Sea (for cereals and wood) and the so-called triangle of trade which ran from the North Sea to North Africa (exchanging goods for slaves) to the plantation economies of the Caribbean and the southern USA, returning with colonial produce (e.g., sugar, cotton, and tobacco) to the North Sea area.

Those trading patterns reflected a crucial feature of early capitalism: the central role of water as a surface for the cheap transportation of goods. Journeys by sea for heavy goods were quicker and cheaper than over land[42]. Good port facilities, either natural (London) or manmade (Amsterdam) were important, as were the existence of inland waterways. Amsterdam gained a pivotal role in relation to the Rhine (cf. Antwerp) and the whole of England was well endowed with navigable coastal ports and inland rivers. However, the exploitation of those potential sources of comparative advantage depended to a very high degree on the possession of certain kinds of knowledge (as Chap. 4). It was those kinds of knowledge which influenced the changing lo-

[42] Braudel (1983), demonstrates that water transport was the cheapest form until the midnineteenth century

cus of economic growth and innovation, especially in the gradual shift of the locus of innovation from regions within Britain to regions within the USA. In this chapter it is relevant to briefly sketch how the theory of four longwaves interprets those shifts of location and to outline what the consequences might be if a fifth longwave were to be unfolding in the 1980s. Longwave theory, especially the interpretation of Freeman and colleagues, has considerable implications for understanding the contemporary success of the Japanese as well as the future positions of Britain and the USA.

The first Kondratiev (Kondratiev 1976) longwave occurred in Britain between 1790 and 1840. Its scope was concentrated in coal, iron and steel, cotton cloth and hardware. At that time Britain was the largest coherent domestic market, was free from costly continental warfare and the USA was a new nation. It has been argued that its types of knowledge were relatively simple, but that judgement by W.A. Lewis understates how problems would have been perceived and experienced at the time. Most of the innovations were concerned with the processes of manufacture and with transportation. As has already been demonstrated with respect to cotton textiles and the USA (see Chap. 6, Sect. 2), it was very feasible for certain other countries to appropriate the new kinds of knowledge and equipment by adapting them to local circumstances. In the absence of serious competition Britain, already successful in the international role of the City, became a major centre for capital accumulation. The regional benefits spread from London to embrace Manchester, Liverpool and its hinterland of cotton mills as well as the industrialising cities such as Birmingham. Britain was able to create its own world-economy with its colonies, except for the USA, which broke away.

The second longwave, commencing in the 1840s, was also based on British innovations in areas like railways and all its related facets. By the mid-1850s Britain produced half the world's iron and steel, two-thirds of the coal and half the cotton. In these areas she remained ahead until the late nineteenth century at which time much of the world's shipbuilding was done in Britain. Yet after the 1860s other nations – Germany, Italy and the USA – were beginning to challenge British supremecy during the downswing. Even in the core industries new processes were introduced in Germany and the USA rather than in Britain.

The third longwave was based on innovations developed outside Britain, including electricity, chemicals, the internal combustion engine, the automobile. These innovations tended to be product rather than process innovations, e.g., automobiles, radios. Freeman contends that they were easier for the innovating country to appropriate. In each of these, British entrepreneurs lagged behind and new regions outside Britain (e.g., Detroit) forged ahead. In certain respects Britain seems to have missed the start of the third wave and to have joined in its unfolding at a later less advantageous stage. Also, overseas firms began to penetrate the British home market, especially automobile manufacturers (notably Ford) and German chemical firms. Several factors affected British performance, including the thinness of the 'filiere' between scientific institutions and the firms and the absence of linking mechanisms through education, the poor level of technical education, the low ratio of domestic investment, and the dominance of the City as an institution and its preoccupations with the gold standard. After 1900 capital was exported through the City, rather than invested in the new industries.

It was the third longwave which marked the emergence of the USA as a major economy. The product innovations were more closely linked with scientific knowledge (e.g., Edison's first research laboratory of Menlo Park, 1879) and they were complicated to manufacture. In the USA the links between science and capital were relatively good, and those between corporate education and the new analytical sciences of management were very close indeed.

This longwave generated a vast demand for commodities such as nitrates, copper, rubber, and oil. Consequently numerous peripheral supply areas were opened up, but all suffered badly during the downswing of the 1920s and 1930s.

The fourth longwave saw the USA become the absolute leader in technology, although Germany did very well in plastics and other areas. The wave took off in the USA around 1940, later in Europe and Japan. The USA opened up a massive technological gap and benefitted from its geographical distance from the centres of wartime destruction. In this period the locus of innovation shifted to the west coast areas. In Europe there was a narrowing of the 'knowledge gap' after 1950 through the import of American technology and knowledge. That occurred through American firms establishing plants in Europe, initially in Britain, and through the creation of technology agreements and the like. Japan entered the fourth longwave.

In the third and fourth longwaves science played a very significant role and that was the area in which the USA, after the 1920s, performed very well, whilst Britain performed very poorly.

8 British Systems of Organizing: Contexts and Directions into the First Divide

8.1 Introduction

The position of Britain as a declining economic and political power is unexplored in the literature of organization studies, but it has been carefully examined by economic and social historians in considerable depth. It is very clear that they consider that the explanation of the recent decline has to be understood in the context of the causes of Britain's rise to pre-eminence. This chapter therefore presents a structured narrative which is driven by the analysis of the capacities to organize. How did the British conceive organization? What were the templates of organizing which became sedimented and taken for granted?

Examining the position of Britain requires a brief references to the general inheritance of western Europe and to the role of the Netherlands. It is important to identify the North Atlantic as a collection of several world-economies (Braudel 1984: 21-86) each emanating from a different core region: Amsterdam, London, New York. The North Atlantic remained a focal area until well into the late twentieth century, by which time the rise of the Pacific Rim linking the west-coast USA to Japan and other smaller centres was surfacing.

The core of the chapter concerns Britain, its position in the North Atlantic economy, the development of its industries, the situation in 1851 at the Crystal Palace Exhibition and at the commencement of the competitive struggle between existing craft forms of organizing and the emergent forms which I shall refer to as the American System of Organizing (ASO). The following two chapters will trace the emergence, development and institutionalization of the ASO.

The notion of structuration as a filiere whose many levels, channels and strands embrace, incorporate and reflect what is happening within work organizations, is quite central. The remaining part of the section outlines the European context. Why did western Europe take such an enormous lead in technological innovation and economic growth from the early sixteenth century? Applying the approach set out earlier in the framework of Figure Fig. 7-1 suggests how the question should be handled (Chirot 1985).

Geographically western Europe was a very small segment of the worlds land surface bounded between the Pyrenees of southern France, various seas (e.g., Mediterranean, Atlantic, Baltic, North) and the Vistula Plains of Poland. Its climate was cold enough in the winters to kill off many of the germs which ravaged tropical climates, yet not so cold as to inhibit the flourishing of a rich diversity of cereals, horticulture, viniculture and other produce. There was abundant rainfall for crops and grasslands and for water power. There were many small valleys with their own rivers, some of which could be used for transport (e.g., the Rhine) and others for water power. The water systems were all independent and so were not as vulnerable to the ravages of warfare or to natural disasters as were the irrigation systems of the Middle East and Asia. Nor did they require complex, oppressive regimes to manage their flows as in Egypt and China (Wittfogel 1957). The land surface was not difficult to traverse and the coastal areas permitted movement by ship, then the cheapest means of shifting bulky goods. There were plentiful rivers which provided linkages from the coast to the inland areas. Politically, western Europe was off the routes taken by the predatory tribes of Asia, Arabia and Africa. Its geophysical characteristics were favourable to growing food and to animal husbandry. The latter was both a source of food and, equally important, a crucial source of power in agriculture. In the medieval period animals were the focus for many technological innovations designed to increase productivity whilst also extending the areas available for cultivation (Wallerstein 1974; 1980).

In that geophysical context, western European structuration evolved very slowly in the period from the end of the Roman Empire to the economic take-off which began in the sixteenth century. In terms of structuration there was a curious feature: the heritage of cultural homogeneity left by Christianity and the political homogeneity from the former Roman Empire were combined with few centres of power and considerable development of powers to feudal lordships. In that context there was a slow, but very significant, accumulation of wealth over the millenium to the sixteenth century. In the eleventh century there was a revival of trade associated with a diversity of innovations, many of them being social innovations. To an increasing extent these innovations occurred outside the rural setting of agriculture and inside the growing number of small towns, particularly those situated in Italy and facing into the Mediterranean. In the towns new occupations arose - artisans and merchants - who were both a new interest grouping and also developed thinking habits and practices which differed from those in the rural areas[1]. They were less tied to the natural base of seasons and the material environment of agriculture.

---

[1] Giddens (1981, Chap. 6); see also the development of towns in Braudel (1984)

The growth of urban areas, initially very slight and very small, played a crucial part in the emergence of a distinctly European form of political innovation based on a compromise between three major factions: feudal lords, kings and towns. The compromise was a state of three estates (Chirot 1985: 185f), a partition of powers, which was introduced in the thirteenth and fourteenth centuries. Although introduced across Europe (e.g., Magna Carta in England), the partition of powers was always fragile and continuously contested, sometimes in intense civil wars. Eventually the compromise broke down in certain regions, e.g., Prussia, Spain, Austria and also France. However, in the Netherlands, Sweden and England the compromise survived. As a political compromise it provided for a degree of flexibility in the political struggles which also aided the generation of another crucial feature, namely, institutionalized calculability and rationality. The consequence was that the state apparatus was flexible and conducive to commercial innovations. Europe's growth rested heavily on commercial innovations in patterns of making contracts, in finance and in modes of transportation (e.g., type of ships).

The general development of calculability seems to have been distinctly European. Landes, in a careful analysis of the development of measurements of time, identifies the peculiarly European concern with regularity and control over the day through its division into equal parts (Landes 1983). He shows that time pieces were valued for practical purposes in everday town life in Europe, whilst in China, for example, they were regarded more as decoration for the rulers.

The emphasis upon calculation was part of an overall development of rationality in the analysis of the linkages between actions and outcomes. Although the development of the sciences is usually given heavy attention, of equal importance were the developments of rationality in legal practice, urban life and commerce. The development of systematic procedures, whose outcomes gained acceptance and legitimacy, was central to the complex state apparatus which required political compromise for its survival. The rules of the state had to be known and had to be separated from other interests of the church and commerce whilst also becoming embedded throughout society. The growth of the legal profession later facilitated the making of commercial contracts and the settlement of disputes. Moreover, the legal profession played an important role in the settling of issues within an increasingly divergent and segmented society (Chirot 1985; North 1981).

Urban centres and small towns became the locations for this new thinking and the intentional manipulation of materials and ideas were core features. Yet, these centres were more like islands within agricultural societies. They only became of greater consequence from the sixteenth century onward.

Europeans were excellent sailors who developed a range of specialized forms of shipping. The Portuguese initially led the way, but during the period from 1480-1620 the Dutch gradually overtook them and developed special ships for fishing and commerce (Wallerstein 1980). In the sixteenth century ports developed outside the Mediterranean, such as, Bordeaux, Amsterdam, Lisbon and Antwerp. European trade developed on two levels. Within Europe there was an increased division of labour and of land usage based on comparative advantage articulated by the merchants in the key ports. These ports were tied to a network of trade routes throughout the Atlantic and more widely, whereby surplus value could be extracted from peripheral areas (e.g., West Indies) through the control of tropical primary goods (e.g., sugar, tobacco) which were brought into Europe.

The innovations were based on commerce, the buying and reselling of goods with their original state remaining largely unchanged. However, in England an important further innovation occurred. England began to import raw materials and agricultural products which were then manufactured so their original state was changed. These new goods were then re-exported as manufactured products.

In the so-called long sixteenth century it was the entrepot of Amsterdam which was the centre of major innovations in commerce. In the eighteenth century those innovations were carried through into England where they were *appropriated* and *refined.*

8.2 The Netherlands: Patterns of Innovation

The case of the Netherlands and its contribution to the development of the commercial and the political institutions of the North Atlantic (including the founding of the USA) is highly important, but can only be dealt with briefly (Wallerstein 1980, Chap. 2+3; Braudel 1984, Chap. 3; Goudsblom 1967). The relevant period is from the early sixteenth century to the late seventeenth century during which the areas around Amsterdam were transformed from a periphery of the Austro-Hungarian Empire into a politically independent core region with the highly developed commercial institutions and the most sophisticated sea-based freight industry.

In the sixteenth century there were two key entrepots, Amsterdam and Groningen, whose merchants created a network of trading which reached from the Baltic into the Mediterranean. They gradually outwitted the Portuguese and obtained their trade. The two entrepots became the centres of urban growth to which streams of merchants and skilled refugees gravitated. Some of these groups were already wealthy (e.g., Sephardic Jews) and others became wealthy. The dominant activity was commerce and the political structure of the towns reflected that dominance, most especially Amsterdam. The considerable commercial skills of its inhabitants were used to link a network of tropical products into the growing division of labour activity within Europe. Dutch commerce in the East Indies involved colonisation and the control of a substantial part of the production flow and was an immense source of capital accumulation. In the New World (e.g., West Indies) the approach was different and Braudel observes that their:

mistake was to erect a commercial superstructure without gaining control of production, without colonising the territory in the modern sense of the word (Braudel 1984: 232).

By 1585 the Dutch held 70% of the Baltic trade (i.e., woods, cereals). The citizens of Amsterdam were able both to create the general shift of trade to the North Sea ports[2] and to benefit from it. Amsterdam was close to the mouth of a major river system which could be utilised for the transportation of freight. Moreover, the Netherlands had a peculiar land formation whereby there were rich sediments appropriate to intensive cultivation. The land was continuously cut by waterways and the control of flooding was a central, ongoing problem.

By 1560 the ports of Amsterdam and Groningen were the centre of the most rapidly growing trade in the world. The merchants soon placed the nobility in their debt and their collective institutions, which were in disharmony with those of Philip II, began to be the focus of resistance to Spanish influence and rule. That resistance grew and it was faciliated by the areas of water and marsh which were ideal for guerilla warfare. The political elites of Amsterdam with their *new* values and institutions, undertook negotiations with the elites of nearby Leiden and Groningen with whom they shared many interests. Initially their alliance was essentially defensive and very successful. The failed siege of Leiden by Philip II (ended 1572) typified the situation. Leiden, although ten miles inland was surrounded by waterways and the land was flooded to drive off the invading troops. Philip II was forced to retreat. Success brought a new opportunity to transform a defensive alliance into new forms of state institutions. The period of 1578/1579 was a crucial turning point when the political elites of Amsterdam arranged a meeting with representatives of eleven provinces. The meeting contained the large, now very prosperous area of Amsterdam, and ten much less prosperous provinces. In 1579 it was agreed, after much debate, that there would be a federation of *equal* provinces amongst whom only unanimous decisions could be taken. Braudel contends that this form of government became known for its weakness and inconsistency because there were endless struggles (Braudel 1984). However, that observation understates the significance of the new political form and its flexibility in the handling of conflicts between rivals.

---

2 Note the role of agency in structuring the economy

The new political institutions of the United Provinces, later described as the social laboratory of Europe (Goudsblom 1967), introduced a new kind of tolerance which decreed that no person could be investigated or prosecuted for their religious beliefs. These values, coupled to the social practices of resolving competition between rival interests, provided a uniquely orderly means of reflection and discussion (Elias 1981). Moreover, the state institutions, sometimes labelled the 'commonwealth of burghers', provided strong, though decentralized support for commerce. Within that context the merchants and wealthy from Amsterdam became very influential. They were preoccupied with the constitution of the new ways of perceiving the world[3] and in searching for those external laws of nature which were pragmatically relevant (as they defined these) to their problems of navigation and controlling the inroads of nature on the Dutch landscape. Rational calculation and the self-controlled anticipation of the future became central values which shaped personality structure through self-imposition (Elias 1981). The joking relationship became a routinised mode for handling the nuances of conflict and for indicating tensions[4].

In the seventeenth century the United Provinces prospered. The earlier dominance of the Baltic trade was complemented by the dominance of the Atlantic and by entering the Indian Ocean. By 1675 Amsterdam had grown to almost a million inhabitants. As a core zone its influence extended into its own hinterland and into regions of England known as East Anglia and London.

These commercial practices are often labelled mercantile capitalism because they involved the state in policies of economic nationalism revolving around the creation of favourable balances of trade and the movement of bullion to the Netherlands from where debts could be resolved and capital accumulated. Gradually the circulation of commodities, as the essential underpinning of exchange, was replaced by paper money - a development which required a greater faith in the survival of the state institutions than the Dutch were able to sustain after 1675.

By that time the Dutch fleet equalled that of all Europe. Its fishing fleets were highly developed to cope with the different demands of inland, coastal and North Sea fishing, as well as distant whaling. The complex structure of fisheries was matched by an extensive division of activities for the commercial sale of the product, its curing, and for the making of ships. Part of the profits were used to develop sophisticated means of handling the drainage of

---

[3] Dutch science and art were heavily involved in the describing of shapes, contours and in the exploration of lighting; for example, Alpers (1983)

[4] These observations are based on a year living and working in The Netherlands and in discussions with Dutch historians and sociologists whilst a Fellow of The Netherlands Institute for Advanced Studies (NIAS) in the Humanities and Social Sciences, Wassenaar, The Netherlands

the land. The rational thinking used by the water engineers gave them, and other professionalised occupations, a core position in Dutch decision-making at all levels[5].

The success of the Dutch is a useful illustration of the relevance of the concept of the filiere to the understanding of innovations because of the array of innovations in social institutions, in commercial practices of making and enforcing contracts, in the many raw materials, and in the institutions of making and using knowledge. Yet the Dutch faced problems. They were involved in land and sea wars which, although often fought by mercenaries, were costly and disruptive. It was the off-shore island of the United Kingdom which became a beneficiary of the Dutch misfortune. Although the British lacked a shipping industry which was as cheap as the Dutch in carrying freight, Britain did possess a favourable constellation of location, resources and structuration which permitted its development at the expense of the Dutch. That transfer of the core of capitalism from Amsterdam to England was signalled in 1651 by the British Navigation Act which required British ships to carry any freight which was to be re-exported. That Act had other effects, most obviously on the New World colonies.

8.3 Britain: An Offshore Island

There are certain contextual and inherited features which contributed to Britain's rise to economic pre-eminence by the time of the Crystal Palace Exhibition of 1851 (Wiener 1981; Rosenberg 1969). These features are examined in this section. In the following sections two points of focus are used as key illustrations: a cameo of the rise and organization of the cotton textile industry in East Lancashire and the templates of organizing which British entrepreneurs utilised.

The geopolitical position of Britain was as an offshore island, close to the continent, yet separated by seas in which the British navy managed to frustrate attempts at invasion. The North Sea is characterised by the kinds of fierce squalls which played havoc with invasion fleets, though as the Dutch demonstrated it was possible to penetrate up the Thames as far as London. The coastlines of Britain were very conducive to shipping and the inland river systems were often navigable so that barges could penetrate inland. Any deficiencies were rectified by the construction of canals in the eigh-

[5] I have made the link between Goudsblom (1967: 140–141) to the analysis of the role of water in Dutch history by Geyl (1964), and the general position of water engineers in the development of The Netherlands; that position may be contrasted with the account of the control of water in China by Wittfogel (1957)

teenth century. There were sufficient natural resources in the land and sufficient rainfall for grasslands and cereals as well as for water power. There was a great shortage of wood, but that was rectified by the discovery of coal.

The economic tendencies which Dutch initiatives had stimulated also favoured Britain as part of the region of expansion. There is clear evidence of close commercial liaisons between Scottish and English merchants and Dutch merchants[6]. Further, Dutch entrepreneurs and bankers came to London and East Anglia where they established their own commercial practices and introduced innovations in agriculture to add to those being developed indigenously. English ports like London, Southampton and Bristol were ideally situated to take part in the triangular trade between the North Sea, the African coast, the plantations of the New World and back to the North Sea (Wallerstein 1980). It is, however, features of structuration and of adaptation to the new commercial possibilities which should be highlighted.

After the Civil War and the interregnum of the seventeenth century the state played a central role in the rise of Britain. Through its actions the British domestic market was unified and became the largest single market for products manufactured in small workshops and similar locations. However, the key development was that of the City,[7] that is the area of London around Westminster within which the Bank of England and a myriad of bankers, insurers and the like became concentrated. The value of the pound sterling had been set and guaranteed in 1560 and that basis was a massive attraction to international bankers and entrepreneurs. Likewise the National Debt was always there and always honoured. In the 1770s, operating from this secure legal protection of market transactions, paper money was issued. From the core of the City and the network of connected agencies in shipping and finance the region of London grew in wealth. Moreover, London increasingly became an entrepot for European trade such as that originating from India, as well as the central pivot of the coastal and inland trade within the domestic market. These significant developments were enhanced by the conflicts between the French and the Dutch which were costly for them as well as distracting to their trading activities. London became the core zone of what was in effect, a southern coalition of powers. Outside London the major eighteenth century developments were in capitalist agriculture. Estate management (Pollard 1965, Chap. 6) was highly developed and the enclosure movement swept the poor, small farmers from the land and left large tracts in the hands of the gentry and aristocracy. By the 1760s Britain was already wealthy and capital

---

6 The Church at Hindeloopen on the now - enclosed Ijsselmeer provides clear and fascinating illustration of the connections

7 The role of the City is examined in Braudel (1984), see also Ingham (1984); Coakley, and Harris (1983)

accumulation was developed. The strength of Britain's economic position lay in its well developed, highly lucrative commerce, initially in the North Sea and the North Atlantic, but then with India and parts of the Far East. The nature of this trading affected the internal economy leading to the export of semimanufactured agricultural products (e.g., wool pieces) and the importing of foodstuffs. Britain produced the cheapest woollen goods in Europe from a cheap rural labour force.

Into that already developed context came the new industries: cotton, iron and steel and manufactured goods. Each of these developments occurred outside the London area, in the Birmingham and Manchester regions, for example.

8.4 Cotton: Clans and Markets

The cotton textile industry which surfaced in East Lancashire and in the corridor linking Manchester and Liverpool developed a distinctive form of industrial orgaization best described as vertical specialization and horizontal competition (Lazonick 1983). The sequences in the production process are shown in Fig. 8-1. The new sector was founded in the second half of the eighteenth century when it displaced the woollen industry as the major form of textile manufacturing[8]. The founding of the sector was a clear example of many diverse entrepreneurs entering a new area and managing to make a successful business. The overall coordination of the sector was achieved through various market mechanisms, most obviously the various exchanges at which contracts were made. The cotton industry was an economic pacemaker with important multiplier effects on ironwork, coal and chemicals. In the period from 1815-1840 the sector contributed about eight percent of the national income and expanded at an annual rate of six percent (Crouzet 1982, Chap. 7). Its output increased by more than 40% whilst wages increased by about five percent. The city of Manchester increased its population tenfold and used three times the steam power of that used in Birmingham.

As a raw material cotton possessed certain advantages over wool. Its toughness and homogeneity meant that it was resistant to the inherent jerkiness of the machinery then being used. The lightness of cotton meant that it was readily transported by horsepack to outlying households. Its supply was elastic so that increases in demand could be matched by expanding the acreage of plantations in the next year, whilst wool was confined to the reproductive cycle of sheep. As demand for cotton rose, so did its production, and the

[8] On the woollen industry see Gregory (1982)

price of yarn fell by twenty-fold between 1760 and 1850. Consequently the price of the final cotton cloth and cotton goods fell to the point where Lancashire cottons undercut the indigenous Indian textile industry. Lancashire received its main supplies from the colonies and from the USA and it re-exported almost 80% to the colonies and less-developed countries. There was also a large domestic market which was supplied with a diversity of products. Overall there was an heterogeneity of tastes in the markets served by the industry and that feature contributed to its overall industrial organization.

Figure 8-1 shows the vertical flow and organization of major production stages. Producing a range of cloths as an end product was a complex process with five main stages and several subsidiary stages: transforming the raw cotton into thread, then yarn; weaving the cloth; finishing processes; and end product. In the founding period all of these diverse processes were undertaken separately on a small-scale basis, very largely located inside domestic households. Gradually a degree of horizontal specialization evolved and certain stages became the focus for radical technological innovations, e.g., spinning (1770s) and weaving (1840s). Most analysts emphasise that these activities were relatively simple and that little scientific knowledge was required. However, that point totally fails to recognise the historical context of that knowledge at that time, and, as already demonstrated, the transfer of that knowledge was a highly complex process. We must therefore assume that the owners of firms required at least two kinds of specific knowledge: the markets for contracts and the kinds of recipe knowledge for their stage in the overall sequence. Because of the complex interdependencies and different routings required by different end-products, the producers would also require certain kinds of knowledge of the overall cause and effect relationships of the different types of processing.

Gradually the industrial organization of cotton was transformed through several stages (Smelser 1959):

1. The cloth was produced from a domestic system in which the householders owned the equipment and purchased their own raw materials; merchants purchased the finished cloths.
2. The merchants began to intervene in the production process by purchasing and distributing the raw materials which were then 'put out' to the existing domestic system. Gradually the merchants took over the ownership of the capital equipment. They in turn were gradually drawn into the longer decision-making chains and into the consequences of the massive expansion of the markets for cotton goods. The merchants began to exert a pressure on the domestic system for regularity in order to enter the relevant markets at the best moment and in order to increase the velocity of capital accumulation. They exerted pressure for regular production. Soon the domes-

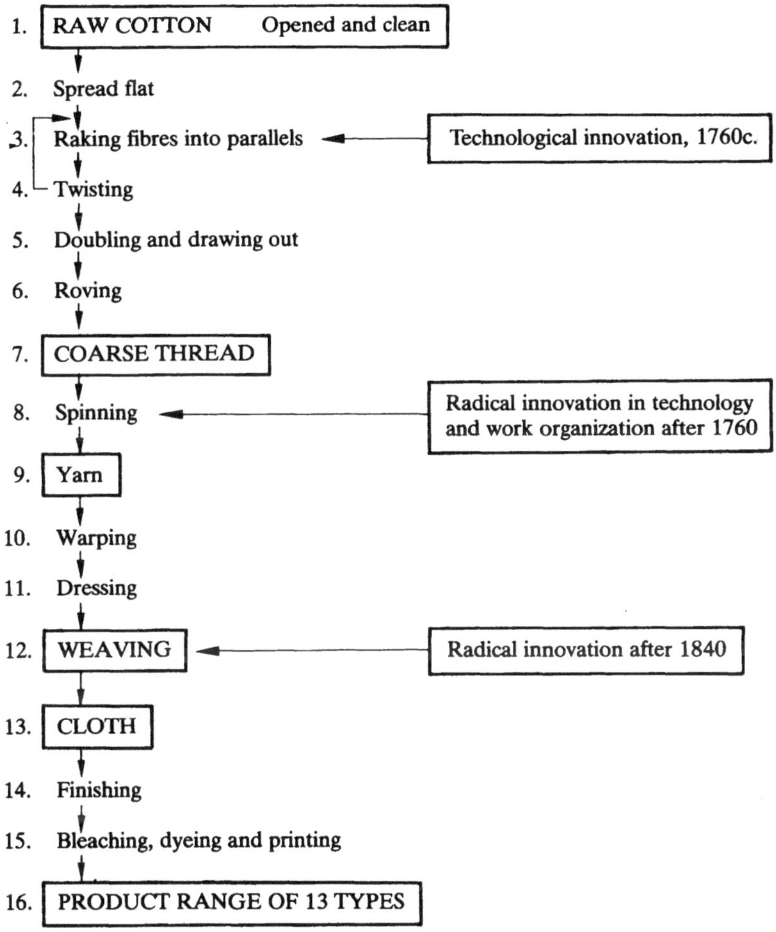

Figure 8-1. Cotton: Production process, 1760-1840

tic system became a bottleneck which was removed by the innovation of the spinning jenny. From then on the domestic system was progressively squeezed and replaced by workshops and factories.

3. The spinning jenny removed the blockage on weft, but created a new blockage on the warp. That was resolved (in 1775/1785) by linking the water frame and the factory system after which there were enormous increases in productivity. In that period the Arkwright model mills were introduced and some forty were erected and operated.
4. Steam power was introduced.

The unfolding of these phases altered the industrial structure of the sector into a complex of medium-sized, specialized forms which were highly localised and connected through market mechanisms. The whole filiere was flexible and appropriate to the early life course of the new process technology.

The introduction of radical technological innovations in spinning resulted in that part of the overall sequence shifting its location from the many households to the more compacted spaces of the early factories. At that stage (post-1770) many new capitals entered and invested in factory production. Factory location was influenced by the developing infrastructure of transportation, especially the new canals in East Lancashire. Great wealth was obtained by some entrepreneurs. Sir Robert Peel, the legendary founder of the police force, left one million sterling in 1830, much of it made with his calico-printing business. The new factories tended to be concentrated and in Oldham 42 were built between 1776 and 1811 by some 50 families from diverse backgrounds: bankers, coal-owners and wholesalers, for example (Foster 1974). These families controlled credit and the supply of raw materials as well as the new places of work, the factories.

8.5 Entering the First Divide: Templates of Organizing

In the USA the three decades between 1850 and 1880 were characterised by the emergence of a new form of organizing the means of production, in which the content of thinking about the flows and coordination of work was the subject of a paradigmatic change (Piore/Sabel 1985). The new form has normally been referred to as the American System of Manufacturing and has been equated with the manufacture of interchangeable parts. What was emerging in America was a new form of organizing, a filiere, which not only included the means of production, but also included the areas of design, marketing and distribution, the physical, technical and social infrastructure of society, and the role of the state. Organizing has to be understood as a feature of both the internal forms within the enterprise and of the relationships between enterprises. Industrial organization and internal processes within the firm are both important. It is therefore useful to refer to an *American System of Organizing* (ASO). These new forms were well suited to the structuring of the American markets and eventually to the gradual dominance of global markets for more than a century after 1880.

What was happening in Britain? What was the emerging British System of Organizing? In what ways was the BSO like and unlike the ASO? This section argues that the differences between the ASO and the BSO should be the focus of attention. In particular attention should be given to the ways in which pre-existing traditional forms of organizing were reproduced within the new industrial sectors situated outside the south eastern region of Britain.

Before examining the BSO three points should be made. First, the 1851 Exhibition at Crystal Palace is a convenient benchmark for comparing BSO and ASO (Rosenberg 1969). The Exhibition was intended to demonstrate British achievements and prowess; the Crystal Palace itself was a remarkable early example of design and planning done according to the principles of functional designing, and with strict attention to the cheap, quick assembly of interchangeable parts manufactured according to strict specifications determined by the exhibition architects (Rosenberg 1969). It might therefore be stated that Crystal Palace is clear proof that the British were 'into' manufacturing based on synoptic, functional designing and on the assembly of standardized components. Some historians do emphasise the 'firstness' of British innovations in manufacturing (see e.g., Musson 1969). However, the real issue is the extent to which that early example became a widely accepted template for the civil engineering and construction industry and for industrial manufacturing. The issue may be resolved by examining the British exhibits within the building. These were *not* based on the principles of interchangeable parts (Rosenberg 1969). The case of British gunmaking is a clear example of where very traditional, craft- and community-based methods of organizing were predominant and very difficult to transform.

Second, Britain created its own world-economy as a subset of the totality of North Atlantic economies. In the British case it was grounded in international commerce and the role of the City. These core activities were complemented by the successful commercial exploitation of the plantation economies of the West Indies, and by raw material supplies from other areas. The Indian subcontinent became a major market for the cotton goods produced in Lancashire and this totally undercut the indigenous Indian textile industry. Third, there was a large domestic market of urban consumers who were serviced by the cotton industry and by goods manufactured from metals of various kinds. The new iron and steel industry was important for those consumer markets, but its crucial contribution was to the infrastructure of bridges and railways.

From these three points it may be seen that the material base of the British economy was still pivotted around the City, around capitalist agriculture and around trading. The image of the 'first industrial nation' is deceptive because, as Crafts' analysis persuasively demonstrates, it overstates the case (Crafts 1984).

Crafts contrasts the character of Britain's early industrial development with the early development of industry amongst leading European economies in the later nineteenth century and provides an account which differs rather sharply from the impressions conveyed by many of the standard texts. He shows that these European countries did not follow the path pioneered by Britain. The classic notion of a British industrial revolution suggests that after 1790 there was a dramatic rise in industrial output based on widespread innovations in technology which in turn facilitated enormous increases in productivity. Crafts contends that development began more than a century earlier and was much more gradual, halting and painful so that the growth in productivity was modest until well into the nineteenth century. Moreover, Crafts concluded that innovative methods of organizing and the use of technology were confined to a very narrow band of staple industries, e.g., refining iron ore and cotton textiles. In other areas any improvements in productivity were only achieved by adding extra labour and by introducing further and new capital equipment. Rather than industry being the focus it was improvements in the efficiency of agriculture combined with the high importation of foodstuffs which released labour for the new manufacturing enterprises. The result was that Britain became increasingly dependent on overseas trade and on a dangerously narrow range of exportable manufactured commodities.

According to this interpretation the British entry into industrial manufacturing was characterised by:

1. Unusually modest levels of productivity and investment in industry
2. A high relative efficiency in agriculture which facilitated the movement of manpower resources from the land
3. An inability to generate high increases in average incomes
4. An international market for a very restricted range of goods
5. An entrepreneurial elite in manufacturing who were less effective and innovative than had been thought

It may be concluded that Britain's inheritance remained locked in traditional habits which lacked the flexibility to respond to overseas competition when that developed after 1880.

How did the new manufacturing sectors organize themselves internally and in terms of industrial organization? It might be assumed that the emergence of the factory as a building with four walls, a roof and high chimney was coterminus with new forms of internal organizing, but that was not the case. To clarify the point it is necessary to examine the extent to which the three basic mechanisms of organizing were utilised: clans, markets and hierarchies.

The traditional forms of work organization utilised by factors[9] was the market mechanism of putting-out, which means the household of the independent operative received raw materials from a factor. The household trans-

[9] Factors were the middlemen and were normally small-scale entrepeneurs

formed the raw material from one state to the next, after which the factor collected the end-product and handled the next stages. For example, the whole of the Birmingham 'toy' industry (e.g., buckles) was based on a community of workshops within the household. The household was paid on a form of piecework and the factors were the coordinating mechanism. As a mechanism it was based in market exchanges with some clan-like features added. Putting-out was a form of subcontracting in which the raw materials were controlled by the factor. Another form of subcontracting in which the subcontractor supplied the whole service, including the raw materials, was found in the construction industry. The basis of organizing was also embedded in the market mechanism. There are clear examples (from the 1790s onward) of how the work of subcontractors was monitored in terms of the time taken to do the job[10]. Putting-out and subcontracting the whole task were the widely recognised templates of organizing.

To varying degrees the worker was organized into craft associations or similar protective alliances[11]. These practices were generally carried into the sites where large-scale activity was undertaken. In coal mining, for example, the colliery owners had the coal seam marked out into areas which were allocated to self-regulating gangs of miners organized around a 'boss' who negotiated the prices of the contract[12]. Thus, gangs of miners were placed in market competition with each other and the miners' lodge emerged as a place for regulating both the forming of gangs and the overall structure of prices. A strong element of the clan dimension was based in the lodge. In the actual coal mine the detailed coordination of activity was largely left to these self-organizing gangs. Therefore, the ownership of the time-space framework of work activity remained in the control of the workforce who also controlled methods of recruitment and the training and licensing of competences based on time served.

Internal subcontracting within factories was the initial and powerful template of factory work. In other words the walls of the factory were used to achieve certain nonorganizational objectives. There were three forms of internal subcontracting: the autonomous gang of workers as in coal mining, the internally employed employees on a form of negotiated piece work, and the time-employed person under a supervisor. In the popular imagination it has been assumed that the time-employed person was typical of the new factories, but that situation was less extensive than might be assumed (c.f. Thompson 1967). There is no doubt that there were many factories run on a regime of strict timekeeping in the manner reported by Thompson, yet his

---

[10] Price (1980); see the manuscripts on Boulton and Watt in the collection edited by Tann (1981)
[11] Turner (1962); see also, Sandberg (1974); Lazonick (1983)
[12] Similar forms of organization were to be found in the coalmining and quarrying sectors; see Samual (Ed.) (1977)

account exaggerates their extensiveness[13]. The case of the pottery firm, Wedgewood, is frequently cited but its exacting practices of timekeeping were atypical. The employer rarely thought in terms of an even flow of production. There were many interruptions arising from the type of products being made, the vagueries of the weather, seasonal fluctuations in demand and human errors.

The internal development of hierarchies required certain kinds of thinking about the market and about the internal organization of work (Chandler, 1977). Moreover, a great deal of specific information about the costs of activities was required. To what extent were these practices of accounting well established in Britain? There is no doubting that estate management had been the target of incremental innovations in accounting practices (Pollard 1965) but, it was the annual round of historical accounting which dominated. Even in cases where a crude form of standard costing had been established, as in brick-making, it was based on past trends (Jones 1982). It was not based on the establishment of a norm within which a given output should be achieved. Nor were attempts made to measure the costs of managerial activities. Instead they were often viewed as an undesirable cost. The British factory evolved slowly and unevenly from the system of inside contracting to one of entrepreneurial control based on personalised decision-making and the employed supervisor. Only in areas like the railways did a large cadre of employed administrators appear. Overall the approach to organizing within the firm tended to reject the forms of planning which became so evident in the USA. In the external relationships between firms, there was a similar tendency to use market relationships supplemented by clan relationships as described in the Lancashire cotton industry. These tendencies became embodied and sedimented in the small educational sector.

The preceding points have emphasised the use of market mechanism as a template to achieve the coordination between small firms and from the firm to its various contractors. The argument has been that the market mechanism was also a major template for internal organization within the new enlarged workplaces of the colliery, the quarry and the factory. Clan relationships were also important.

Three forms of clan relationship need to be considered when identifying the BSO.

First, in labour organizations and in the early trade unions there was a strong ideology of clan relationships, but that always faced the corrosive impacts of the economic mechanisms as shaped by the agency of employers. The degree and form of bargaining between employers and organized employees varied

[13] Thompson (1967); see critique by Clark, P. A. (1986a); Smith (1986: esp. 194-195)

considerably between, for example, cotton, coal and construction[14]. In cotton the employers permitted various trade associations to develop and used their existence to regulate the fixing of prices in the labour contract. Hence a whole array of associations and occupations developed and became institutionalized. Their existence complemented similar forms amongst the owners. To varying degrees the factory owners sought to impose a new discipline on these groups, but its rigour was always contested, and the circumstances of variability in demand sometimes provided the labour organizations with a space to determine their degree of impact on the space-time framework of work. The slow incremental developments in production technique and in production knowledge developed their own roadblocks; the structures became ossified. Yet, those features did not become serious inhibitions until the later nineteenth century when the nature of innovation became more strongly shaped by developments in science. In coal mining the butty system and the miners' lodges served a similar function, though the contest and struggle with colliery owners was harsher. In the economic downturns the prices for work were pushed heavily downward and there was little attempt by the owners to achieve lower prices through more effective organization in either the colliery or in the industrial sector as a whole. In construction there was a subtle balance. Large-scale projects of a civil kind were becoming more commonplace and a small number of builders had begun the practice of speculative building. Also, the growth of general contracting was beginning to undermine the large areas of autonomy enjoyed by the various subcontracting groups in the industry. Second, in the struggles within the workplace the processes which later became labelled the 'politics of consent' were in their embryonic stage. Few employers were yet skilful in choreographing the ideological context. Paternalism was a forceful strategy for obtaining a degree of stability in the effort bargain. Third, the main influence of the clan templates was amongst the owners of capital themselves.

Wiener has implied that clan templates were developed in the English public schools and the ancient universities from where they became the base for the so called anti-industrial subculture (Wiener 1981). In Chap. 6 it was argued that the British middle class did develop distinctive templates of being in a team. However, Wiener failed to consider the role of such templates in the sectors of the City where close personal relationships were highly important. Nor did Wiener consider the functionality of those templates to service in colonial administration or in the military. The City, the colonies and the military were three major features of British society.

When considering BSO it may be useful to return to the case of Philip II and to consider the extent to which there was a parallel between the British con-

[14] See the studies of the development of brickmaking in Britain by Janssen (1986: 165-181)

cern for styles of authority fitted to their mercantile capitalist empire and their capacity to recognise new forms that were emerging in the New World. The implicit theme of this chapter has been that the configuration of phenomena in the British context favoured one form of economic success - mercantile capitalism - without providing the filiere for the First Divide. The British developed forms of flexible structuring based in a market nexus and in community-like associations which were highly adapted to the variability of output levels faced within its economy, but the attempts to shift to alternative forms did not gain sufficient momentum.

8.6 Institutions of Work

In the period to the midnineteenth century there were important transformations in the institutions of work (Crafts 1984). The pre-industrial heritage from the seventeenth and eighteenth centuries contained two central features: paternalism and market individualism. British paternalism, more accurately English paternalism, was rooted in the medieval practices and statutes regulating the operation of the guilds. Within those bounds there was strong clan-like protection for the members within their overall subordination to the care of the master craftsmen and those of similar status. There was a highly developed filiere of rights and obligations. Market individualism co-existed with paternalism, yet was a very different system and was one in which labour was already treated as a commodity recognised in Common Law. An important feature of those medieval institutions was the concept of 'liberty' and 'liberties' (Fox, 1985).

Individualism has a complex place in British and American national ideologies[15]. In the British medieval context 'liberties' referred to rights to pursue one's own sectional concepts, whilst 'liberty' referred to the ways in which the gentry protected themselves from the aristocracy through Parliament and the Common Law. On the whole Parliament was sympathetic to employed labour whilst the Common Law emphasised individualism and was negative in its stance towards collective forms of action. In the eighteenth century there had been a conflict between economic interests and issues, that is, liberty with alliances based on liberty. Out of that conflict came the notion of liberal individualism which was so important in the nineteenth century. Through those values the artisan aspired to receive respect, while enmeshed in potential conflicts in the economic sphere through the dynamics of horizontal cohesion and sectional interests.

The eighteenth century was characterised by the decline of the previously salient institutions of paternalism and the strengthening of the individualist

[15] McFarlane (1978); O'Connor (1984); Riesman (1954)

creed, especially in the political economy associated with Adam Smith. In that context the labour market was depicted as a necessary and 'natural' instrument whose coercive consequences should be taken for granted. The labour market became an important feature of the control strategy adopted by English capitalism. As Dore observes (Dore 1973) it was the cash nexus which came to dominate and that predisposition contrasts very sharply with certain Japanese practices and institutions of work. The emerging values and institutions emphasised plebeian, artisan independence and left them (after the demise of paternalism) with the minimum of supports. Thus, the artisan strata were thrown into a competitive market situation in which they became aware of the vertical structure of society. Out of that configuration and conjuncture there developed the foundations of an historical alliance and an historic conception of workplace conflict which was the craft heritage in Britain. In the first half of the nineteenth century that heritage came under strain, especially in the face of state repression (e.g., Combination Acts, 1799-1800) and the 1820s and 1830s were characterised by intense struggles leading to the repeal of the Combination Acts in 1824-1825 and the Reform Act of 1832. Then came the debate on the 'new' social order, the growing strength of the trade unions, and a renewed interest in paternalism. After 1850 there was the emergence of a new framework of institutions governing work.

9 American Systems of Organizing: The Early Foundations

9.1 Introduction

This chapter provides an analytically structured narrative of the period from the founding of the New World to the American exposition of 1853 which, like that held two years earlier at the Crystal Palace, provided a clear demonstration of the novel character of American agriculture and production[1]. Those exhibits are normally explained as outcomes of the American system of manufactures, a label which owes a great deal to the visit by three leading British experts to the USA in 1853[2]. They were sent by a special commission from the House of Commons in London to investigate and report back on the methods of manufacture which were being used in the USA[3]. The three experts concluded that the American system of gun-making was based on an approach to design and on workshop practices which were quite different from those being used in the craft based gun-making community of Birmingham[4]. They recommended that the British government should attempt to appropriate those methods through their imitation at the specially created Enfield Armoury near London. Historians have used that visit and the attempt by the British to imitate American methods as the turning point in Anglo-American approaches to organizing and to innovation. It is a convenient conjecture, but certain revisions and extensions are required.

First, the notion of an American system of manufactures has been the subject of intense debate and the most recent interpretations emphasise that its defining features (e.g., the interchangeability of parts) were not as evident in the 1850s as they became in the twentieth century[5]. Moreover, it has become

---

1 A variety of sources were utilised. The work of T.C. Cochran provides a useful and extensive introduction from the viewpoint of cultural history, Cochran (1981)

2 The journey and report by Anderson, Wallis and Whitworth is introduced and reported in Rosenberg, (1969)

3 The reports were extremely detailed and included charts of the times taken to manufacture specific items such as the gunstock; as such the report provided a sharp contrast with some of those from individuals and teams visiting the USA in the past four decades

4 Birmingham as a craft-based community has been characterised by Briggs (1963). See also the approach taken to industrial communities by Sabel (1982) and in Piore, and Sabel (1985), and by Rosenberg (1969) on Birmingham in the nineteenth century

5 See the collection of articles in Mayr and Post (eds) (1981); Hounshell (1984)

clear that manufacturing cannot be understood on its own because manufacturing practices were the (unintended) outcome of a configuration of factors and of the agency of a variety of factions each pursuing sectional concerns. It is, as has already been noted, necessary to think in terms of the notion of filiere: a network of channels and structures whose interdependencies played a crucial role in the final phenomena. It is more sensible to refer to an American System of Organizing (ASO).

Second, in order to understand the American approach to organizing and to the role of innovation it is necessary to examine certain selected features of the founding culture and structuration of the USA in order to compare its predispositions and capacities with those of Britain.

This chapter will therefore examine two areas.

9.2 The American System of Manufactures Reconsidered

The notion of the American system of manufactures occupies a central position for Americans explaining the rise to economic pre-eminence of the USA and for all those historians and analysts seeking to explain the relative decline of Britain (Rosenberg 1969; Smith 1981: 63-101). It is therefore necessary to examine this notion and to consider just how far it provides an adequate basis for the enormous weight of explanation which is built upon its simple theme. So the first task is to account for the centrality of the notion, then the next set of tasks is to show why that notion requires reconsideration. I shall argue that the American system of manufactures was an important *facet* within a series and sequence of events, many of which were contextually shaped, rather than the linear outcome of human intentions. The American system of manufactures has to be understood within its antecedents, with respect to the agency of key individuals and certain entrepreneurs and part of the overall structuring, the filiere of American approaches to innovation.

At the Crystal Palace Exhibition of 1851 the American display was disparagingly referred to as the 'prairie ground' by the British press because its exhibits were spread about and the stands had the appearance of desolation (Rosenberg 1969). Also the exhibits were designed in a very functional style and were rather utilitarian; e.g., icemaking machinery for the household; fireproof safes, soap for removing grease, locks, mechanical reapers, and guns. During the Exhibition the Americans demonstrated the high performance of their reaping machines (David 1975) in field trials and by the close of the Exhibition their contributions had attracted considerable interest, especially the firearms. Two years later, when the American Exposition was held, a deputation of three experts – Wallis, Anderson and Whitworth – were sent by the House of Commons to investigate American methods of production and to provide a detailed report.

Wallis was head of the Birmingham College of Art and Design and Whitworth was an admirer of American machine tools for high quality grinding. Their very detailed reports described several novel features of the American practices. First, American machinery was flimsy and poorly constructed and the three men were critical of that feature. Also, equipment was typically adapted for specialized uses on a single operation. That particular point was noted, but its significance was not fully understood. The machinery was designed to handle wood (e.g., gun stocks) as well as metal. Second, the ingenuity of the exhibits was noted including two machines, a sewing machine and a woodworking machine which was excellent on labour saving. Third, they noted that the products were designed for a short life because durability was not so important. The products were designed for the cheap end of the market and so they were rather homogeneous and could be produced at high speed. Fourth, the three visitors found that American education was very interesting and described its approach as "adaptive versatility". Fifth, they found that the craft ownership of skill was very much weaker than in Europe. Finally, they found that steam power was used more on boats than to provide motive power in the factories and they concluded that this was because of the fast-flowing rivers in the valleys providing the power for the mills. One common element which seemed to underpin those features was that products were assembled from what *appeared to be interchangeable parts.* The principal and most obvious example was in gun-making at the Colt factory, a firm which had attracted great attention at the London Exhibition.

Wallis, Whitworth and Anderson reported back to the House of Commons in 1854 and their report had two major consequences for the people of Birmingham and the government. The manufacturers of arms from Birmingham ridiculed the report, arguing that the three experts knew nothing about gunmaking and that the American methods were inferior. Birmingham had been the focus for manufacturers of guns since Tudor times and during the eighteenth century Birmingham had been the world's major centre of manufacturing. During the European wars of the early nineteenth century it had produced six million guns in 11 years. The Birmingham approach to manufacturing was based on extensive and varied craft skills and a highly diversified metal-working industry arranged around the master gun-makers. Rosenberg notes that, though output had been considerable, in 1811 more than two hundred thousand guns were rejected and from 1815–1850 the industry was in a state of torpor (Rosenberg 1969). However, the ridicule by Birmingham arms-makers did not discourage the government (see Chap. 10).

In the USA the firearms industry was a key area for the gradual development of a very significant form of organizational innovation (Smith 1981; Hounshell 1984, Chap. 1). It all began at the turn of the nineteenth century when the government offered contracts to Eli Whitney for the production of arms based on the promise that the weapons would be produced from parts which were sufficiently standardized to be interchangeable. The reason for that decision was that many conventionally made firearms became defective on the battlefield where their repair was a highly skilled process undertaken by armourors. There was a bottleneck. The American government became interested in the possibility of having firearms which were sufficiently standardized for their repair to be a matter of the simple replacement of the defective component by semiskilled men. In fact Whitney was unable to meet the terms of the contract, but the government persisted with the basic design principle, *thereby taking the American industry on a different learning route to that previously established in Europe.* The new methods of design and manufacture were easier to establish in the USA because there was no organized resistance from manufacturers or from craft groups. The government continued to press for standardization with the interchangeability of components. Moreover, the manufacturers established a system of mutual support amongst themselves and with the separate government establishments. The military contracts became the mechanism through which slow progress was made towards the target of standardized products which were interchangeable. That target became a template of *future practice*[6]. By the 1850s firms like Colt had achieved a measure of success. Certainly, the basic problems were known and they formed the agenda for future developments.

In 1854 the British labelled the approach they had seen, the American system of manufacturers. In the next section I examine the overall structuring, controls and cultures in which it was embedded.

9.3 Location and Resources

The geopolitical location and resources of North America were very different from those experienced in Europe and they contained a number of very favourable features as well as certain distinctive challenges to the existing European technologies of agricultural equipment and firearms[7]. The natural

[6] The extent of future orientation is indicated in a variety of ways; for example, Pulos reports how museums and expositions are milieus for presenting images of the future, Pulos (1983); see also Segal (1984); Brookeman (1984). Likewise the reconstruction of the evolvement of the vision of a workerless factory by Noble (1984)

[7] Based on a wide variety of sources including the studies of Cochran and Chandler already cited

resources were very favourable, but their utilisation was shaped by the distinctive sets of values and practices which were developed in the New World.

During the seventeenth century the east coast of North America became an increasingly important part of the growing North Atlantic trading pattern. The coastal strip facing the Atlantic was rich in potential. Generally speaking it was accessible to ocean-going shipping and there were deep, navigable riverways providing entrance to the hinterland. The coastal waters were useful for transportation and were well stocked with fish. The deep bays and the penetrating river systems reduced the time and costs of transportation, especially after the development of the canals through the Great Lakes, to bring agricultural products to Europe in the 1830s. The coastal plain contained sufficient areas of fertile soils to grow a diversity of crops ranging from cotton and tobacco in the south to cereals in the middle areas.

Wood was a crucial clue to the early American economy and to the distinctive development of a range of wooden products (e.g., textile machinery) and to a whole array of woodcutting machinery which played a significant role in the general trajectory of technological development[8]. There were enormous amounts of wood which were either readily available or easily transported. Moreover, there was enormous versatility in the range of woods with soft woods like cedar and pine which could be used as building materials to clad the sides and roofs of houses, as well as tougher woods which could be used as a substitute for masonry (Pulos 1983). The tougher woods were used by the Pennsylvania Dutch to construct large wooden barns. There were also very strong woods such as maple, cherry and walnut which were used in cabinet-making and in joinery and tough sinewy woods like ash and oak which were essential for constructing wagons, schooners and larger vessels. The abundance of wood meant that there was an opportunity to experiment. Later, in the nineteenth century the abundance of woods probably facilitated the development of high-speed sawing mills which were very productive whilst also being highly wasteful through the creation of enormous quantities of sawdust (Rosenberg 1976, Chap. 1 + 2). Wood was also a useful material for experimenting with in models (e.g., Oliver Evans). The diversity of woods meant that it was possible to imagine their use in different combinations. For example, the development in the 1840s of the balloon frame house as a kit of struts and side pieces was made possible by the combination of distant hard woods for the frame and soft local woods for the cladding (Gideon 1967). The American homestead could be readily expanded, replaced, reshaped in wood. Also, many products could be made in wood.

---

[8] Pulos (1983: 10ff.); Rosenberg (1981; 1975); Rosenberg (1975: 32–49); Rosenberg (1963)

The fertility of the coastal areas and the immediate hinterlands meant that it was possible to grow surplus crops, some of which could be exported to Europe. Moreover, behind the coastal plains in the inland areas there were the regions of the mid-West which would become the richest cereal-growing areas in the world. Beneath the land there were rich resources which were waiting to be discovered; including coal, gold, iron ore, oil and other materials.

The climate was diverse, but there were large tracts of temperate zones which, like Europe, provided sufficient ranges of temperature to kill off tropical germs whilst still being adequate for crops and livestock. The precipitation levels were generally good and that combined with the steep-sided valleys of the east coast provided fast-running streams in the valleys which could be used to drive waterpowered mills.

The geopolitical position of the New World is of interest. It was far enough away from Europe to avoid being drawn into the most destructive events of European conflicts, especially in the twentieth century when the USA was able to manufacture large amounts of war materials without the expensive and disruptive problems of bombing. Yet, the east coast of the USA soon became an integral part of the various world-economies centring in the Netherlands, Britain, France and Germany.

The immense amounts of space available and the absence of powerful indigenous populations permitted the occupants of the New World to undertake almost two centuries of frontier expansion (Cochran 1981) and thereby to sustain a distinctive set of values, norms and institutions well into the nineteenth century. The different regions of the USA provided the opportunity for an extensive social division of labour based on the comparative advantage of different regions.

9.4 Founding Cultures

There were several founding cultures of which two major branches became very important: the cultures established in the northern corridor between New York and Philadelphia and the southern plantation cultures associated with cotton and tobacco. Each of those was influenced by the frontier mentality. Beyond those founding cultures it is important to focus upon the struggles to establish certain core values and practices with respect to the position of the individual and to the position to be given to capitalist agency and institutions (Riesman et al. 1975; Riesman 1954; O'Connor 1984). The outcomes of those struggles were strongly shaped by a relatively small proportion of the total population, that is, by those in the middle and upper classes who played a highly active role in choreographing the new culture. It was members of those strata whose competition with one another, and whose modernising tendencies provided the most important inputs into the making of American rituals and institutions[9].

One important area of debate about the content of American culture, institutions and rituals has been around the extent to which there is individualism or conformity (Riesman 1954). A recent analysis of the problems of adaptation to the changing forms of economic competition has highlighted the complex and contradictory dimensions to American individualism (O'Connor 1984). Other studies emphasise conformity. The attention which is given to individualism in the analysis of American culture has to be balanced by attention to the capacity of certain institutions to become 'taste markers' (Pulos 1983: 133 ff.). As Riesman observes the character of American individualism has to be set in the context of the very remarkable capacities for conformity (Riesman 1954).

The founding cultures and forms of early structuration of the colonies established in the northeastern areas were different from those found in any European regions. There were two major sources of immigrants: those who wished to leave Europe in order to establish new freedoms and liberties and those who were involved in the commercial exploitation of the New World. These two factions were in competition, yet the needs of the former were sometimes achieved through the means provided by the latter. In the new setting there was an absence of the privileged European elites with their traditional values and so a social structure was constituted which was much more truncated than that found in Europe. Without that sedimented overlayer of elite mentalities the new sets of values were crystalised and most importantly the new colonists were quickly able to alter European values and practices in an adaptation to one another and to the exigencies of their new setting where European artisan skills were absent.

Five major areas of transformation in key values and practices may be observed. First, there were important differences in the approach to human planning which affected family life and personality structures. The socialisation of the young was much less rigorous than in Europe. Consequently more permissive values were facilitated within which there were more loosely defined role prescriptions combined with greater pressures for conformity amongst a set of interacting colonists. So individuals were given an opportunity to develop their own intentions and trajectories whilst simultaneously having to learn codes of mutual adaptation which the colonists believed were essential for survival. Human planning became an individual and social feature. The colonists undermined the spatial restrictiveness of many Euro-

[9] There has been little recent analysis of American rituals; it would be relevant to have an equivalent to Lane (1981)

pean families and permitted younger marriages. The family became *the* organizational institution and the *household* became a key economic unit which in the nineteenth century had an important consequence for the development of the American system of manufacture. The household was prime user of the functional products of nineteenth-century manufacturers (e.g., cloth and icemaking equipment) and within the household the mass-produced goods were customised in various ways, for example, by the use of sewing machines.

Second, the northeastern colonies were typified by a strong emphasis on a regularised space-time framework in their lives and in their worldviews[10]. With respect to space the early settlers espoused values of rationality and precision which included the defining of America as a territorial map and wilderness which *should* be transformed. The spatial aspects were cognitised in a maplike manner around the notion of grids. The grid template was applied to street systems thereby providing a core framework from which later extensions could radiate. Space was sometimes the focus for 'model layouts' which created an endless repetition of rectangles, yet also permitted the development of varied wooden housing. These practices of treating space fitted very neatly with the attachment of commercial values to all spaces so that they could be readily sold. Space was commodified at a very early stage.

The colonists also possessed distinctive conceptions of time (Caroll/Noble 1980):

1. There was a strong emphasis on the blocking of time into segments, particularly in the organization of their daily lives. The concern with the calendar was less than in Europe, possibly because of the absence of traditional institutions like the European churches, and because of the predominance of agricultural rhythms of working.
2. Time was much more oriented towards secular, future-oriented concepts and towards a more infinite time than in Europe. Therefore concepts of time were much less dominated by notions of God's time.
3. There was a strong orientation towards thinking that time was money so that economic transactions should be evaluated against segments of time. The orientation probably shaped their approach to commercial transactions.

Thus the space-time framework of the northern colonists possessed a European inheritance whilst shedding certain restrictions which inhibited commercial practices.

Third, particular conceptions of knowledge and skill were developed. The colonists and their successors set out to create an 'American tongue' (Pulos

[10] Landes (1984) The treatment of time and space is particularly interesting in Carroll and Noble (1980)

1983), a language of their own in which there was a new calendar of events and a distinctive recognition of American rituals. An important constituent of the language was the notion that knowledge should be useful, that is, a highly pragmatic conception of language which has been described as the 'how-to-do-it' approach[11]. Europeans seem to have underrated the importance of pragmatic knowledge in American culture, yet that language is often a crucial embedding feature of American innovations, and equally so in planning systems[12]. For example, the development of knowledge about chemical processes in the period 1890-1920 was arranged around the crucial concept of unit operations, yet that organizing principle – which became part of the architecture of the consultant A. D. Little – was barely understood in Britain until well into the twentieth century[13]. The concern to organize knowledge lies at the basis of so much of American culture, including the development of agricultural research (see Chap. 3). By the eighteenth century there was a notion of best practice as the single most effective method of operating[14]. Notions of skill differed from those established in Europe, especially from the highly specialized and segmented skills found in Britain. In the colonies the stress was on becoming skilful in whatever range of activities were relevant to a particular setting. Given that many settings contained relatively isolated homesteads there was less opportunity for either the travelling skilled craftsman or the urban labour forces (e.g., construction). The concept of the skilled worker therefore gave greater attention to the possession of a range of skills: the generalist. Moreover, skill was taken to be an attribute which was subject to revision according to the circumstance. In the frontier context that circumstance was often changing. The artisan trades of Europe were judged to be unsuited and the colonists soon learned that their own survival depended on generalised crafts which could be flexibly deployed like lumbering, farming and carpentry. These differences from Europe had considerable implications later for the establishment of new occupations and for the locus of the ownership of skills. Pragmatic knowledge and generalist skills were loosely intertwined in the restless approach to finding best practice and to constantly searching for better practices. Great attention was given to the recording of methods of working and then using such accounts as the basis for learning and reflective revision.

Fourth, the early colonies developed as agricultural communities spread over wide areas. The ready availability of land was soon seen to require an

---

[11] The how-to-do-it theme also appears in American Films

[12] There are some interesting questions to be asked about the actual level of contemporary utilisation of computer aids by American management; see research by Harvey Brooks and Maryellen Kelly, Harvard University

[13] For example, McCabe, Smith, Harriott (1984) for the notion of unit operations

[14] Best practice is a distinctly American approach to organizing knowledge; see Chaps. 3 and 7

approach which emphasised extensive farming rather than intensive farming. Hence, it was important to develop some means for handling the quantities of farm produce and output. Consequently European implements, such as the scythe and the axe were modified. Much later, in the nineteenth century, specialized farming equipment was developed. The fertility of the land meant that food was produced much more cheaply than in Europe. Later, that would permit Americans to have a larger disposable income and to use that surplus to buy goods of simple, functional design which had good performance and were easy to repair (Rosenberg 1981). Those attributes were requirements for homesteads which were some distance from repair facilities. The early demand for simple, functional products became an important facilitator of machine-produced goods in the nineteenth century. Labour-saving activities and devices were of great concern in the founding period and they had a special value to the immigrant populations, in part because there were few domestics.

Fifth, as noted earlier in the examination of the role of the state in the transaction-costs approach (North 1981, Chap. 14), the new colonies developed a highly codified conception of property rights which was administered by an influential legal profession and with state support. Because many of the early colonies were founded as commercial enterprises the property rights gave much more favourable opportunities to economic development than in Europe. In the founding period the state institutions protected the property rights of individuals, though these were later appropriated by corporations.

The colonial period contained a whole set of enabling conditions which included a highly literate population; an egalitarian, adaptive and utilitarian culture with a relatively open social stratification system; a central government formally committed to economic growth; sufficient old world crafts to form a basis for their appropriation to new requirements; and a widespread home handicrafts sector (Uselding 1980). America contained a rural middle class anchored in the frontier regions and frequently occupying relatively isolated locations in which there was little opportunity for ostentatious display. That rural strata would play an important role as the demand side of the American economy developed.

Throughout the founding period the Americans were great borrowers of European technologies and practices. They continued to be borrowers until well after 1850, yet they were also rapid adopters and most significantly they were very skilful in handling the appropriation gap: they modified and re-invented the European inheritance and adapted it to their values and contexts. In so doing, they found themselves taking *directions* of innovation in work organization and technology which were of a different character to those found in Europe. It was the directions as much as the pace which was to have such an impact in the nineteenth century.

9.5 North Atlantic Economy

The Americans learned a great deal from the disjuncture between the European practices and the exigencies of adapting their wilderness concept of North America to events as they occurred (Rosenberg 1969). The previous section has outlined some of the ways of behaving and the techniques for problem solving which came to serve as templates for individual Americans. These provided a framework for interpretation around a model culture of the generalist craftsmen and utilitarian ways of organizing. The framework provided a foundation for interpreting experiences, as well as transmitting that learning from one generation to the next.

The colonies became an integral part of the North Atlantic economies. The southern states supplied raw materials like cotton and tobacco to Europe, whilst the northern states became self-sufficient in farming and in commerce, but Europeans like Adam Smith never anticipated the potential for manufacturing. After Independence the USA commenced a slow, but sure, development of agricultural capitalism and commercial capitalism within an infrastructure of state support and provision. From Independence onwards there was considerable evolution in methods of production and in distribution. The extent of evolution can be examined by comparing the last decade of the eighteenth century with the midnineteenth century (Chandler 1977). Initially, distribution was handled by general merchants who were confined to local markets where they bought and sold all types of products and carried out the basic commercial functions. There were few specialist merchants, but there was then an enormous growth of middlemen who handled thousands of transactions without having much acquaintance with one another. A market-based system sustained by formalized rules gradually evolved. Their commercial practices and procedures relied heavily on forms developed earlier by the Italians, Dutch and British, though it seems likely that the speed of transactions may have been quickened.

A considerable degree of the commercial activity was focused in New York and in the corridor from there to Philadelphia. Thus, by the midnineteenth century the specialist merchant had emerged along with the brokerage houses in the ports and there were long chains of connections creating a large network of credit based on bills of exchange and promissory notes.

Throughout this period the pervasive unit was the family farm and only a small percentage of Americans lived in towns of more than 2500 persons. The northern farms grew crops and there was a limited amount of lumbering. On the southern plantations cotton brought commercial agriculture and after 1800 its growth was greatly stimulated. Once established cotton moved westward. It was on the cotton plantations that some of the early management tasks emerged, but those settings did not provide templates for best practice in the USA. In general the crops grown in the period of the 1790s moved along different routes to the incoming goods, especially in the developing midwestern areas. However, with the opening of the Erie Canal in the 1820s there was an enormous burst of trading which was completed by the development of specialist distributers. In that period most freight was moved on water and in the 1830s steam power was applied on the internal waterways.

Gradually there was the development of a commercial and informational infrastructure. Even the largest enterprises – the finance houses – were relatively small and most enterprises consisted of the owner and a few male clerks. Basic forms of accounting were utilised and there was little effort to analyse the sources of costs. In fact the fluctuations in the price of commodities would have made such an analysis extremely awkward.

There was only the slightest development of manufacturing and little use of machinery outside lumbering. However, in lumbering there was an increasing use of machinery including various specialized attachments. As indicated earlier, the textile industry grew through many small establishments and the demand for cloth was met by mechanisation. Although the textile industry did not become a template for the American system of manufactures it did, as Chandler observes, provide some early evidence of the American predisposition towards the use of specialized equipment which could be readily modified. In that respect it was similar to the sawmilling industry and to the inventive strain represented by the varied designs of Oliver Evans.

It was in the area of gun-making that the roots of the American system of manufactures were slowly sedimented, largely through the agency of the government and its requirements for particular attributes in its weapons. Figure 9-1 presents schematically the theme of this next chapter with respect to ASM. The intention is to show that ASM evolved in the context of particular structure of consumer demand and a particular predisposition by suppliers. Most importantly, it evolved as slowly as did the game of rugby union examined in Chap. 6. By the time of the 1853 Exposition the early failure of Whitney to produce interchangeable, standardized components from specialized equipment was being met in the government arsenals and the private armsmakers (e.g., Colt). Yet, not all components were interchangeable and the specialist equipment was applied to the wooden parts of the guns as much as to the metal components. The point is the direction taken rather than the pace. In fact, as noted, the cost of producing guns by these methods was considerable. Also, various accounts show that the overall coordination of the various workshops in the armouries was very rudimentary. As yet there were few signs of the emergence of the *visible hand* of American management.

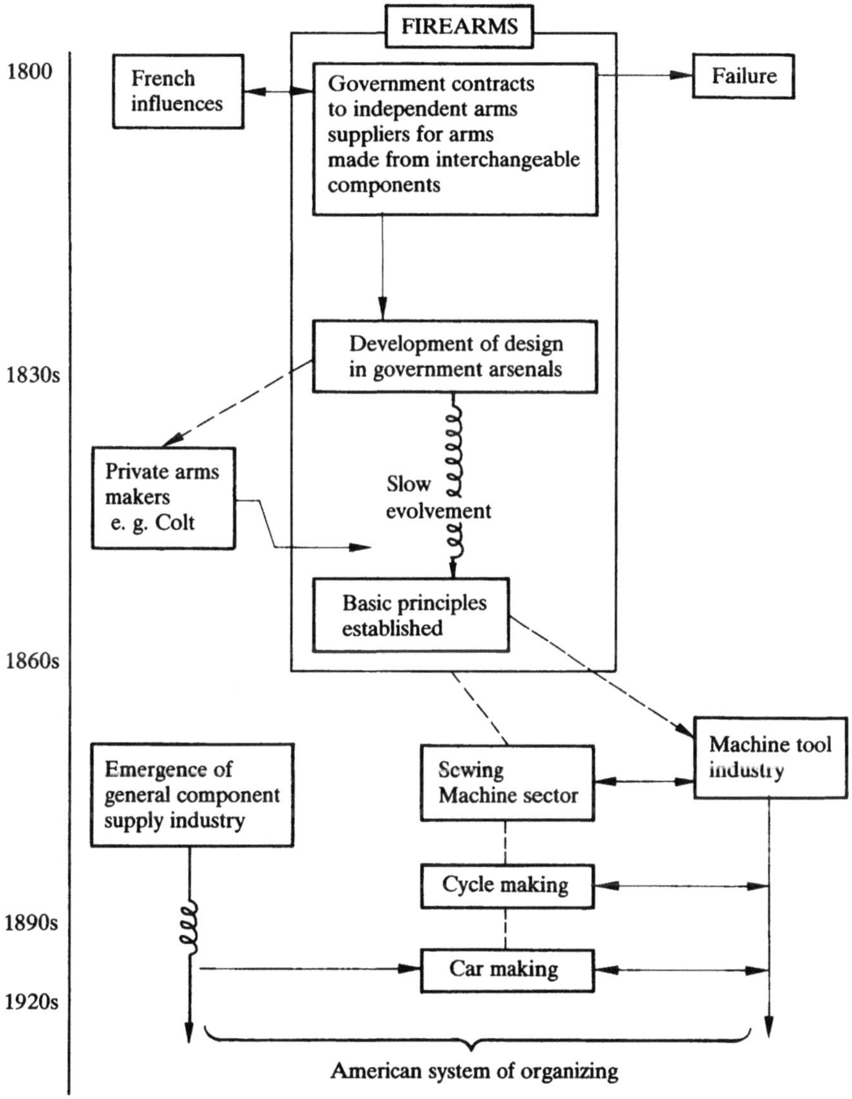

Figure 9-1. Evolvement of the American system of manufactures

9.6 American System of Manufactures in Context

In Europe the factory system had been developed in accompaniment with various forms of the division of labour where the division of labour on certain tasks had been minute. For example, in the Dutch fishing industry and most especially in the case of manufacturing in Britain. That inheritance was altered in the American context.

The Americans chose a different manufacturing style from the Europeans and although that choice of style was facilitated by the rich endowment of natural resources it has been persuasively argued by Sawyers that American social attitudes predisposed the choice in certain directions (Sawyer 1954). It is evident that there was a fortunate degree of interaction between those predispositions and the solutions chosen for certain practical problems associated with warfare.

There were three major influences on the position taken in the very early nineteenth century by the American government[15]. First, they had recruited French advisers on military technique who had brought with them a penchant and commitment to the encyclopaedic construction of details on all aspects of the military from the broad sweep of strategy to the detailed performance of the rifle, to the drilling of troops and to the synchronization of firepower from heavy guns. American generals and senior officers appropriated those ideas and allied their basic architecture to American concepts of useful knowledge. The French influences are evident in engineering treatises, testing procedures, the design of arms and most important, in educational techniques. Although reference is sometimes made to the French system, it is important to observe that the French inheritance was gradually transformed. By the 1830s there were a variety of American texts on military knowledge which emphasised the importance of simplicity in the production of armaments. Second, the experience of the war finishing in 1812 had made the government keenly aware of the limitations of American gun-making and repair facilities. Third, the requirements of a frontier society probably influenced attention toward the efficiency of armsmaking and their operational performance. The three influences set in motion the search for an alternative approach to armoury practice and to the heavy involvement – in financial and institutional terms – of the government. That government involvement in the military took diverse forms, such as map-making, gun-making and major civil engineering projects (e.g., Erie Canal). The national armouries located in the New England states became the region where those nineteenth century influences took root.

[15] See the collection articles edited by Mayr and Post (1981)

In the national armouries, including that at Harper's Ferry in Virginia, certain individuals undertook a search for a system of uniformity in the production of arms which would insure that the problems of gun maintenance were reduced (Smith 1981). The search for uniformity focused on the *anticipation* that it would become possible to manufacture guns from standardized wooden and metal components. The attempts to create standardized uniform components required the parallel development and high performance of several activities, namely, gun design, production scheduling, measurement of products, monitoring of quality and corrective procedures. Also, it required the creation of bureaucratic systems of accounting for inventory and costs. The chosen solution was to achieve uniformity in the production process by developing specialized machinery and associated fixtures (e.g., jigs and gauges) on which craftsmen could produce components which were so sufficiently similar that they required only very slight filing at the stage of assembly. That solution took more than four decades to develop in the face of considerable opposition.

The attempt to implement principles of uniformity and specialized machinery commenced after 1816, but there was a constant stream of failures of various kinds. At the Springfield armoury the craft-oriented employees objected to methods which required them to undertake a smaller segment of the overall process. In the 1820s there were chronic problems and sharply conflicting opinions about the need for mechanisation. The early pioneering of standardized production (e.g., Wadsworth) developed a network of exchanges of information and of men which was devised to enhance the chances of success. By 1834 they had achieved a degree of success and there was a major attempt at the codification of practices. In all those developments it was the consulting engineers who gradually enlarged their own role to include the coordination of the overall process. Their tactic of having mobile mechanics transformed the search for uniformity into a quasi-educational phenomenon. By the midcentury their followers were appearing in the emerging machine-tool industry where, as Rosenberg shows, the capital goods sector contributed very significantly, to the development of large-scale equipment in the second half of the century (Rosenberg 1981). The machine-tool industry became a national reservoir of learning about technical problems. Returning to the armaments industry the search for uniformity led to close analysis of many aspects of military production, for example, to the investigation of cast iron for cannons, the analysis of gunpowder, and the establishment of chemical laboratories in 1849.

By the time of the American Civil War the development of armoury practices had created a *template* of production techniques *in which design was used to standardize components* and in which specially designed equipment and fixtures were used to produce components, a good number of which could be interchangeable once some slight filing had been undertaken. No one would have expected to save money by using such production techniques yet even partial interchangeability of components was very useful. For example, at the Colt factory, which was privately owned, there was interchangeability of parts for many of the components, but hand fitting of the most crucial parts.

Armoury practices have to be seen as a very important tributary in the development of an American System of Organizing after the Civil War. Moreover, it needs to be observed that the new production techniques were increasingly set in the context of increasing control of the content of workers' jobs and of their performance. That required new kinds of managerial activity in coordination.

Parallel to the development of armoury practices there were other significant developments which Hounshell has described as 'system thinking' (Hounshell 1984). Yet, as Hounshell demonstrates certain important areas of major corporate growth in the USA were initiated by using European methods of production. The most salient example is in the manufacture of sewing machines between 1850 and 1880. At the founding of the new sector the typical firm such as Wheeler and Wheeler built 300 unique machines and only one firm, Brown and Sharp, employed men using the template of armoury practice to the design of the entire sewing machine along with its special tools. They found that very difficult to implement. The most successful firm was the Singer Company which (until 1863) used European methods of production, including the employment of internal contracting. In that period they built up an enormous market share based on *marketing innovations*. After, 1863 there was a decade of piecemeal introduction of 'system thinking' during which period the internal contracting systems were removed. By 1880 the firm had achieved system thinking and was producing a very high proportion of components which were manufactured to standardized designs on specially designed equipment and jigs.

The three decades after 1850 have been identified as the period within which there were an array of simultaneous developments which resulted in the emergence of an American capacity in mass production methods of manufacture and distribution. During that period the established systems of internal contracting were progressively removed and the pace of removal quickened after 1880. Then for four decades there was a radical change in the coordination of factories. The role of the foreman was emasculated and planning departments staffed by college-trained men were interposed. These made a very selective use of Taylor's thinking and practices to construct a new body of knowledge about the design and control of work. Equally important was the revolution in marketing and distribution and the missionary zeal with which American entrepreneurs invaded the European and world markets.

10 The American Market: A Key Base from 1870 to the 1960s

10.1 Introduction

This chapter extends the thesis advanced by Piore and Sabel, who contend that in the three decades after 1850 there was a struggle between two alternative systems of manufacturing (Piore/Sabel 1985):

1. One was the British template of a well-developed division of labour, sometimes regulated by craft groups, in which the linkages between the many small workshops were typically achieved through mechanisms which combined market relationships between scores of middlemen with horizontal clan relations; the Manchester Cotton Exchange was a good example
2. The other system eliminated the coordinating role of the middlemen and replaced them by large corporations within which coordination was achieved through the hierarchy; in that system there was an extension of the minute division of labour into the utilisation of equipment so that there was a strong development of technological solutions to the problems of obtaining efficiency in transactions

Piore and Sabel concluded that the second system, which is an American rather than a British or European form of organizing, came to dominate the American scene by 1880 and remained dominant until the 1970s. The American system of organizing also became the single most salient international template for the organizing of production. The Piore-Sabel thesis provides a useful point of entry into the analysis, yet it requires certain modifications.

Here it will be shown that the period from 1850 to 1890 witnessed the discovery by certain American entrepreneurs of mass markets. These markets were created through the development of advertising, financial credit, selling techniques and their integration into the infrastructure of transport and telegraph. The manufacturer entered distribution thereby obtaining the profits previously accruing to the many layers of middlemen who linked small-scale production and the final consumer. These were marketing rather than production innovations.

Within the large scale enterprises which emerged in specific sectors new forms of handling production and the internal coordination of the enterprise were constituted. In large part these new forms relied heavily on the defining of procedures, especially for distribution, and in the production of specialized information which was interpreted centrally by the entrepreneur and by the central staff functions attached to the head office. The standardization of procedures was coupled to high formalization and often coincided with the allocation of tasks to specialized functional areas within each of which there was further specialization of tasks. By the 1880s there were several clear examples of entrepreneurs who had managed to achieve a dominant market position. Thus the template of market domination became more salient.

The rationalization of the production system tended to follow on from the development of a secure market linkage as for example in the case of the sewing machine industry (Hounshell 1984, Chap. 2). However, competition on price induced a search for internal economies of scale and it was this search which provided the opportunity for the bearers of the approach known as the American system of manufactures (ASM) to shift their attention from working in the government-run armouries to applications in the private sector. Key experts moved from the armouries into sewing machine firms, later to the bicycle industry and then into electricals and automobiles. With them they brought an attempt to introduce the *design* of the product and of the production process as a totality which was linked to the market and to forms of work organization. It proved difficult to introduce the thinking behind ASM into production units, but there were successes. By the 1890s the template of uniformity, standardization and design was salient. Also, a cadre of experts with missionary zeal arose to spread the new ideas.

From the 1890s through to the 1920s the approach of mass distribution and the templates connected to ASM were increasingly diffused and refined. Further, the basic thinking was adopted by certain professional groups – the corporate engineers – and was institutionalised in both higher education and in the internal training facilities which certain firms came to establish. The control by the corporation over training was a crucial step in the appropriation of skill and in providing the opportunity to specify how new skills would be allied with new equipment in the constitution of new occupational roles. From the 1920s into the 1960s the basic template for mass distribution and mass production was refined, articulated, packaged and diffused. In that process the stress was given to continuous incremental innovation, particularly to the design of products from common components. Design began to focus on the frequent restyling of the 'outer skin' of the product around a slowly changing array of unseen components. The car is the prime example, and as Abernathy observed, productivity came to dominate innovation as American firms competed on price (Abernathy 1978). Thus, the senior executives of American firms learned how to be active agents in the shaping of the market and in moulding new sectors of tastes to the existing capacities for incremental innovation (Pulos 1983). Within that framework there developed highly formalized systems of production organization, mainly through the application of mathematics to the problems of scheduling production efficiently and in defining suitable batch sizes. The American approach emphasised downward, forward-oriented control by the experts in the technostructure. They had appropriated the levers of control and in order to sustain those levers they facilitated the development of systems like materials resource planning (MRP) which, in accompaniment with master production scheduling (MPS), defined the allocation of resources within the space-time framework of the firm. As we now know, these systems tended to create inventories in order to protect the plans from immediate decay[1]. However, that limitation did not become apparent and widely recognised until the 1970s.

Innovation had become part of a filiere in which considerable skills had been developed in monitoring and learning from existing practices. So much so, that a whole array of consulting agencies developed to facilitate the channelling of best practice technology and administration to the new users. It seemed that every American innovation found some agencies who claimed a skill in its marketing and application. Consequently, when new administrative approaches to training like Organization Development (OD) surfaced in the 1960s there were specialized agencies to market them. In effect Americans developed a remarkable skill in the transfer of best practice based on the continuity of mass distribution and mass production. The transfer of innovations was so developed that *imitation* became relatively straightforward (DiMaggio/Powell 1983).

In 1960 the American economy was benefitting from its early entry into the fourth longwave and its technological leadership had not yet been appropriated by the Europeans and by the Japanese. There was a form of structuration in place which was quite different to that of a century earlier and was now fully articulated as a filiere connecting many channels into the same direction. Those channels included the family-education linkages leading through to the massive surge in demand for business education and for the postgraduate MBA. By 1960 the three major areas of managerial activity had been differentiated (Kaplinsky 1984) – production, coordination and design – and recipes had been established for their linking together. Those recipes were underpinned by research on administration (e.g., PIMs) and reinforced through external higher education and through the massively growing sector of training within corporations[2]. It seemed that with 'mature markets'

[1] Hayes and Wheelwright (1984) The rate of decay of corporate planning was a neglected theme of organization of design; however, it was dealt with in Galbraith (1977, Chap. 4 + 5)

[2] The approach, which was developed by the Strategic Planning Institute in Boston, examines

and ageing 'life cycles' that future profits would rest on the capacity to diversify into related activites and then into unrelated activities. The business schools rapidly contributed to this move by introducing the ideas of portfolio management and the analytic manager.

In 1960 the enormous American market, with its distinctive structure of tastes and with their distinctive pace of change, had not been penetrated to any great degree, least of all by British firms providing products. Rather, the whole of the twentieth century seemed to be a case of the American invasion of Europe and Asia, at least that was the appearance in 1960, but beneath this surface changes were taking place. America's rivals, or at least some of them in Germany, France and Japan were learning to appropriate American best practice to such an exent that they were managing to build their own significant market bases. Also the structures of markets were being altered by the agency of new suppliers (e.g., from Japan) whose domestic markets provided a more competitive and flexible array of learning paths than found in the USA.

10.2 An Integrated World-Economy

10.2.1 World-Economy

The notion of a world-economy was introduced earlier to describe the networks of commercial relationships which the Netherlands, Britain and certain other European countries created as more or less self-contained subsets of international capitalism (Braudel 1984: 21-86; Wallerstein 1974; 1980). In each case a metropolitan region of the country (e.g., Amsterdam, London) was the core zone which mediated and to a degree orchestrated a complex process of capital accumulation. Those subsets shared the larger region of the North Atlantic and that region continued to be the major network of international capital until the 1980s.

The founding of the American colonies and their growth into the USA can be fruitfully examined by applying the notion of world-economy within American territory. In effect, given the relatively high disposable income which cheap foodstuffs and highly favourable natural resources provided, the American economy became a consumer of manufactured goods at an early stage. At first these goods were manufactured in Europe by European

the rule-like patterns found in corporate ratios for specific market conditions. PIMS stands for the Profit Impact of Market Strategy and the objective is to compare one firm with a similar firm to uncover strengths and weaknesses. Its origins are probably in the approaches developed in GEC in the early 1960s. The rules which form the basis of the program are used creatively to explore the 'what if?' type of strategic intent

methods and according to European concepts of style. Although, as Pulos shows, those goods continued to be attractive to a significant segment of the wealthier American population, other very large segments (e.g., the agricultural homestead) purchased goods which were simple, functionally designed, easily repaired or discarded, homogeneous and produced in ever-increasing quantities and marketed through major distributers (Pulos 1983). By 1880 the USA was both a significant market and also one which had a distinctive structure of tastes that were well suited to the key attributes of the American System of Organizing. Put simply, too simply, the growing American market provided a key base from which its entrepreneurs gained the potential to achieve economic dominance in Europe and in the rest of the world. By 1880 that process of domination was well underway and the Britain of Queen Victoria, rather like the Spain of Philip II, was looking in another direction. The full strength of the American capacity for capital accumulation was formidable. The growing population of the nineteenth century ensured that new households were being continually established and that propensity was reinforced by high fertility rates and by extensive immigration to the USA. American entrepreneurship evolved in the context of a constantly expanding potential which it was also an agent in creating through the policies of high wages and through the shaping of consumer desires. The period of a century after 1880 is one in which, through a series of transitions, an American System of Organizing provided a distinctive place for innovation as a routine activity and for its key institution of education.

As the third and fourth longwaves of sectoral development unfolded the USA contained new core regions for their evolvement (e.g., cars in Detroit, aircraft in California). Whilst two major international conflicts destroyed the accumulated wealth of Britain and finally disrupted the world-economies of Britain, France and the Netherlands, the geographical location of the USA proved to be a beneficial feature. Within its own world-economy the major corporations developed in a distinctly American style, a style that was very much shaped by and reinforced through a body of ideas and institutions associated with the management sciences. As a subset of ASO these provided the bodies of thought on top of which more tactical recipes were added (e.g., experience curves, life cycle models) (Conley 1970). From that key economic base American corporations and thinking were diffused throughout the world. The Japanese learned about Taylorism in the 1920s and about quality standards in the 1950s; the French vigorously adopted and adapted American thinking on mass production in the late 1940s. The British? I shall look at their experiences more closely in the following chapter. Here the objectives are to show that ASO took a distinctive shape and gained a strong hold on key institutions. The issue of the adaptability of ASO will be addressed in Part IV.

10.2.2 Infrastructure: Transport and Information

The American infrastructure of transport and information became highly developed after the second half of the nineteenth century. In three successive waves there occurred new forms of transportation: railways, roadways and airways. With respect to information there was the development of the press as the carrier of information, then telegraph, followed by telephones and the whole twentieth-century communications industry. The case of the railways has been given high importance by Chandler's suggestion that their evolvement provided a major template for how to administer the business corporation (cf., the civil service) (Chandler 1977). Roadways were constructed around the development of the automobile. Airways have been less examined, yet their development as land-based linkages across the USA provided the domestic industry with certain potentials not present in Europe.

In the 1850s the railroads became the first modern business enterprise. Sometime earlier they had been established by private capital in competition with water-based transport. The investors owned the rights of way on the largely single-track lines and other enterprises were permitted to use the railroads, but that proved costly to administer. Railroads soon overcame the competition from waterways and were able to capture the market in heavy freight because of their greater speed and low construction costs over rugged terrain. For instance, the New York to Chicago trip was reduced to three days and Chicago became a rail centre which displaced the waterways centre of St. Louis and became a major midwestern city. The railroads provided the user with dependable, precisely scheduled transit of goods in all kinds of weather, That, coupled with the intensive use of the same track and equipment brought costs down even further.

The railroads were the first enterprises to separate ownership and administration, thereby permitting the incoming salariat to develop careers as administrators. The operation of railroads required the coordination of several contiguous areas under a headquarters staff who devised precise timetables and accompanying rules to prevent accidents on the single-line track. These were administered by superintendants and the American railways became the embodiment – at that time – of a precise spatial and temporal framework of thinking which seems to have impinged very directly on the consciousness of employees.

Considerable data about the operating parameters was collected, but not collated or analysed. There were daily reports on the system which were consolidated into monthly statements. However, that data were unused until after the 1860s when the costs of operation began to escalate. By that stage the large internal structures of middle and senior managers were operating within clearly defined areas of responsibility and through standard operating procedures. Through the development of collective tickets and the like it was possible for a multiplicity of separate companies to extract their share from the users' payments. In the 1870s the competition over freight increased both the cooperation between railway companies and also the extent to which the whole nationwide system became standardized on common rules and principles. All these procedures were developed by middle and senior managers, whilst the very senior managers engaged in deciding when to make alliances and takeovers with rivals. By then there were well-defined career patterns, defined practices and an increasing use of committees to review performance. According to Chandler's interpretation these managers were the visible hand which was replacing coordination through the market. By the 1880s the Pennsylvania State Railways employed over one-hundred thousand persons and possessed an internal network of corporate education.

Thus, the railways provided a key means of linking together the American market so that goods could be transported between regions, each of which was better able to pursue its comparative advantage. Moreover, the railways became part of a precisely scheduled, fast moving, temporal system of transporting inventories. By this stage their own internal sources of data were being collated and used as the basis for discovering the main sources of high revenue. By the 1890s, with their accumulated wealth, the few, large and surviving companies had begun to diversify into various different activities. Overall the railways provided an important example of how to use bureaucratic structure to achieve high financial performance and in that respect they were a striking contrast to governmental bureaucracies being developed during the same period in Britain.

The development of the railways was paralleled and closely intertwined with the development of the telegraph as a means of conveying information between far-distant markets. The telegraph was also a means of ensuring that the railways ran according to their schedules. Together they provided a first step in the direction of fast, inexpensive linkages in the market. In the twentieth century subsequent developments of roadways and airways extended and eventually overtook the position of the railways.

10.3 Design, Administrative Structures and Corporate Education

10.3.1 Design

Pulos states that an American design ethic was established and influential by the 1830s and Noble argues that in the universities the syllabus for engineering had adopted design principles by the 1870s in preference to the laws of nature (Pulos 1983; Noble 1977). To unravel the context of innovation inside American corporations, it is necessary to reconstruct the emergence of design thinking in the USA and explore how its core tenets broke away from European influences.

At Independence sufficient Americans were wealthy enough to import household and other manufactured products from Europe and to rely on the imported pattern books as major influences on their concepts of the home and its contents. However, at an early stage the imported pattern books were copied and then adapted to create an indigenous form of furnishings known as stick furniture which included the rocking chair. By the midnineteenth century the middle-class Americans were looking to the outputs of the new mechanisation to provide them with an array of household goods designed and produced in America. The tastes of that market segment were articulated with propensities for mass production.

The middle-class customers expected to purchase household articles which would last for a short time and they were unwilling to pay for durability. Compared to Britain their tastes were relatively homegeneous and rapidly changing and they were unwilling to pursue marginal increases in efficiency unless these were known to provide certain financial savings. These features of consumer demand provided a neat fit with the capacities which American producers were able to satisfy. The link between demand and supply was articulated by the American design ethic.

In the 1830s leading American thinkers on design (e.g., Greenough) sought to establish schools of design based on European concepts, but their efforts were opposed and undermined by indigenous developments. The consumers sought products which were utilitarian and yet handsome, the so-called democratised elegance sometimes associated with the influence of the Shakers. Suppliers met that demand with American-made glassware, furniture and household coverings all produced in much larger volumes than in Europe. A 'game of style' emerged between the suppliers and the consumers in which the suppliers introduced a constant tendency towards change, including the careful development of seasonal variations. Fortunately for the suppliers the American middle class often sought guidance on style through 'taste makers' whose writings were highly influential. The suppliers were able to harness that potential.

Design as a deliberate process of the prior modelling of some future item or activity had been a widely spread predisposition of the founding northeastern culture. By the 1830s distinctive American recipes and heuristics had become established. The recipes required that form and functions of products were clear and simple. This can be observed in two of the basic household items: the cast-iron cooking stove and the sewing machine. Each of these were eventually sold in great volume through innovations in merchandising and both became significant in reducing the household drudgery of cooking and of making clothes whilst also increasing the potential for variety in the home. These products were produced through means which were designed to be economical and efficient. Their customers provided a growing market of new households owned by people who wanted a degree of styling (through change) combined with labour-saving capacities in their household goods. The concept of the home was addressed by the suppliers through newspapers and advertising (after 1870).

After the 1880s a further segment was added to the consumer market because the American lower middle class and upper working class households also had net disposable incomes which could be spent on consumer durables. These segments purchased simply designed products. The Beaux-Arts style replaced the colonial design style and received enormous publicity through key exhibitions and through the museums where the public were *oriented towards the future* more than to the past. The Arts and Crafts movement was associated with a shift of interest from the livingrooms to the kitchens, laundry and bathroom. Time and motion studies were used to design products and to give those products an image of scientific usefulness to eager consumers. In the early twentieth century a whole array of new household electrical goods were developed such as toasters, cleaners, telephones and lighting, as well as fans and electric irons. All these were the products of American concepts of design.

It was central to the activity of designing that the *end product should be anonymous* and that no single person should dominate the creation of style. Thus design as a general capacity was directly connected to and integrated into mass production and it was clearly recognised that the design of mass-produced goods was different to the design of goods in small batches. Moreover, the design of the final product was often undertaken with consideration given to the problems of manufacture and to the requirements for special equipment and fixtures. The American taste makers ignored the European modern movement whilst appropriating some modern features into industrial design. That development was particularly influential on the department stores (e.g., Macy's). One of the interesting developments was in the recruitment of Harley Earl from California to design car shapes for the General Motors Company in 1926. In the following decade design became even more important with the advent of plastics and by that time designers had come to occupy a key place between the new materials and volatile public taste. In the American context the designing functions were closely allied to the market dimension and to the merchandising of products as well as having close connections with the organization of the productions system. The cementing of those linkages was greatly facilitated by the development of professional engineering.

In the USA the professional engineer emerged as an expert in applying science through a design perspective, and the conception of science included the social sciences (Noble 1977). Production engineers, from the 1870s onwards, gradually gained a monopoly of knowledge about the coordination of the production techniques and production institutions. The scope of that knowledge included the design of work organization and it was embodied in the education received by engineers. In important respects it was the engineering profession who, inside corporations, led the formulation of responses to changing levels in capital accumulation. Noble argues that they played a central role in the survival of capitalism through their capacity to design solutions to each set of new constraints and contradictions within the capitalist process. During the three decades after 1850 American higher education was partially transformed. The liberal arts colleges and foundations were overtaken by the parallel development of technical education and an important segment of the original arts syllabus was transformed into the social sciences. The arts became more scientific, but science became more empirical and design oriented, especially in foundations like M. I. T. The trend was to make heavy use of laboratories and experimentation. Moreover, by the 1890s the training of engineers, already infused with design principles, had started the process of incorporating the new social sciences.

Once inside the corporate hierarchies these graduate trained engineers had no real competitors, particularly in the new industries of electronics and chemicals, nor were they encumbered by the 'rules of thumb' which were established in the mechanical engineering industries. In their own settings the engineers developed new forms of business analysis and a small number of them established consultancies (e. g., A. D. Little in 1915). Soon all branches of engineering identified professional engineering with business leadership. Moreover, a significant number of corporate engineers became involved in social reform movements in the early parts of the twentieth century. From the 1890s onward the corporate engineers, no doubt influenced by social Darwinism, became the progenitors of corporate reforms, especially in the science-based industries. They viewed themselves as laying the foundations for successful business and introduced extensive standardization in all areas of activity. Their reforming zeal drew in the American Society of Mechanical Engineers (ASME) and together they had a strong influence in standardizing the operating procedures of firms in the automobile industry. During the First World War they developed methods of assisting mass production in the American armaments industries. Their actions did much to create production techniques which, in effect, reduced the transaction costs of internal operations. Through their actions and through the active intervention of capital in higher education corporate engineers were drawn into the industrial processes and the process of integrating higher education to the analytic and design approach of the engineers had the indirect consequence of undermining the traditional, works-trained managers with careers based on local knowledge. Within the corporations where they worked they introduced extensive internal, formal training.

After the 1920s their attention was drawn onto a new locus of problems, work organization. There they achieved what Noble regards as one of the fundamental innovations of American capitalism "the transformation of human labour into an abstracted means of commodity production and capital accumulation" (Noble 1977). Thus, the problem of *how* to fashion the structures of flexible bureaucracy that were essential for the efficient administration of newly won business empires was solved by the analytically oriented engineers who developed means of achieving economies of scale in equipment coupled to forms of discipline, work organization and motivation. They pioneered the formulation of rational procedures across the whole corporation. It was they who led the way in establishing planning and designing departments and in the introduction of ever more horsepower per worker through electrification.

10.3.2 Administrative Sciences

The notion of a science of administration received a strong impetus in the USA, much stronger than any support to be found in Britain (Child 1969). That said, it must be observed that the administrative sciences, in the plural, contain a partially competing array of bodies of thinking about the best means of running corporations and making large profits. There are some common underlays such as the idea of design (e.g., job and organization design).

Three notable steps in the evolvement of the administrative sciences may be seen in Fig. 10-1. The initial step began with the Systematic Management movement between 1880 and the 1920s. Then there was a generalised commitment to the discovering of a body of principles to replace the 'rule of thumb' thinking and internal contracting found in most American firms. Several strands developed and Taylor attempted to package these within a single philosophy and a single set of practices, but that attempt was rejected. In the second stage there was extensive differentiation in types of thinking: the development of accounting practices; the founding of The Institute of Management Sciences (TIMS) and the post-1945 growth of Organization Design and of Organization Development. These were partially competing bodies of thinking which fitted neatly with the segmented and functional specialization of large-scale corporations (Lazonick 1983). In the third, contemporary period, those divisions have been breaking down and there are

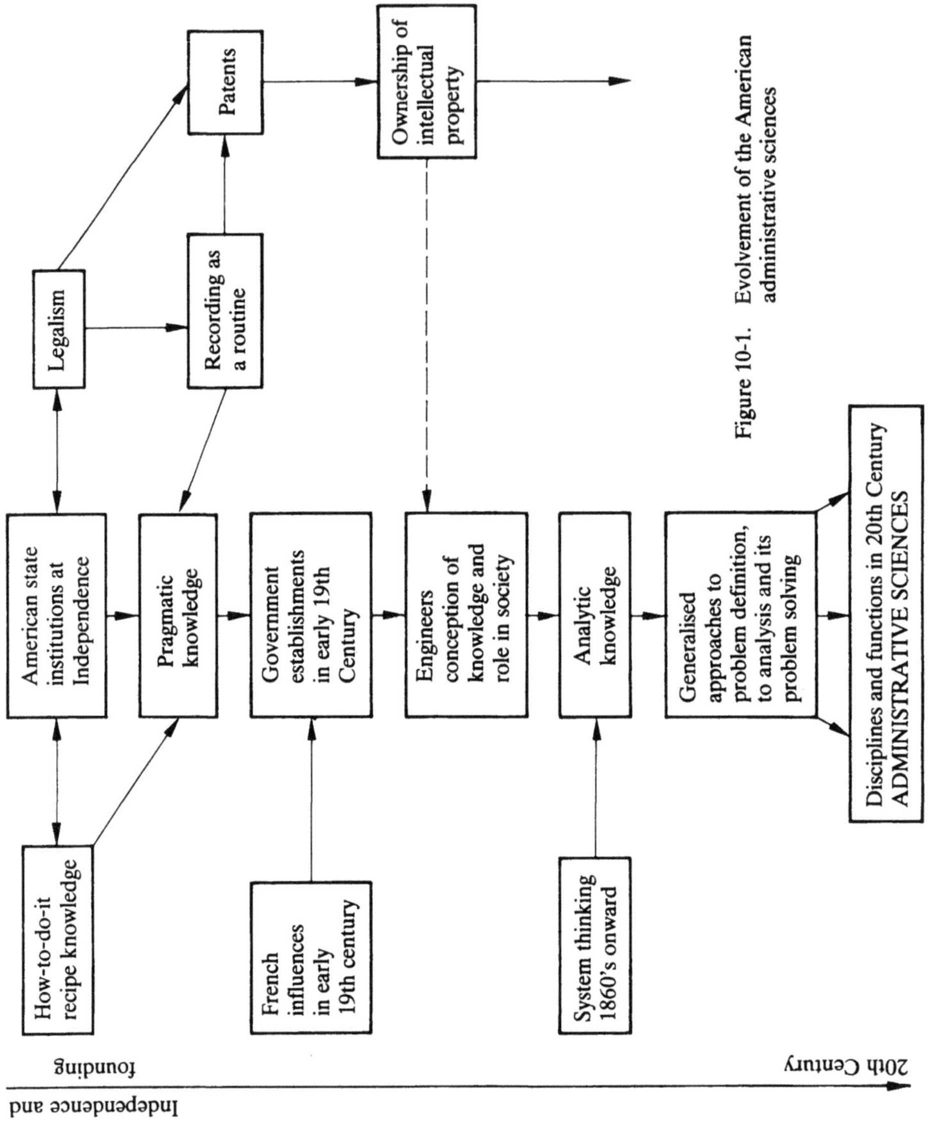

Figure 10-1. Evolvement of the American administrative sciences

clear indications of merging, behavioural accounting, for example (Kanter 1984). Those developments, still in a rudimentary phase, reflected overall changes in the actual structure of managerial work, particularly the integration of design, coordination and operating tasks.

It is useful to track the development of the notion of administrative sciences by noting Uselding's assessment of the position in the midnineteenth century

(Uselding 1980). Uselding observes that up to the 1850s American concepts of knowledge tended to be very concrete and object oriented, very empirical, with a strong preference for trial and error learning precluding the construction of abstract principles. According to his interpretation the use of models was rare until the systematisation of knowledge became a preoccupation in the second half of the nineteenth century. Uselding's view almost certainly understates the actual and potential situation whilst also being a useful corrective to any exaggeration. Uselding seems to have been attempting to emphasise parallels between the USA and postwar Japan, yet his own analysis shows that rudimentary systems for the creation, storage and dissemination of information were being established in the 1830s through at least three agencies supported by government means:

1. Schools of engineering were founded on the back of the Erie Canal project after 1817 and their influence spread to other projects
2. The West Point Military Academy became a centre for the retention and transmission of engineering practices
3. The Ordonance Departments also became centres for the transmission of systematically organized thought

These early examples, often the unintended outcomes of government actions, were enormously reshaped after the Civil War, when social Darwinism made an important contribution to that reshaping.

Social Darwinism was popularised through the serialisation of the writings of Herbert Spencer, an Englishman whose writings were relatively unnoticed in his own country[3]. Hofstadter, in a considered comment on the influence of Spencer in America concluded that:

> American society saw its own image in the tooth-and-claw version of natural selection, and that its dominant groups were therefore able to dramatise this vision of competition as a good thing in itself. Ruthless business rivalry and unprincipled politics seemed to be justifed by the survival philosophy. As long as the dream of personal conquest and individual assertion motivated the middle class, this philosophy seemed tenable, and its critics remained a minority (Hofstadter 1955).

Although the serialisation of Spencer's writings provided a great impetus those ideas were recast into American concerns.

Three features of the influence of social Darwinism on the development of analytic thought may be noted. First, although Spencer depicted man as caught in events and sought to show that there is an inevitable unfolding of processes to which man had to adjust, the American interpretation has given a strong thrust towards a future-oriented and action-oriented approach in

[3] For a clear account, see Peel (1971) Hofstadter (1955)

which man could shape and design events. Consequently, Americans defined the social sciences as a positive, practical instrument for achieving given ends. Second, social Darwinism highlighted the existence of strategic and consequential choices by individuals and firms. The availability of choice was shown to be ever present. Likewise, foolish or unlucky choices of direction might be fateful. It was implied that choice ought to be based on the search for the underlying causes of events. In other words a game plan could be constructed through which targets (e.g., winning) could be achieved. Further, the game plan could be used to monitor events as they unfolded to ascertain whether the chosen directions were actually being taken. Third, social Darwinism became closely associated with analytic knowledge and with the significance of education. In particular the observation of events came to be regarded as the initial step towards their reflective scrutiny after which the best methods could be adopted. The notion of best practice, already established as part of the how-to-do-it knowledge, was transformed into a more analytic approach and allied to research (e.g., as in the government arsenals). These three features were very strong amongst the middle classes who carried their thinking into the higher education system where it was fused with complimentary patterns of thinking in a diversity of areas, including sport (see Chap. 6). Also, social Darwinism has its own concepts and debates about design and these contributed to the general design ethic already noted.

European analysts of American managerial thinking have given an exaggerated and unbalanced attention to Taylorism[4]. The attention is misplaced because Taylorism has to be grasped within the overall trajectory of evolvement of analytic sciences of administration in the USA.

There are several reasons for focusing more broadly and extensively than on the influence of Taylor. First, Chandler observes that Taylor failed to tackle the problems of the coordination of flows in the enterprise or even to recognise the limitations of isolated specialization (Chandler 1977). Second, there is widespread evidence that the bundle of elements (see Fig. 10-2) which Taylor considered to be tightly bound was rarely applied as a totality. Users loosened the bundle and only selected certain elements. Some employers, such as Henry Ford ridiculed Taylor's approach and adopted quite different approaches to the analysis of work. Third, Wrege and Peruni have shown that certain of the claims made on productivity savings are difficult to confirm from the known records of the companies (Wrege/Perroni 1974). Fourth, in an extensive series of investigations Nelson has shown that Tay-

---

[4] Compare for example the European and North American contributions to the First International Workshop on Taylorism held at the CNRS, May, 1983; see Montmollin, and Pastre (Eds.) (1984)

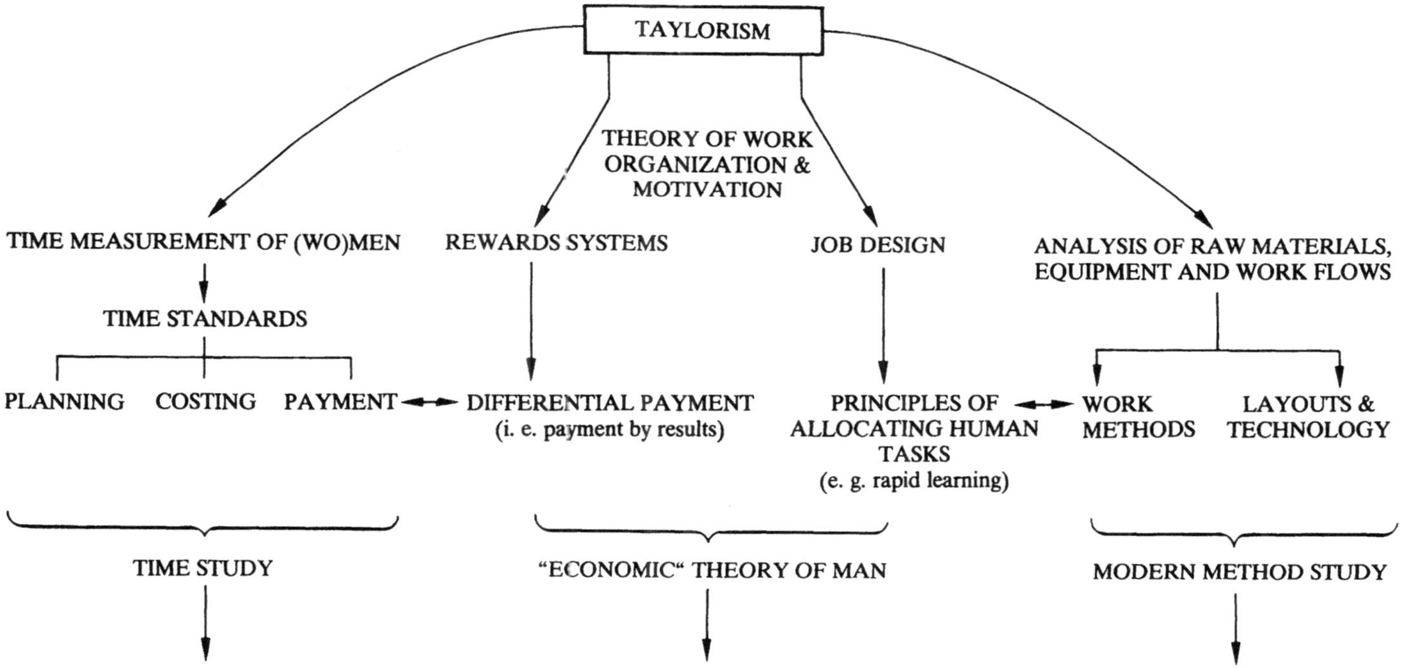

Figure 10-2. Taylorism: The bundle of elements

lor's contribution to the analytic sciences was superceded during the First World War (Nelson 1983: 51-66) Nelson's interpretation, like Noble's account of the rise of the professional engineer, indicates that attention should be focused on the evolvement of knowledge bases in production engineering and in the management sciences, including operations research.

It is more fruitful to examine Taylorism in the context of the wider development, between 1880 and the 1920s, of the Systematic Management movement which was very closely intertwined with the development of the corporate engineer. The movement proclaimed the existence of certain basic problems arising from the rapid growth in the size of firms after 1865. The increases in the specialization of the product and the production process was accompanied by fragmentation in the use of technology and in the organization of work when the internal contracting system came under severe strain. The Systematic Management movement argued that 'analysis' and an administrative science were required to balance the use of technology and labour. Their approach integrated cost accounting with the development of production control and they claimed that managements' role was to cope with the fluctuations in the requirements of production rather than to leave them to the dynamics of internal contracting. Their approach was to construct a paper replica of production which would describe, audit and reflectively examine the performance of the firm.

The approaches which were developed contributed to the increasing bureaucratisation of the operation of the firm by:

1. Increasing the specialization of roles and functions
2. By standardizing practices
3. By enormously increasing the extent to which procedures were written down and records were maintained, that is, formalization

From the early 1900s onward the technostructure of experts and other quasiprofessionals grew in size relative to the direct, production workers. These technocrats removed the mechanisms of internal contracting and replaced them with their own planning systems. Also they were highly active inside corporations and also outside in establishing consultancies.

By the 1920s Americans possessed an increasingly rich body of administrative experience. For the next half-century there evolved parallel administrative sciences for each of the major functional areas. For example, in marketing psychological schemas were developed to depict the choice behaviour of the consumer (e.g., AIDA). Two examples deserve brief attention, management sciences and behavioural sciences.

The management sciences, as represented by TIMS and by its various journals combined a search for empirical heuristics with a problem-oriented approach to theory building[5]. For example, one of the endemic problems in the chemical industry is the so-called four-year cycle. This consists of the almost simultaneous discovery of a new product after which firms design and order new plants, but find that the construction industry is overburdened so that the opening date for the new plant has to be extended. Then soon after the product is ready for delivery the price has to be reduced because of competition from other producers. The problem could be resolved if a formula could be devised and adopted by the competing firms. Rao devised a formula by which the profits of all firms could be increased if they could agree on sequence of priorities: a form of sophisticated collusion[6]. Another area of enquiry examined by the management sciences concerned the theory of teamwork treated mathematically. On the more empirical level research on the patterns of behaviour of firms led to the construction of a marketing heuristic, PIMS (see note 2). Behind these developments, and those in operations research, lies the claim that administration can be simulated, modelled and prescribed by applying mathematical thinking. The classic successes have been in terms of production scheduling and transportation problems.

The behavioural sciences developed in the shadows of the management sciences and somewhat within the patronage of corporate engineering. Many of the early experiments (e.g., Hawthorne) were initiated by production engineers and the whole area of job satisfaction was significantly sponsored by nonbehavioural scientists. After 1945 the behavioural sciences grew greatly in importance when they became part of the applied areas and of the business schools. Throughout the 1960s new areas grew claiming to address the problems of design and change.

The growth of the management and behavioural sciences was paralleled by other functional areas and disciplines. That parallelism began to break down in the early 1970s from two directions: executives in corporations sponsored efforts to achieve bridges between functional areas (e.g., accountancy and behaviour), and the nature of managerial work began to change with the development of information technologies which created common databases and began to provide an image of the synoptic view of the corporation. Moreover, the new technologies enabled the gradual displacement of certain types of managerial activity and the progressive integration of design, coordination and operations management.

---

[5] The Institute of Management Sciences although worldwide in its membership is relatively less influential in Europe than in North America where its extensive organization provides professional channels for current thinking; it promotes a series of journals for the dissemination of thinking at different levels of abstraction

[6] The formulation of the sophisticated collusion model was undertaken by R Rao and D Rutenberg, Queen's University, Kingston, Canada

In the 1980s a rash of books and articles implied that the administrative sciences should integrate functions. That can be clearly seen in the early reactions to the competitive success of the Japanese where it was shown that the whole time framework of the corporation could be tightened up considerably by removing temporal buffers (e.g., inventories) and by reducing the impediments of collective decision-making.

10.3.3 Corporate Education and Training

At the time of the First World War, Europeans observed that American soldiers were trained at all levels and that there was an assumption that any skill could be defined from the observation of operation conditions and then transformed into a specification for training[7]. After the Second World War, when the focus on training was revived and stretched into a major activity, an increasing number of corporations came to regard their internal training programs as an essential part of their future success. Moreover, in a context where changes in the basis of competition *implied* new skills, certain corporations concluded that the investment in human capital was highly valuable for two reasons. First, it left the control over the content of occupations within the zones of influence of the employer rather than to the itinerent employee. Second, training improved the speed at which new employees became productive. In many respects the approach to training paralleled the development of coaching in athletics, particularly of the development of American football. The all-powerful coaches were an early childhood symbol stating that performance was a matter of applying best practice, how-to-do-it thinking and drills. In that sense the potential for appropriation of skill by the corporation was already well developed.

In the past four decades those earlier potentials have been massively developed to the point where in size and content, corporate education challenges the traditional monopoly position of schools and colleges. For example, in the mid-1980s the leading consultancy of A. D. Little began to advertise that students could take their eleven-month MBA and that course simply reflects a more extensive development of overlaps between public education and corporate training. Moreover, as the Eurich Report stated, corporate training now has its own academic legitimacy so that employees can get college credits from their employer's courses to be used in the public system of accreditation[8].

By the 1980s corporate education had grown so much that it was estimated that eight million employees were involved and that expenditures of up-

---

[7] I am indebted to Bernie Bass for this interpretation during a seminar at SUNY, Binghampton

[8] Eurich Report, cited in *The Times Higher Education Supplement*, 1985

wards of 40 million dollars (1985 prices) were being expended. The range of courses being offered was very wide and had been expanded from narrow technical skills at all levels into the general areas of education. A small, but growing number of corporations grant their own degrees. In addition a small number of major employers have indicated that they intend to create Master's level courses in specialized areas (e.g., semiconductors).

From a British perspective the real significance of the American approach is the routine nature of the investment in corporate education at all levels. Thus, for example, the launch of new competitive initiatives by General Motors in the mid-1980s was accompanied by an extensive rejigging and enlargement of shopfloor training.

10.4 Market Control and the Modern Enterprise

10.4.1 Agribusiness

After the midnineteenth century, an increasing number of American firms sought to gain market dominance over their rivals by innovations in marketing and distribution devised to reveal the structure of American tastes whilst also providing a template for articulating a close fit to those markets by the enterprising firm. In that context the development of agribusiness was very important.

Americans had soon discovered that European practices of agriculture required considerable revision in the New World, particularly in the conditions of the midwest. Americans found that they had plentiful fertile land and they discovered, and recorded, successful ways of exploiting the advantages, but they faced problems of the uneven labour supply (David 1975; Rosenberg 1976). The demands of European farming, certainly of British farming, required considerable variation in the supply of labour according to the time of the year, yet in the USA the labour supply was inelastic. Solving that problem was not so difficult in a society with cultural predispositions towards the idea of labour-saving devices and towards the flexible use of equipment for specialized purposes. So agriculture became one very important focus for technological innovations. The suppliers provided increasingly sophisticated equipment which was reliable and readily repaired. The case of the mechanical reaper and the rise of its principal supplier, the McCormack corporation, is an apt illustration (David 1975). The farmers borrowed money to buy the machinery rather than wait to borrow or hire, thereby providing a market for a robust machine. Once the market was established the suppliers moved towards competition on the basis of price with the competitive edges resting on careful design and the gradual, very slow adoption of the principles of standardization and other features associated with the ASM. The explosion of grain prices in the 1850s meant that the farmers could purchase the machines and the gradual introduction of ASM reduced the price of equipment still further and brought in extra demand. The development of an agribusiness drove the development of the manufacturing of machinery and coupled well with the parallel development in the infrastructure of telegraph and railways. In turn the wealth which was produced turned Chicago into a major centre. From the 1860s onward massive food-packaging firms began to emerge.

Rosenberg argues that the capital goods industry occupied a crucial and pivotal position as a location in which the demands for specialized equipment in each sector of application – sawmilling, food packing, farm equipment, household products – were connected into general problem-solving skills (Rosenberg 1976, Chap. 1+2). The machine tool industry became a reservoir of skills which facilitated a high degree of technological convergence across sectors, thereby diffusing embodied knowledge to a high degree. After 1865 there was a massive growth in the total amount of manufacturing and that coincided with a tendency for the concentration of manufacturing within each sector on a small number of very large producers (e.g., Carnegie and United Steel in the 1890s). In the next sections four examples are briefly reviewed: cigarette-making, electricals, chemicals and automobiles. In each sector multinational enterprises had emerged by 1910 and were beginning to enter the international market using the USA as a base from which to develop. All four invaded Britain (see Sect. 5).

The large firms were based on marketing skills and great competence in the coordination of production, but their long-term survival required attention to the institutionalisation of innovation through design. In due course that problem was resolved, for the moment at least, and became routinised into the American system of organizing mutual learning in large enterprises and also facilitated a relatively high degree of the transfer of best practice amongst the total population of firms. Chandler, in his early study of the use of divisional forms of organization indicates that the appropriate forms were diffused and imitated rather quickly (Chandler 1962). Moreover, American firms (e.g., Ford) seemed able to transfer that learning into the British and European markets.

10.4.2 Cigarette Industry 1880–1900

In 1860 the cigarette industry contained many small firms and quasi-independent growers of the leaf (Chandler 1962; 1977). The farmers and plantation owners sold their leaf to local wholesalers who in turn sold their stocks to brokers who cured the leaf and stored it ready to release onto the market at opportune times. At that time most Americans smoked or chewed tabacco and very few smoked cigarettes. The manufacturers of tobacco leaf purchased the leaf from the brokers and then cut the leaf into pieces ready for processing after which it was packed ready to be sold to the wholesalers. In turn the regional wholesalers distributed to the thousands of small retailers in their locality. The industry contained many small firms in a quasimarket system.

In 1880 there were four large firms producing tobacco who held 80% of the 'smoking tobacco market'. Duke, who had decided in 1881 to manufacture cigarettes, was the owner of a very small tobacco firm and his rapid rise to pre-eminence in two decades illustrates how quickly large-scale modern enterprises could surface in the American context. The transformation in the cigarette industry centred on innovations in technology and marketing. The technological innovation was created by the American, Bonsack, who constructed the first successful cigarette-making machines and attempted to sell these in the USA and Europe. Duke decided to enter the cigarette market in 1881 and realised that two machines, each producing about 120000 cigarettes per day, could meet the current low demand. He also realised that each machine could produce more than 40 times the output achieved by any operative, so he began to investigate the total, world market. In 1884 he bought licensing rights from Bonsack for two machines. Simultaneously he:

1. Opened branch offices to buy the leaf directly from the farmers in order to cut out the middlemen
2. Established offices for distribution in all the leading commercial centres on a nationwide basis and staffed these offices with salaried employees

From that basis he sent his emissary, Wright, to investigate the world market. Duke commenced production. Within one year he was one of the five major producers of cigarettes in the USA. However, whilst his four rivals each operated a single-function, production firm, Duke was totally integrated. In that year he finalised the contract with Bonsack.

Fortunately a tax alteration brought a favourable opportunity which Duke was able to exploit to the full by reducing the price of his cigarettes and by increasing the amount of advertising. Within four years, by 1889, Duke was making and distributing 834 million cigarettes and he had forced his competitors to merge and then rationalize all the production into six plants within the new American Tobacco Company. Duke created a tightly organized, wholly integrated firm, specializing in a single product: cigarettes. American Tobacco had a small headquarters staff at the New York office. They received information and monitored the three major segments of the business: buying leaf, manufacturing, and selling. Each was run as a single department with highly developed standardized procedures and cost controls. The leaf purchasing was extended into the control of the farmers and plantations by specifying the type of leaf, when it was to be delivered, and by buying the future production. In that way the middlemen were rapidly eliminated. On the production side, also a single department, the new machines could readily handle the stable leaf, though there were still sources of variability in the prior departments of preparation and the later stage of packing, neither of which was then mechanised. By 1898 Duke had an administratively centralized, tightly coordinated enterprise and had moved into spacious offices on Fifth Avenue, New York. A large part of the space was taken up with leaf purchasing and with selling. By that time his leading advisers were career specialists. The six factories in Rochester, Virginia and North Carolina produced 3.78 billion cigarettes that year, of which 1.2 billion were exported. By 1900 American Tobacco was a worldwide enterprise with a factory in Australia and owning manufacturing companies in Japan and Germany. Duke was ready to invade the British market (see Chap. 11). The growth in the scale of the firm was accompanied by total vertical integration and by the takeover of suppliers of all the basic raw materials (e.g., pasteboard, rum flavouring) and the suppliers of much of the basic equipment.

The control of the company was exercised from Fifth Avenue, where the staff set targets and a tight timetable. The timetable was required to ensure that the cigarettes arrived at the retailer in as fresh a state as possible. That was a difficult task because in the days before cellophane wrapping the leaf could dry very quickly and become bitter in taste. All orders were telegraphed from New York to the factories which had to deliver shipments according to the defined schedule. The factories also had to send daily reports which were analysed in great detail under various headings. From those figures the headquarters staff were able to calculate labour costs to 0.00035 per thousand cigarettes. The relative costs of each of the factories were then compared so that it could be decided where the brands could be made most cheaply. American Tobacco were able to calculate prime costs of production as well as costs of advertising and selling, and their sophisticated system utilised the approach to renewal accounting developed in the railroads. In the 1890s Duke made enormous profits which he intended to deploy in gaining worldwide control. Yet top management remained a small, close-knit group involving one brother and a long-time associate. Their functions were strategic: the promotion of growth.

The growth of American Tobacco between 1880 und 1900 may be seen as an illustration of what was happening across a number of sectors in the closing decades of the nineteenth century.

10.4.3 Electricals, Chemicals and Automobiles

Three major industries emerged in the period between 1890 and 1910: electricals, chemicals and automobiles. They led the growth of the economy from then into the post-1945 period. America gained a pre-eminent global position in electricals and automobiles which it retained until the 1960s. The position of chemicals was more contested, especially by the German companies. In each of these industries major large-scale corporations surfaced, dominated their domestic markets and gained significant shares of markets abroad. I shall take the automobile industry to illustrate the main themes of innovation.

The electrical industry was responsible for the diffusion of new sources of power and new types of powered equipment which greatly altered the American factory and the household. That industry was also an early learner of the importance of innovation and its leading firms had, within two decades, established a rhythm of incremental innovation which permitted them to use the design functions to link changes in consumer taste to alterations in the raw materials and in the production technology. Design was underpinned by a heavy corporate involvement in the routinisation of scientific research and patenting through the establishment of internal laboratories within which new potentials for rapid capital accumulation were appropriated. In the founding period – possibly a new longwave – these firms acquired a definite skill at design-based assembly on a large scale. Their designed products were immensely successful in bridging the marketing and manufacturing requirements. Also, the electrical industry became major employers of the new corporate engineers who in due course (e.g., 1920s) began experiments in different forms of work organization. In concert with German firms (e.g., AEG) the leading American firms, especially the General Electric Company, entered the European markets and established bases in Britain.

The chemical industry also developed. Firms like du Pont began in gunpowder where they merged and rationalized their rivals into a large concern and commenced the unrelenting process of related diversification and of entering global markets. The chemical industry was also the site for developments in the American system of organizing, including the establishment of large corporate laboratories and research bases in the universities. With respect to organizing there is only space to mention one example: the notion of unit operations as developed by A.D. Little. The basic idea is elegant and simple. The whole process of a chemical plant is divided into homogeneous parts and a body of production knowledge is defined for that segment. The segment later became known as a unit operation. The simple principle was accompanied by extensive formalization in which the existing procedures were defined in great detail and recorded in manuals. These became the basis for instructing operatives and for defining the time-space boundaries of their discretion. The notion of unit operations, which may be recognised to be distinctly American, was developed in the 1890s and diffused in the First World War as a general principle. Later, it was incorporated in orthodox production engineering. Its significance is cultural and pragmatic. In many respects there are parallels between the idea of unit operations and the whole approach to the playing of American football. In each case a large phenomenon is broken down into defined elements each of which is studied intensively by applying general principles to create a specialized body of knowledge. The practice of using unit operations, in one form or another, became an everyday feature of American systems of organizing.

The extent to which the electrical (e.g., radio) and chemical industries became key sites for the development of an American system of organizing deserves further attention, yet it is useful to take the automobile industry as a key illustration for the development of innovation.

According to Abernathy the automobile is the "legitimate capstone of the American system" (Abernathy 1978). In Chap. 4 I examined his account of how production and the production technique in the American automobile industry evolved over the period from 1900 to the 1970s. Now it is appropriate to examine how innovation became organized in the automobile industry and why certain kinds of rigidity seemed to become established.

Until the late 1960s competition within the American market was sufficiently complex to make entry very daunting for foreign competitors. Few non-American manufacturers produced cars in sufficient volume to make a dent in the American market. Certain ranges of European speciality cars hold well – Volvo, Jaguar, BMW – but the mass produced cars had a very limited market appeal, except for the VW Beatle. In most respects the structure of American tastes was well satisfied by the indigenous car-makers and they were in a prime position to shape that taste through advertising and through the media of films and television. There were implicit barriers to the national market. Much earlier, in the first two decades of the century, Detroit had become "a home" to the industry and the region was known as Motor City. In 1920 there were some 50 manufacturers still in business, amongst whom Ford (with the T) held about half the market and many firms were very small. The major development in the 1920s centred in General Motors.

In the mid-1920s there was a major transformation in the American car market. The basic demand for a cheap, reliable and relatively robust form of transportation had been satisfied and new market segments were appearing. General Motors was constructed through the *successful merging* of several companies. In its strategic policies GM was an interesting and theoretically puzzling mixture of centralization and decentralization[9]. Unravelling that complexity requires attention to much more than the company's use of the divisional form of organizing in conjunction with top-level lateral committees.

One important feature was the distinctive approach to the styling of the car. Sloan realised that the 'outer skin' and shape of the car was the focus of attention for the American public, whilst also realising that any attempt to base competition on those features could be expensive. However, that cost could be minimised if two conditions were achieved:

1. The firm created an overall style through using design as the linkage between the consumers and the arrangement of production
2. Small annual alterations could be made to the skin of the car sufficiently obvious to indicate the newness of the car, yet often so small that they could easily be accommodated in the final assembly process

Sloan recruited Harley Earl to head up his styling departments in 1926–1927 and Earl produced a series of styling changes which sometimes involved major extensions and most of the time only involved cosmetic alterations (Bayley 1983). The major alterations were timed to coincide with annual changes in the model range. With these two features GM were able to penetrate and create a whole new structure to the American market and it was they rather than Ford who benefitted up to 1940. The other beneficiary was Chrysler whose innovative engineering in the 1930s secured them third position amongst American car-makers. The period from 1925 through the 1930s was followed by another transition in the 1940s.

The all-purpose road cruiser was the dominant design from the 1940s into the 1950s. It was a large closed-in steel-bodied car with water cooled engine, rear wheel drive and automatic transmission: the 'roadster'. The main suppliers were GM and Ford who competed on the basis of styling. In this period Chrysler's fame for innovative engineering attracted less attention and they gained a smaller share of the market. Producing the roadster was a matter of incremental innovation based on the styling of the body. Beneath the body the large size of the American car provided the designers with plenty of room to undertake a slow evolvement of the thousands of basic components. The focus was on getting costs down through the application of transfer machines to link segments of the process together. The skin of the car and its inside cabin were the focus for annual changes. Often these annual changes could be accommodated by simply substituting different, interchangeable features or even simply colours.

---

[9] Compare the analysis in Chandler (1962) with Learned, and Sproat (1966)

In effect the degree of customer variability was reduced by the skilful use of designs which were, relative to the American context, easily stretched. Thus, the American auto industry gradually moved towards a position of increasing certainty and routinisation. Abernathy infers that the content of decision-making became standardized and that the bodies of knowledge required to run the major car firms evolved very slowly within increasingly formalized paradigms. The production of cars seemed to be approximating the earlier situation of cigarette making. It was said that the auto industry had reached maturity. Abernathy observed that American firms had traded off incremental productivity against incremental innovation, and claimed in 1978 that by resolving the dilemma between productivity and innovation in favour of the former the car firms had entered on a learning path which shifted them towards a position where their existing forms of structuration would actually inhibit innovation; there were blockages to innovation. By the late 1960s car firms had come to specialize in marketing and had neglected production. Further analysis of the American industry suggested that its paradigmatic knowledge had evolved towards rigidity since the epochal breakthroughs by Ford and that each step in the evolvement of the industry reinforced the tendency towards preferring productivity to innovation.

10.5 Innovation as a Filiere: Imitation and Rigidities

Several analysts of the contemporary state of the American system of organizing have suggested that there is both a high degree of homogeneity (or isomorphism) amongst enterprises and also that this has arisen through processes of imitation. Yet, it would be hopelessly misleading to suggest that a single, homogeneous collection of practices typified all American enterprises. Instead, it would be preferable to presume that American organizations tend to reproduce a typical variety of responses in particular contexts. One way of expressing the range of the American system of organizing (ASO) relative to the British system of organizing (BSO) would be to compare their approaches on a continuum of choices from whether to adopt direct control or relative autonomy as shown in Fig. 10-3. Direct control would be represented by the tendency to define the future requirements through a highly formalized system of design undertaken by experts and then to define formal procedures through which current performances could be monitored. Responsible autonomy may be defined as the tendency to devolve the content to decison-making down the organization so as to permit the occupants of roles to define the future and to interpret the means through which it might be achieved. Taking Fig. 10-3 and comparing the discussion of American and British predispositions suggests that the typical American range will be towards the direct control pole and the typical British range will be towards the responsible autonomy pole. Clearly there will be many exceptions, yet this approximation provides a useful way of drawing some of the analysis together. American football may be placed at the direct control end of the continuum, whilst rugby union is towards the responsible autonomy pole.

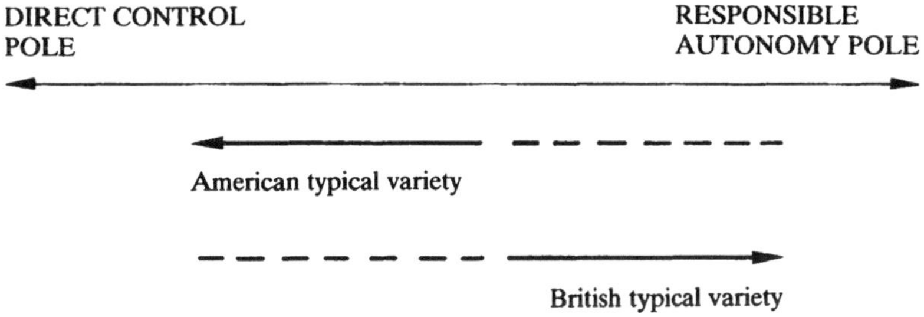

Figure 10-3. Direct control and responsible autonomy in relation to the typical variety in the USA and in Britain

American approaches to organizing at work have been associated with relatively high degrees of formalization in procedures. It is likely that this tendency towards formalization, which is often accompanied by well-developed financial controls explains why American enterprises have been able to grow in size whilst obtaining more of the benefits of scale economies and of the learning curve than British enterprises seem to have been able to achieve. Within an overall tendency towards formalization the ASO has been coupled to a tendency to use analysis in a very systematic way as a deliberate means for dividing larger problems into smaller ones, each of which can be tackled within a synoptic perspective. Again American football illustrates a general principle of factoring problems and does so in a way which reveals how the design of the game is part of the factoring process. The ease of factoring problems is enhanced by the ways in which such a high proportion of the conditions of play are equalised between the teams. However, that cultural capacity can only be fully deployed when the nature of competition permits the time for reflection. In the case of the two most American of sports, baseball and American football, the competition was limited to intra-American football until very recently when amateur Japanese baseball teams defeated the USA. In the case of industrial and commercial activities American corporations competed with one another within their own world-economy until the 1960s. Either foreign corporations were unable to decode the American market or they were unable to match American products on a cost for cost basis until after the early 1960s. Likewise American corporations found some difficulties in the international markets though their advantages of costs and of an early entry to the third and fourth longwaves permitted them to obtain technological rents.

It is now time to return to the debates surrounding the American system of manufacture. Looking back with the advantage of hindsight, it may seem that the ASM began as the deliberate policy of the government to have firearms which could be repaired in battle conditions. The persistent attempt, over several decades, to develop methods of production based on uniformity required an accompanying alteration in structuration and also was an agent in the gradual creation of islands (e.g., armouries) where the ASM was embedded in a system whereby production was factored into a cluster of technologically specialized components each of which was created in a defined subunit populated with specialist equipment. These subunits were part of an overall design so that there were progressive workflows ensuring that each subunit fed into a flowline. The whole process was coordinated by specialized staff who applied principles of discipline and designed specialized equipment. The cascade of that system out of the armouries into sewing machine manufacture then bicycle assembly and after that, automobile production, occurred over a period of more than six decades. Even then the process was incomplete.

The diffusion of ASM did not occur through organized channels, such as exist, to market certain products. Instead the diffusion occurred through the movement of skilled exponents.

11 British Systems of Organizing: A Case of Incomplete Modernization?

11.1 Introduction

This chapter and the next one form an entity. The objectives of this chapter are to examine the most commonly held thesis about the position of Britain relative to that of the USA. The thesis states that Britain and the USA are on the *same trajectory* of evolvement in their forms of organization, but that the British systems of organizing (BSO) *lags behind* those of the USA by one or more decades. It is therefore presumed that novel forms of organizing and specific innovations initially surface in the USA from where they are eventually diffused to Britain. The next chapter examines specific instances of where American innovations were introduced into the British context.

The issue of whether British institutions can be said to be on the same trajectory of evolvement as those found in America is a complex one. The thesis advanced in this book cautions against the assumption of a similar trajectory and replaces that notion with the argument that the structuration constituted by the British actually sets them on a different trajectory[1]. These two points imply that the British systems of organizing may be very 'inefficient' at appropriating American innovations. If so, then indications of adoption may be at a deceptive surface level. For example, there may be cosmetic similarities between the multidivisional form (MDF) in the USA and in Britain, whilst at a fundamental level of the structuration of the British firm the old forms of organizing continue (see Chap. 12, Sect. 2.3).

It has also been assumed that the British adoption of an American innovation is essentially a matter of imitation. There are many specific examples of this thesis and two may be cited. First, Chandler, in a well-known series of studies contends that the multidivisional form of administration surfaced in the USA during the 1920s and was then gradually diffused within the USA until the 1960s when it was diffused and adopted by British firms. The claim

[1] There is still a residual influence of the extreme forms of the convergence thesis which theorized that nations were developing common organizational infrastructures and patterns of social control around work. I have taken the view that insufficient attention had been given to the unlikeness between societies. At the present time that unlikeness still requires further development. However, once achieved it will be necessary to re-assess the extent of alikeness across societies both directly and indirectly through alternative and equivalent structural forms

that MDF was adopted in Britain is one of the best-researched instances of transatlantic diffusion from the USA to Britain. Second, Littler in a neat analysis asserts that the approach of Taylorism was carried to Britain in the late 1920s by the agency of Bedeaux engineers who undertook assignments in some 250 firms (Littler 1982). The firm of Bedeaux later became known as Production Engineering (PE) which became one of the main British consultant firms. These two examples assume that the diffusion occurred through the active agency of American-based consultants (e.g., McKinsey for the MDF). The two studies illustrate the commonly held thesis that British firms lag behind those American ones of a similar type, and that British firms imitate American innovations which they tend to adopt as a package, *albeit imperfectly.* The thesis therefore contains a number of basic assumptions which have already been demonstrated to be untenable and/or requiring refinement.

There are several issues concerned with the transatlantic diffusion of innovations which must be introduced at this point.

First, the central issue in the diffusion of innovations into Britain is one of appropriation. To what extent did the British users manage to narrow the appropriation gap by redesigning the original innovation to fit the intentions and processes in the British context? As noted in the previous chapter an entire filiere of American institutions were involved in their appropriation of European innovations and in their respecification for the American context. To what extent did the British manage to tackle that problem and to develop a filiere within which British requirements and exigencies were influential on the definition of the problems and in the creating of diagnostic approaches and in the constituting of best-practice solutions? Further, there is the question of how the British managed to handle the progressive evolvement of each innovation over long periods of time. As demonstrated earlier, all innovations continue to evolve, thereby changing their shape and uses. Thus appropriation is a matter of taking charge of that envolvement in a manner which permits and encourages the user to be internationally competitive. Appropriation also requires the creation of channels and agencies, including the role of the educational system in all its facets as well as the development of consulting agencies.

Second, it becomes clear that from the late nineteenth century onward the British faced the double problem of maintaining a distinctive variety of forms within the BSO whilst appropriating from the Americans those forms which are relevant and modifying American innovations for the British context. The problem is complicated by the fact that the British and the Americans have been in different world-economies. It is only with the advantage of hindsight that we know that the processes of capital accumulation have tended to favour the USA rather than Britain. It is also apparent that the

American advantage was partly based in their rich resources and in their self-contained, land-based world-economies which permitted them to take advantage of Britain's costly involvement in two major conflicts. Thus, it was not only the American cultural innovations – including technologies – which contributed to American success. The recent success of the Japanese, who lack basic raw materials, demonstrates the point. How, therefore, were the British to cope with the problem of learning from the Americans? In order to explore that question this chapter examines the development of BSO in the twentieth century, and the following chapter examines specific cases of what happened to selected American innovations in Britain.

The chapter contends that British forms of structuration were very different to those found in the USA and so the growing American presence, in its various forms, became a disruptive influence on British learning. American concepts of best practice and the enterprises which embodied them came to Britain and constituted a powerful overlayer of ideology, practice and technology which actively competed with indigeneous developments. The argument is made that the American overlayer was only decoded with difficulty by certain enterprises. Resistance to the influence of the ASO was undertaken for a complex variety of intents which as a configuration often overpowered British channels and agencies. So, although the American templates were often held up as *the* templates they were also held up as templates to be rejected.

11.2 Loose-Coupling and Devolvement in Work Organization

11.2.1 Introduction

Loose-coupling is achieved by permitting a high degree of autonomy between adjacent subunits and by not attempting to preplan and to specify the relationships between them. Devolvement is a situation where the control over the pace of the task and over its sequence is pushed down the hierarchy to the lowest levels. It is important to emphasise that high degrees of loose-coupling and of devolvement can be quite consistent with degrees of paternalism and with strong social control by a particular strata who themselves handle coordination in a personal way.

There are important comparisons between the ASO and the BSO in the extent to which activities are tightly coupled through a network of channels. In the British case there has been a persistent tendency towards the use of loose-coupling between subunits, with both horizontal and vertical subunits. Also, there is a high tendency to devolve certain areas of decision-making. Devolvement in the BSO may be contrasted with the ASO where the tendency was to use direct control through extensive, often formalized, prestructuring of the future by highly educated and expensively trained technocrats who applied technologies to the processes of control.

In order to understand the British templates of devolvement of loose-coupling and their curious, paradoxical relationship with state socialism and with paternalistic management it is useful to clarify the meaning of devolve-

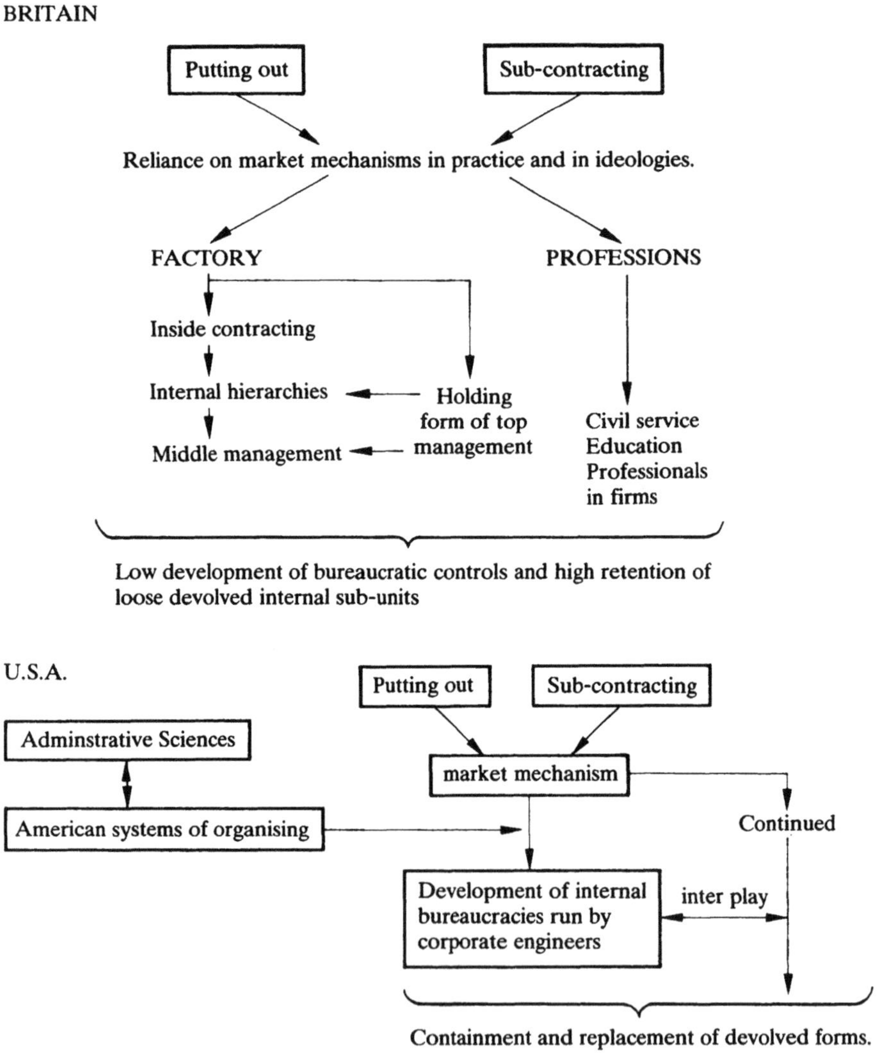

JAPAN

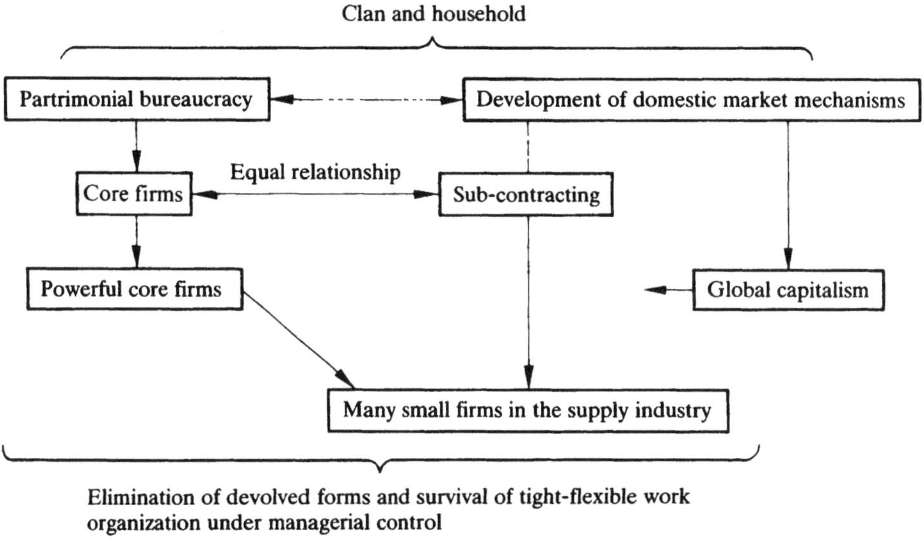

Figure 11-1. Devolved forms of work organization: Britain, America and Japan

ment in terms of the debate over the relative significance of markets, hierarchies and clans as mechanisms for allocating resources between alternative priorities (Ouchi 1980; Williamson/Ouchi 1981: 347-370). According to the current statement of the debate by Chandler, the ASO relies very heavily on the development of contract-based hierarchical relationships within firms followed by the monitoring of contracted performance through the use of managerial accounting as a proxy for the operation of the market (Chandler 1977). From Chandler's perspective, which is very Weberian in its emphasis on the efficiency of the impersonal dimension of bureaucratic structures, it would be theoretically consistent to observe that the BSO makes a heavier reliance on market mechanisms and on clan-like relationships amongst horizontal stratas of the owners of capital and the owners of labour time. That interpretation has been the thrust of Chandler's more recent explorations of the BSO (Chandler/Daems 1980; Chandler 1980; 1984). However it should be recalled that in those analyses he has attempted to examine the BSO as if it were on the same trajectory of evolvement as the ASO and was simply lagging behind because of the capacity of clans to impede the development of the visible hand. Also, Chandler asserts that in Britain the markets were a more effective mechanism than in the USA. There are three problems with that aspect of Chandler's thesis:

1. The persistent tendency to invoke a convergence explanation of advanced industrial societies
2. The evidence on the relative efficiency of markets in the USA and Britain is highly tentative and hotly contested
3. It obscures from investigation the question of whether there has been and is a typically British form of organizing and of cognitising innovation

The preferable alternative to Chandler's position is to balance the characterisation of the visible hand by distinguishing between direct control and responsible autonomy (i.e., devolvement) in the manner introduced in the previous chapter.

In this section it will be argued that the BSO has been persistently angled towards the use of responsible autonomy devolvement as shown in Fig. 11-1 which compares Britain and America in terms of their reliance on devolved forms of work organization. Japan is included to provide a link forward to the discussion in Part IV.

It is pertinent to observe that the position in Britain of the template of devolved, loosely coupled work organization can be explored in a very revealing way by considering which templates of work organization the government of Thatcher (in the 1980s) has turned to in its attempts to insert managerial controls into the state-run National Health Service. The NHS is the largest employer in Western Europe with more than one million employees! The Thatcher government has sought advice on re-organizing the NHS from the retail industry[2]. It is relevant to note that competition amongst retailers was engendered by the government in the 1950s and 1960s by legislation which removed retail price maintenance. That is to say, the legislation increased the level of market competition and altered the cosy relationship of collusion between manufactures and retailers. The retail sector is now amongst the most competitive in Britain and is the one area in which the commodification of space, time and inventory has been most fully articulated. Retail is a marked exception, but an important one, and retail is the template which has been chosen for introducing higher efficiency into the state health service. From the retail sector two firms were selected as outstanding examples of commercial acumen, skilful innovation and for directly controlled work organization. John Sainsbury is Britain's most successful chain

[2] In Britain the retailers have moved into a powerful position and in some cases operate as the equivalent of final assemblers through the design policies of their buying departments. Certain retailers have also been perceived as successful in the financial sections of the press. Moreover, as the French observers constantly note, Mrs. Thatcher's own biography includes a small-scale retail background; For a remarkable comment see *Paris Match*, 12 August 1986, pp 48-49: Magie/Elisabeth Rein Ne Va Plus

of supermarkets with twice the rate of return per square metre of its nearest competitor; Marks and Spencer occupy an enviable position as a major retail chain with more than 250 stores in Britain and new stores abroad. Key executives from these stores have been involved in major attempts at removing loose-coupling and at reducing devolvement.

11.2.2 Colonies, Trading Companies and the Professions

Devolvement takes many different forms. In the administration of the African colonies the British ploy was to achieve a degree of hegemony through the political co-option of the local elites, thereby reinforcing their often precarious power, whilst having them act as agents for inserting British commercial enterprises into the local barter-based economies. A classic instance of this policy was the appropriation of cocoa products in the western regions of Africa. The use of poll taxes and similar devices ensured that the indigenous farmers had to enter monetary exchanges rather than persist with local barter in the domestic markets. The general result was that the British established areas of regional influence with the least costs of administration. In the case of cocoa the local farmers were left with the freedom of manoeuvre with respect to the means of production, but were forced to operate through the British-controlled market mechanisms. Yet, the British were often able to present themselves as the neutral party refereeing local conflicts. In the colonial approach considerable autonomy was left to the locality concerned (e.g., regions of India).

That degree of autonomy was also reflected in the role played in colonial trading firms by local merchants and by agents of British trading concerns. A similar set of practices governed the activities of the British civil servants who were sent out to administer the protected areas. It is therefore no accident that much later it would be discovered that British corporations tended to adopt a mother-daughter relationship to their subsidiaries.

In a similar way the relationship in Britain between manufacturers and distributers typically left the distributer as the key informant on the state of the customer. That was just another example of the British tendency to combine a division of labour with the use of loosely articulated connections between the segments. The reliance on that approach by British automobile manufacturers in their relations with the distributers left the latter in a very powerful position which will be further examined in Sect. 11.3.3.

The devolvement of responsibility and the practice of loose-coupling was highly developed in the professions of law, medicine, education and engineering. The legal profession in Britain occupied a monopoly position in which, until this decade, they were not permitted to advertise their services and their fees. In medicine the consultant not only occupied a position of immense status, but also held a much higher degree of control over the daily coordination of a hospital than does an American equivalent[3]. Coordination was devolved down to each subunit where a particular 'firm' of medical specialists elected their own leadership. The firm handled their own group of theatres and beds in the wards. In other words, the coordination of flows through the hospital was not planned centrally by cost-conscious administrators, as Zerubavel reports is the case in American hospitals (Zerubavel 1979), but was the outcome of subunit decision-making.

In education there were similar practices. In schools, for instance, the legal control of the curriculum lay with the head of the school and its interpretation normally devolved to the teachers. The principle of teacher autonomy was strongly protected. The same features were to be found in the training and occupational identities of the engineering professions. The point to be emphasised is that in all these cases – and they cover a significant segment of the salaried middle-class occupations – the professional was expected to write the organization script rather than, as in America, to use his discretion within a script which contains some predispositions towards systemically coordinated actions. The point refers to the nature of the linkages between the professionals. Their interstructuring has exhibited a very high degree of loose-coupling.

The loose coordination of professional activity in Britain became the target for change programs in the 1960s. Particular attention has been given to the very large area of the medical profession. In hospitals a succession of state-instigated changes have been introduced to increase the throughput of patients and reduce the length of stay of patients in the hospital[4]. A major area of change has been in the development of means and mechanisations to achieve coordination between the subunits, an area in which the administrators were able to insert some degree of measurement of performance by comparing levels of output between similar subunits from different hospitals. For the administrators, however, entering the operational areas of medicine has proved much more difficult. There, established and new practices have remained firmly in the hands of the professionals who also control the choice of new technologies. A very recent investigation of the introduction of new medical technologies into British hospitals strongly indicated that the choices were made at the subunit level (MESS program; see note 2). That posed only slight problems for stand-alone technologies and data-processing

---

[3] These observations are derived from the studies of ESRC, WORC already mentioned and from the related MESS program conducted by John Child and Ray Loveridge

[4] Sir Keith Joseph, a leading conservative politician and thinker was closely associated with the attempts to restructure two major areas of state employment in Britain: the health service in the 1970s and education in the eighties.

systems, though it did create an unnecessary lack of uniformity in the choice of equipment which could disrupt collective learning and retard the maintenance of the equipment. The major problem area was that many of these new technologies should have had a systemic, integrative role in the hospital. For instance, computerised patient records ought to have contained details of the different drugs prescribed to a specific patient by different doctors in order that incompatible drugs could be identified. However, that system has proved somewhat difficult to establish. Likewise with many other systemic technologies.

One of the major areas of reduction in the degree of loose-coupling can be observed in the professions associated with construction (Clark, P.A. 1986b: 130–143). Unlike education and medicine, they are in a direct-market context, and major construction firms have elected to work in areas where they encounter international competition. A report on the construction industry produced in the mid-1960s detailed the antagonistic relationships that existed amongst the multiple parties involved in construction, namely, architects, building suppliers, suppliers of services, construction firms, quantity surveyors and civil engineers (Higgin/Jessop 1965). Since then the various bodies have sought to tackle the problem and their actions have been enabled by the increased sophistication of the clients who have imposed exacting contractual specifications (Clark, P.A. 1986b).

11.2.3 Rationalisation and Bureaucracy

The characteristic features of the British situation, relative to what occurred in the USA, are twofold: the lack of rationalisation of companies after mergers and the less extensive use of bureaucratic mechanisms.

There are many and extensive examples – some of which seem amusing or even absurd – of British mergers being created from several, possibly scores of small companies, most of which retained some degree of autonomy through the occupation of a seat on the board of the new company (Hannah 1983). Moreover, after the merger a considerable amount of the previous productive capacity would be retained in its existing form. This predisposition has been admirably summarised by Hannah in his account of the growth in size of British firms (Hannah 1983). The point will be illustrated in Sect. 11.4 through a comparison between the creation of the Imperial Tobacco Corporation (I.T.Co) and the earlier discussion of how Duke merged American firms. In practice, I.T.Co. was, relative to many British firms, quite strongly rationalised.

The tendency in the British case was to attempt to use the form of a holding company rather than to apply corporate rationalisation or to impose and insert the multidivisional form of organizing (Hannah 1983, Chap. 3+6). In the holding company, the many subsidiaries operated with a high degree of independence and in sectors like the carpet industry, managements in the same firm were sometimes totally unfamiliar with other similar firms within the same corporation. Throughout much of the twentieth century, until the later 1960s, this strategy seemed to have been highly successful. For example, certain major firms in the industrial midlands of Britain operated as holding companies – GKN, for instance, in the metal components and supply industry. Likewise in the automobile industry, the merger of 1951 between the two largest British firms, Austin (from Birmingham) and Morris (from Oxford) created the British Motor Holdings (BMH), yet the two companies retained a high degree of almost competitive independence. Later, between 1968–1972, when further mergers occurred and were accompanied by attempts at rationalisation, the internal management essentially failed to handle the process.

One of the important areas of weakness within merged and nonmerged enterprises was the development of structured, bureaucratic mechanisms of operation. In practice British firms embraced the bureaucratic mechanism much less extensively than did their American counterparts. The issue is one of the processes of managerial coordination and of the tightness with which the core values and procedures of the firm are devised and applied on a routine, daily basis. A superficial glance at any comparative data on the structuring of British and American organizations would merely indicate that the American organizations are slightly more structured for each unit of size[5]. However, a closer examination of the coordination process would reveal that the use of planning was less systematic, particularly in the routine application of cost controls. For example, in the 1960s it was common for medium-sized firms outside the petrochemical sectors to rely heavily on historical accounting rather than use managerial accounting and cost standards. In the public sector the use of financial controls was more concerned with balancing the books than with the cost effectiveness of the service. It was only in the 1970s that modern financial controls were systematically inserted and used regularly. Consequently, British attempts to adopt American plantwide

[5] The two original Aston studies examined about 140 establishments of diverse kinds in Birmingham and nationally. The scores for the scales measuring the dimensions of bureaucracy were used in conjunction with size to establish a profile which was then used as an index for British establishments. Applying the same instruments in the USA through a collection of small-scale, partial replications, suggested that the 'architecture' of bureaucracy of US establishments was very similar to that found in Britain, but that American establishments tended to have higher levels of structuring of activities through formal procedures and formal role specialization. The full program of researches have been edited and reproduced in a series known as the Aston Programme, published in London by Saxon House; for example, vol. 1 is edited by Derek Pugh and David Hickson

financial incentive schemes like the Rucker Plan and the Scanlon Plan imposed severe strains on the routine inventory method and on existing procedures for monitoring costs. Often the American innovations had to be considerably modified for the British firm (Clark, P.A. 1986b: 130-143).

The formalization of information which did exist in British firms often failed to encompass the heartlands of innovative activity. Here again there are important exceptions such as the large petrochemical firms and the major firms in the food, drink and tobacco sectors. The overall tendancy was for the designing process to be thinly formalized through mechanisms which were given ritual status, but were not readily available for reflective development. The use in the car firm, Rover, of the Blue Book is an excellent example. The Blue Book was the benchmark of procedures to be followed, yet its successful utilisation depended on a highly personal set of relationship amongst senior managements and the successful transmission of their implicit culture of designing from one generation to the next[6]. Formalization was present, but its detailing and openness to reflective editing was quite limited.

The coal-mining industry has been a good example of the absence of rationalisation and the limited use of hierarchical means of the visible hand[7]. When the industry was privately owned (up to 1947) the rise and fall in the market price for coal affected the colliery owners in a fairly immediate way. The employers tended to match these fluctuations by lowering or increasing the piece rates. In the collieries there was a relatively slight development of labour-saving devices through technology, which was explained by claiming that the variability of the British coal seams rendered the power machines unsuitable. However, power machines were developed in Germany where there were also small and variable seams. During the Second World War, when the industry was brought within the policy of state planning, the coalition government initiated a major review of coal-mining productivity and safety to investigate how technology could be used to increase productivity. The Reid Committee noted that technologies had to be developed to cope with the three stages of coal mining: getting the coal from the underground face, transporting the coal to the surface for grading and packing, and transferring the coal to users and distributers. After 1947, when the industry was taken under state control, the immediate pressure of the market mechanism was removed and specialized centres were established to develop the requi-

---

[6] The use of the Blue Book by the Rover Car firm illustrates the degree and type of formalization of the design activity in that period. The Blue Book was treated as the prime reference, but it was not formalized into a series of specialized procedures as in North American car firms, such as Ford; rather the Book was used in conjunction with traditional and paternal forms of control

[7] I am indebted to Joe Townend (SPRU), Richard Whipp (SIBS, Warwick University) and the WORC project on the National Coal Board for these observations

site technologies. In the 1960s the newly developed coal-getting machine was successfully introduced in all pits irrespective of their costs of production (Townsend 1980). The expensive coal seams of South Wales were not rationalised. Instead the image of cost saving was mobilised by pointing to the large number of men who voluntarily left the industry. That claim totally obscured the costs of retaining expensive pits. In fact the attempt to rationalise the number of pits was not undertaken with any vigour until the 1980s.

11.2.4 Forms of Payment and Skill Ownership

If bureaucratic controls were used sparingly in the British context how was work coordinated and how were new skills introduced? An examination of the basis of the employment contract shows that the employers tended to use informal standards of effort control and to couple these with a reluctance to challenge the employees over the ownership of skills, especially with respect to established skills.

In terms of the system of payment, British employers showed a strong preference for applying informal standards to both time wages and to piecework. Fig. 11-2 is based on two dimensions[8]. The horizontal axis contrasts the use of informal and formal standards for monitoring effort, whilst the vertical axis distinguishes between payment by the time period (e.g., the hour) or the piece. The argument here is that the British tended to move *up and down* the A-B line[9]. Some observers have claimed that the British employers have moved from A to B. The evidence on employment shows that the majority of employees from all sectors of work have been on time wages, but that the proportion on piecework does fluctuate upward and downward. There was a burst of piecework applications in the engineering trades, manufacturing and assembly, from the 1890s right through to the 1960s and a wide variety of payment systems were utilised. The engineering employer's preference for piecework was so strong in the twentieth century that the use of piecework was a basic requirement for membership of the Engineering Employers Federation (EEF). Piecework was also used very widely in the textile industry. In the main these applications of piecework were done on the basis of estimating the work content rather than through the systematic application of work study (see Chap. 12). A similar informal process was used for calculating the prices and allowances to be paid in the coal-mining industry which employed a million men at the turn of the century. Forms of piecework were also characteristic of the construction industry. Yet, many employees were paid on a time basis.

---

[8] Littler (1982) suggests this framework

[9] Compare with Littler (1982) who indicates that the direction of evolvement was more unilinear

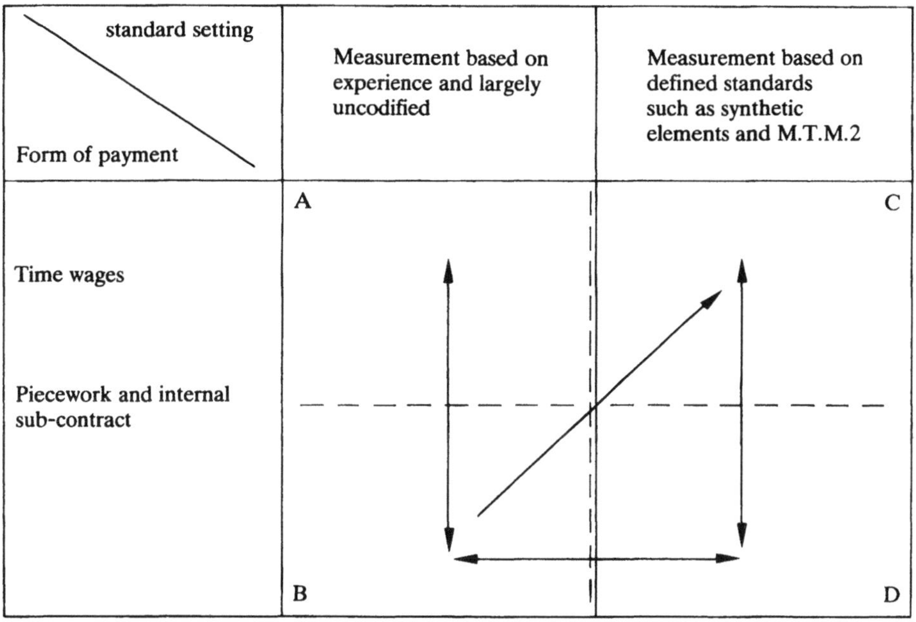

Figure 11-2. Time and wages

With reference to Fig. 11-2 it may be noted that in the USA a significant number of the medium- and large-scale corporations shifted from informal standards to formal standards in the early part of the century. The example of Ford and the five dollar day is merely the most salient and an early example of a switch in that sector[10]. After World War II there was a further switch in the USA away from piecework altogether to a concentration on specified time work. By the 1960s certain employers in Britain were moving in that direction. Significantly, an early leader was in the subsidiaries of the Dutch-owned electronics firm, Philips, where the Premium Payment Plan was used to specify work standards[11]. The shift out of informal piecework into formal time standards was hotly contested by the employees and the trade unions, especially by the local level of the trade unions. The shop stewards were keenly aware of the consequences of the transition for the political power of the shopfloor and for their own roles as representatives of labour. Many employers were hesitant.

[10] The removal of piecework by Ford was a remarkable move at the time
[11] Premium Payment Plan, *Metal Working Production*, 19 February 1964, pp 5-7

In Britain the shift away from implicit piecework was very much influenced by perceptions of what was thought to be happening in the USA. Sometimes British employers misunderstood the lengthy, incremental and cumulative process which had led to the position of American firms in the 1960s. Consequently, the problems of introducing the transition in the British context were dealt with in a very simplistic and naive manner, most obviously in certain parts of the automobile industry.

There is clear evidence that the British have also shifted from informal towards more formal standard setting, but that the pace and content of the shift has varied greatly between sectors (e.g., retail compared with textiles). The position of training and of skill ownership has been an important ingredient affecting the pace and content of the shift.

Compared to American employers, British employers have not appropriated the ownership of skill, nor have they been so prepared to invest in internal training and education. To unravel this predisposition it is necessary to consider the position of the trade unions with respect to the ownership of skill. In the nineteenth century one of the most important associations for labour was the Amalgamated Society of Engineers (ASE). Their rules stated the following in the 1860s:

It is our duty do exercise the same care and restraint over that in which we have a vested interest – i.e., our skill – as the physician who holds a diploma[12].

By the 1870s certain key, persisting features of the modern British labour movement were already in position: a section of the male working class already held voting rights in national elections; the Trades Union Congress (TUC) had been established in 1868 to create a collective forum for the labour movement; the unions had established local trades councils in a cross-union basis to liaise in specific localities and to attempt to regulate the influence of the employers; the form of unions was already crystalised in sectors like construction, engineering and the craft industries; collective bargaining had been established in coal, cotton and steel making (Hinton 1983). In the founding period trade union actions had a great deal to do with preserving job autonomy as well as maintaining levels of pay. Their disposition was centred in the craft unions, especially in printing, engineering and construction. The most skilled workers in these areas were typically recruited through portals which were controlled by the unions who handled the rates of entry and also the apprenticeship for the skilled occupations. Their position paralleled that of the modern professions like medicine and the role of the universities.

---

[12] Contained in the book of rules for the Amalgamated Society of Engineers in the 1860s; cited by the Webbs and used as an illustration of a basic value system by Hinton (1983)

Hinton observes, that in the founding period, skill was treated as the cultivation of an aura of mystery around the activities of one's occupation (Hinton 1983). Moreover, it was successfully claimed that the skills of the most skilled workers could only be transmitted through the experience of the tasks under the supervision of a master. Thus, the coalminer learned to tell the difference between an innocent creaking of roof of the coal-seam and a creaking which presaged disaster. Knowledge belonged to the workers rather than to the employers and it was the workers who had to conceive the tasks and to execute them. The coalminer's knowledge covered the whole process from the preparation of the tools, through the setting up of the simple equipment, shoring the roof and transferring the hewn coal. Hinton states that the miner controlled his own pace and "bossed every minute of the working day".

That position of control was put under pressure from the employers, yet the counter-pressure of the most skilled occupations was also influential and the outcome was that the major alterations occurred in the roles of the helpers and the semiskilled workers. The employers were particularly active in trying to loosen the unions' control over the ports of entry during the conjuncture of an upward movement in demand of the business cycles. Employers pressed for extra production by extending the number of employees recognised by the union. As a result even the most powerful unions, like that representing the typographical workers, abandoned whole sections of the industry to nonunionism in order to retain control over its own apprenticeship systems.

In practice Victorian trade unionism rested on the activity of relatively small groups of privileged workers who defended their status against the encroachment of the semiskilled. There was a structural division in the workplace between the skilled and the rest which was reproduced in wage differentials. Outside the workplace, in the community, one section of the working class self-consciously differentiated themselves and played a central role in the formation of a stable trade union movement. It was they who embraced the language of 'self-help' and promoted the idea of a 'workshop community'.

The rise to eminence of such groupings and strata coincided with a period when some considered capitalism to be a passing aberration, while others were more than prepared to support groups who competed with industrial capital and when the middle-class bourgeoisie passed through moments when they feared that the whole fabric of British society could be overthrown. Jones has demonstrated that such moments occurred in the winter of 1886 when the innocent march of the London casual labourers through the fog was rumoured to be an insurrection (Jones 1972). Later, in the period from 1917 to 1921 similar fears about labour were evoked.

The consequence of these actions was to inhibit employers from acquiring the degree of control over tasks, jobs and skill which key American companies had acquired from 1890 to 1917. In the British case even when the employers broke the challenge of the Amalgamated Society of Engineers in 1893 there was a relatively slight attempt to gain control over skill by the employers. Moreover, in the British context there was no rising strata of salaried administrators nor of corporate engineers who allied their careers to capital and developed the aura of the professional control of work through notions like best practice.

In certain respects the control of labour over skill was quite precarious. For example, in cotton it was the employers who permitted the spinners to sustain an image of skill. When it suited the employers they tended to reposition the frontier of control between the spinners and the semiskilled piecers. The strength of the trade union position lay in its overall size and its central role in the collective bargaining over piece rates. Also, the employers used the spinners as a supervisory elite to intensify the work of the piecers. That pattern was also to be found in other settings.

In the industries where new technologies were being introduced, such as foodstuffs, breweries and cigarette making, the employers tended to incorporate skilled sections into supervision and to embody the new knowledge of the process in the technology. In those areas there was a much stronger interest by employers in the analysis of skill, yet compared to American tendencies, there was only a slight attempt to appropriate the skill.

The result was that by the 1960s only a small segment of the labour force received systematic formal training; under seven percent.

11.3 Markets and Sectors

11.3.1 Commerce and Shipping

What were the characteristics in the twentieth century of the markets which the British occupied and the sectors which its business leaders entered? Assessing the British patterns of innovation is a delicate subject which British historians have approached gingerly. Individual business histories are often revealing, but the synoptic overviews have tended to concentrate narrowly on the size of British firms rather than on their capacities and international competitiveness[13].

At the start of the twentieth century some of the British were already noting that the international competitiveness of British goods and services was being challenged from the USA and Europe. Yet, the British had created a

[13] An interesting exception is Williams, Williams, and Thomas (1983)

world-economy with the City of London as a major pivot[14]. In 1900 the two most profitable sectors were shipping and the merchant banks. In shipping Britain held a virtual world monopoly of freight transportation as the trade routes coincided with British naval strategy. One-third of all the world's tonnage was carried in British-owned ships, the very great majority of which were built in Britain. In effect Britain was a maritime colonial power. Merchant banks came second in earning capacity (e.g., Barings) and they played a key role, as international capital, in the transfer of capital in technology-embedded goods to certain areas of the world. Often these investments were in thinly populated countries and that reflected both a lack of competitiveness and also an encouragement, because those markets were the least likely to appropriate the technology. Even so, Britain's initial technological lead was progressively narrowed, sometimes by innovations (e.g. in steel-making) which assisted rivals more than British firms. From the City the services of banking, credit, advice, investment and insurance spanned the world. British capital was exported to invest in various parts of the world (e.g., Argentina).

The particular world-economy in which Britain was embedded combined these trading services with roles as a merchant of raw materials from the colonies and from areas in which Britain sought to establish its influence (e.g., the oil-rich state of Iran). In terms of manufacturing, Britain concentrated on its staple industries (e.g., textiles) and its 'domestic' industries (e.g., food, drink and tobacco) rather than in entering the new sectors of electricals, chemicals and automobiles/transport. Although the British government has been described as noninterventionist, there can be little doubt that the few sustained major initiatives to create new departures owed a good deal to government-inspired actions.

The failure to enter the new sectors before the First World War and again after the Second World War strikes at the very heart of the British self-image as the workshop of the world. Yet that very self-image is revealing because the third longwave was based on the laboratory rather than the workshop. The British are supposed to have had a distinctive competence at invention, yet as events demonstrate, if they were good at invention then they seem neither to have been able to appropriate the benefits for any length of time, nor to turn the claimed inventions into commercial innovations (Freeman 1974). There are exceptions of course and it is sensible to note two examples: rayon, the manmade filament, and float glass.

The history of Courtaulds and the discovery of rayon is a fascinating story of entrepreneurship and chance (Coleman 1969; 1977). The discovery of rayon

[14] Ingham (1984) Coakley, and Harris (1983) Turnover of foreign exchange in London is now about 90 billion dollars daily whilst New York is 50 billion and Tokyo is 40 billion (*The Times*. 20 August 1986; Report of the Bank of England Survey)

was largely a matter of scientific push. The chemistry of cellulose was investigated in the midnineteenth century in order to discover a satisfactory filament to include in lamps and other products. In 1905 the firm known as the Viscose Spinning Syndicate became the British licensee of inventions already patented by Muller in Germany, but the firm had considerable difficulty in making the filament in sufficient quantities for commercial exploitation because some production problems had not been resolved. A coincidental visit to the plant from a senior member of the silk company, Courtaulds, led to the latter agreeing to fund commercial development. However, the owner of the patent challenged the authenticity of the British work in the courts. Fortunately for Courtaulds the decision of the court only blocked a relatively unimportant segment of the patent. That provided Courtaulds with an opportunity. Between 1907 and 1912 they developed the successful production of rayon filament in Britain so that financial losses of ten thousand sterling were soon turned into profits of more than three hundred thousand. Courtaulds were able to sell rayon as a substitute for silk and therefore obtain high profits and the profits were used to force down prices and to defeat rivals. With these profits and the successful learning of the process the directors of Courtaulds were able to enter the American market by acquiring the license and patents of an insolvent American firm and establishing a factory in Pennsylvania. By 1912 they held the dominant position in the American and British markets. However, during the Second World War those advantages were 'traded' in return for American supplies so that the key advantages of subsequent commercial exploitation occurred in the USA. In the case of float glass its developers, the Pilkington Company, were able, after a fascinating and instructive process of development, to obtain commercial advantage over a relatively long period through the 1960s and into the 1980s (Barker 1977). These two examples are important because they refer to products which are essentially international rather than domestic in the market base.

The importance of the market in which a population of firms have elected to compete has been shown to be important (See Chap. 3, 5 + 7). In the British case there are two features of the elected markets which should be acknowledged. First, there are many items whose production and successful marketing required the decoding of a distinctly national cultural taste. These were domestic markets and in the first half of the twentieth century British firms were often able to hold their domestic markets, especially in the areas of foodstuffs, including confectionery, drinks and tobacco. It is therefore not surprising that with the evolvement of production technologies for mass production that British firms, and joint British-Dutch firms, managed to maintain an adequate market position. There were, though to a decreasing extent, distinctly British tastes for cigarettes, beers, bread, confectionery/sweets and basic foodstuffs. In those sectors British firms were already large by 1905 and they tended to survive into the last quarter of the twentieth century, though they now face more exacting competition (Hannah 1983).

The domestic angle also assisted many other industries; construction, for example, in both the civil and housing sectors. In those areas foreign firms had to overcome British tastes and regulations. In some cases certain British firms played a central role in the designing of connections between the domestic market and the capacity of indigenous producers. In retail, for example, the firm of Marks and Spencer was, in effect, a design-oriented organization which assembled components of clothing (and later foods) into a marketable form. Marks and Spencer's policies with their textile suppliers included a high degree of structured-encouragement towards higher productivity and towards sustained, incremental innovation[15]. Although the firm concentrated upon the design of the product, their executives were keenly aware of the problems of production. Moreover, the scientific advisory committee on raw materials within Marks and Spencer included the introduction of a very distinguished external adviser on the new raw materials and on improvements to existing raw materials. In this way, the company acted as a channel coordinating developments in the chemical industry which had a considerable consequence for the survival of parts of the textile industry. Second, the other feature of the market is the British proclivity to remain within the world-economy associated with its former trading connections (e. g., China) and with its so-called Empire, especially with India. This world-economy was as enclosed as the Mediterranean worldview of Philip II in the seventeenth century. The results of failing to enter and to learn how to compete in the more prosperous markets of Europe and North America became more and more consequential.

It is useful to look at some examples to illustrate these themes. The cotton industry provides a neat illustration of how the choice of India as a market between 1900-1918 contained the seeds of later, virtually permanent decline. A good indication of the staple domestic industries is provided by the cigarette industry. With respect to the new, twentieth century industries of the third longwave, examples may be given from chemicals, automobiles and electricals. Before examining these a brief reference should be made to the military sector because it has occupied a central position in the consumption of financial resources and in the employment of that scarce resource, the British graduate. Unlike the American armaments industry there is little indication that it has been as productive of innovations in products or in management techniques as has been the American industry (Kaldor 1980: 100-121).

---

[15] The approach of Marks and Spencer to the control of suppliers dates back into the early 1930s, possibly earlier, and was grounded in the special relationship with the Leicester-based firm, N Corah (St. Margaret), who at that time were one of the largest in Europe; by the 1960s the locus of initiative in the relationship was firmly in the hands of Marks and Spencer

11.3.2 Cotton and Cigarettes

The fate of the cotton industry in Britain during the twentieth century provides a revealing probe which can illuminate how market differences, combined with differences in entrepreneurial strategies, led the British down a different learning path to the USA. The direction taken was beneficial in the short run, but contained the seeds of disaster for the industry in the 1930s.

In the latter part of the nineteenth century the American textile industry introduced radical innovations in its technology for spinning and for weaving, but these innovations were only adopted in a narrow segment of the Lancashire cotton industry (Lazonick/Mass 1983). Why did the major part of the British industry still continue with the old technologies and why did the industry head into permanent decline in the 1920s and 1930s? The analysis of these questions has to be undertaken at two levels: the industrial structure of the sector, and the choices made by entrepreneurs in the period 1900 to 1914 of markets and production techniques. The existing structure of the industry contained numerous features which were impediments to a shift towards higher mechanisation, and the choice of markets heavily reinforced pre-existing tendencies towards inertia. British entrepreneurs responded to international competition by choosing the types of markets which could be served by making incremental modifications to the existing forms of technology and retaining much of the forms of labour organization already established.

In the first part of the twentieth century there was a burst of growth in the Lancashire cotton industry. Growth of output and the retention of relative profitability were achieved by choosing to enter the massive markets of India and China and not to enter the markets of Europe and North America. The demand side of the Indian market was distinctive and it accounted for around half of the exports. The Indian market consisted of quickly changing styles of fashion-oriented cloths which were highly finished, yet of low quality. The Chinese market, which consisted of the wealthier classes in the towns, was smaller and different from that of India, accounting for about ten percent of exports. In addition to these two markets the other areas of high sales were all in the former 'white colonies' and in the poorer developing countries, not in Europe or North America.

The choice of markets provided the opportunity for the renewal and reproduction of the existing industrial structure of the cotton industry rather than its transformation. The most significant features of this were the very extensive degree of both vertical specialization by the stage of the process (e.g., spinning, weaving, finishing) and the horizontal specialization within each stage by firms who focused their production capacity on a narrow segment of the market. Moreover, the different stages of the process tended to be located together in towns which were sometimes 30 miles apart. This industrial structure created many layers, in each of which were a plethora of small family-owned firms, with the industry being coordinated through market mechanisms centred in Manchester. These were well developed and also consisted of many small, family-owned firms acting as shippers or merchants. In each case the shippers sent samples to the agents in the buying countries and obtained orders. In order for the suppliers at each different stage to obtain longish production runs they tended to specialize and therefore to obtain a monopsony.

The producers captured the low-quality markets by cutting the costs of production. First, they used cheap, short staple cotton in conjunction with the introduction of a high degree of substitution of even cheaper noncotton raw materials in a process known as sizing. The extent of sizing in Britain was approximately four times higher than in the American industry. These decisions reduced the cost of the basic input, though the new input did create problems for the operatives because of breakages. Second, the cheaper raw materials were subject to breakages in the yarn during processing. The rate of breakages would have been very high if used with the continuous motion of the new partially automated technologies and hence the costs of production would have been increased. Moreover, there were not the long production runs to justify the new technologies. Therefore the producers chose to continue with the incremental modification to the existing technologies whose performance was increased very slightly by speeding up and by small-scale modfications. The old technology, for example, mule spinning had an intermittent action which caused less breakages to the low-quality yarns. Further, the old technologies were readily adapted to the week by week planning of workloads. Third, the traditional forms of work organization were retained along with established, local piece rates which remained fairly stable. It has been argued that this form of work organization was at its peak in the 1890s, but that its efficacy was reduced in the twentieth century. Fourth, the new technologies were only adopted in a small segment of industry serving the higher quality market segments. The configuration of these four factors permitted the owners of Lancashire firms to maintain a reduced level of profitability between 1900–1918, but to do so by accepting lower returns for their managerial labour and hence having less reserves for later investment.

The burst of economic activity in selling to these low-quality markets reinforced and deepened the vertical and horizontal specialization of the industry, thereby heightening factors which would inhibit the introduction of the new technologies because these required longer production runs of higher quality cloths through more vertically connected processes. By 1914 the American textile firms, who were supplying different market segments, had become familiar with the new technologies so that 90% of their spinning capacity was in the new ring spindles compared to less than 25% in Britain. The British industry relied on old, already amortised technologies to retain a large industrial structure and way of life. In the 1920s their position declined and that continued into the 1930s.

The case of the *cigarette industry* provides an interesting illustration of Anglo-American similarities and differences (Alford 1973; 1976a, 1976b, Chap. 3). The story hinges around events between 1899 and 1902 when the American Tobacco Company entered the British market by trying to buy John Players and eventually acquiring Ogdens in Liverpool. Yet, by 1902 Duke had withdrawn. Why was Duke unable to dominate the British market? In large part because of the peculiar entrepreneurial, administrative and engineering flair of the Wills family and their associates whose political skills had been instrumental in putting together a defensive alliance of British producers to create, in 1901, the largest firm in Britain, the Imperial Tobacco Company.

Wills were a long-established firm based in Bristol. In the 1840s they had innovated in the branding of the tobaccos and this feature gave them an increasingly strong presence as the railways drew the pattern of distribution into a single market. At that time tobacco was chewed and smoked by the working class and was sold through some quarter of a million retail outlets. The innovation of the briar pipe coupled to the introduction of milder Virginian leaf stimulated consumption. Wills were pioneers in marketing and in distribution who, by 1860, had constructed a national network in which their salesmen were also sources of intelligence.

The strategy of the company was to maximise the sales by measuring weight. In other words, to increase their market share. Their accounting procedures defined profits as the residual reward to management and therefore the directors searched for sources of profit and found it difficult to envisage any great increases from tobacco sales, So, in 1871, Wills introduced an expensive hand-rolled cigarette into the luxury market of London. Nine years later Harry Wills, who was responsible for engineering developments, acquired the UK exclusive use of the Rose packing machine and the following year acquired the exclusive UK rights to the Bonsack cigarette-making machine for 4,000 pounds. Wills commissioned a very thorough testing of the early model from which an outstanding report revealed both faults and their remedies. The latter were skilfully woven into a clever definition of the UK patent for the cigarette-making machinery. Machine-made cigarettes were introduced in 1883-1884. Then in the 1890s the very cheap, five for a penny, Woodbines were introduced even though there was no obvious evidence of a large market. In four years the Woodbines' share of the total market increased from one-fifth to two-thirds and the size of market increased fivefold. Moreover, cigarettes were more than five times as profitable as tobacco.

By the 1890s Wills, previously a large firm amongst several, had obtained 60% of the market.

The threat of the American invasion acted as a catalyst. Wills were active in putting together a defensive alliance in which they were the dominant partner. It was agreed that the members would share technical information and draw profits according to a defined formula. The new company, Imperial Tobacco Co. (I.T.Co.) confronted the American invasion with a countermove of entering the USA market. After a very short encounter the two firms agreed to a division of interests. I.T.Co. were to have the British market and to withdraw from the USA and they jointly were to form British-American Tobacco (BAT) with Duke as its head. By 1902 that arrangement left I.T.Co. as the largest firm in Britain and they continued to dominate the UK market from then until the late 1970s. The process of rationalising factories proceeded slowly and consisted of concentrating production at the Players factories in Nottingham and Wills factories in Bristol. I.T.Co. was the object of several investigations by the Monopolies Commissions, but was declared to be "in the public interest". In 1967 the firm began to plan for the introduction of the new integrated technologies in Nottingham and Bristol[16]. Then the throughput time of the process was more than 100 hours and each segment of process was buffered from the previous section by almost ten hours work in progress. The new plans reduced the throughput time to a few hours and tightened up the connections between the stages. In certain respects the approach to operation of the old plants and to the design of the new plants was at the frontier of administrative practice. When in 1972 Britain entered the Common Market the 1902 trading agreement had to be scrapped because it was contrary to the Treaty of Rome. Since then the opportunities for foreign firms have improved and the Imperial Group gradually lost market share. Moreover, the very belated attempts to diversify, for example, into American hotels (i.e., 'HoJo'), were unsuccessful and the company was taken over by the Hanson Trust in 1986.

It is relevant to note that the Imperial Group was still a major British company in the 1960s. Its market share had been protected during wartime and the postwar period and after external consultants had warned the directors about the threat to their market share (in 1960) the company moved very quickly to recruit brand managers and to launch new cigarettes like the very successful 'No.6'. Later, around 1970, the firm began to diversify into foodstuffs by acquiring firms making crisps and supplying chickens, but their entrance onto the fish sector was more awkward. The early acquisitions reflected the capacity of the firm as a marketing enterprise. Likewise with the acquisition of a major brewery chain. However, the entrance into the Ameri-

[16] The case study of John Players is reported in Clark P.A. (1972a, Chap.3-6, 8+9)

can market proved more difficult, particularly because the firm shifted into the hotel business where it did not possess any core skills.

During the 1970s the whole sector of food, drink and tobacco became much more international. Public tastes for foreign goods increased markedly and the economies of scale required that British firms entered foreign markets. Those changing demands placed considerable strains on British firms in these sectors.

11.3.3 Cars, Chemicals and Electricals

The *wholescale* development of these three sectors, each of which contains considerable diversity, was much less evident in Britain than in the USA. In practice the British chemical industry has proved to have been the most robust, but that was significantly helped by state actions.

The automobile sector was very slow to develop (Whipp/Clark 1986). Initially French cars were imported and the German firm Daimler opened a plant in the West Midlands. As late as 1912 Ford entered the British markets with assembled kits in Manchester and became the largest single car-maker with 30% of the market. Rover, the third largest producer, only supplied 1200 cars per annum at that time. The small British-owned firms tended to be both manufacturers of components and assemblers. Lanchester, a firm which attempted to buy in specified components from West Midland suppliers, found that they could not get metal components in the requisite state for interchangeable assembly (Saul 1967). Even in the 1920s it was necessary for carworkers to undertake adjustment to many components in order to achieve the required degree of fit. British car-makers remained small into the 1920s when, following the recession of 1921, a new very small car, the Austin 7, became the state of the art design and was followed by the small car from Morris. These two firms gained a high market share and the structure of the market was altered. Many small firms survived the 1920s and new firms like Rootes and Jaguar entered in the 1930s.

The dealers, who handled the sale of the cars to the public, were autonomous, independent agents (cf. USA) who controlled the channel of information from the consumer back to the car-maker. Power in the automobile industry was finely balanced between the dealers, the car-makers and the suppliers of components and equipment. For small firms like Rover the power of the dealers was such that at one stage their nominees were placed on the board of the company. In the case of Rover the firm relied on the dealers for continuous market intelligence and depended on reactions of the trade at the annual car show for the planning of the forthcoming production.

Rover only institutionalised innovation and design after 1932 and then did so in a way which was only really effective for a small, paternally run firm.

British car firms primarily produced for the domestic market and then developed outlets in Australasia, South Africa and the colonies. They did not penetrate the European market to any extent. Nor did they establish a secure toehold in the American market, except for the Jaguar range, and it is important to note that Jaguar has been outsold in the American market by the principal German speciality car-makers. That market orientation remained into the late 1980s.

In the early 1960s the banks began to warn their clients that the leading British carmakers were poor prospects for long-term income. British car firms tended to be viewed with suspicion by the investors who expected to receive high rates of return. The warning was prophetic. By the mid-1960s the leading firm, BMH, was moribund in its product range. The famous Mini was not making money and the engineering innovations incorporated into other cars created difficulties in assembly which were not recouped by high sales and profits from the salesrooms. The earlier looseness of linkages between the manufacturers and the dealers opened into a chasm, especially when the dealers were left to cope with a much higher degree of presales warranty work than was the case for incoming foreign cars and from the increasingly successful Ford plants whose cars had a much better record with fleet sales. The Labour government, concerned about foreign competition and seeming to believe that mergers could provide a new life for ailing managements, manoeuvred the commerically successful Leyland into a merger with BMH/BMC to create the largest British grouping. Two studies of that merger done from quite different perspectives reveal the lack of integration which occurred[17]. At the level of design the crude attempts to reconfigure existing design capacity produced confusion in the design-translation-commissioning process. In the mid-1970s the firm had to be taken into state ownership and its market share fell steadily from over 30% to around 16% in 1987. The British-owned auto industry tended to recruit its managerial staff from within and not to recruit graduates. At the shopfloor level the unions were important channels of recruitment, particularly for the skilled workers.

The position of the chemical industry also reflects government intervention, though much earlier, in the mid-1920s (Reader 1977). The experience of the First World War had revealed to the cabinet that Britain's explosives and chemical industries were very unsatisfactory in their performance. In the 1920s the government encouraged a merger of several firms, into a new company, the Imperial Chemical Industries (ICI). Its early leadership gave consideration to American concepts of organization and so from an early stage

---

[17] Detailed studies of the mergers are rare

it was organized on a divisional basis with a multidivisional form. The headquarters staff was located in London at Millbank and the divisions were located around the country, with a heavy concentration in Teeside and near to Manchester. The ICI company followed a distinctive trajectory for the British context. As a science-led company, it recruited graduates.

From the mid-1920s onward the development of ICI became very closely intertwined with the concerns of successive governments, particularly during the Second World War. As a company, ICI became one of the world's major chemical concerns, and as such was one of the front-runners in the development of scientific research laboratories.

In electricals, British-owned firms were established in the early twentieth century, yet by the 1920s they had been overpowered by foreign, especially American entrants (e. g., Westinghouse). Of the early firms those who sought to compete in the consumer goods markets did very poorly until the 1960s when the holding company of GEC combined them with large, financially lucrative defence contracts.

There is an interesting comparison to be made between the British firm of Lucas (Nockolds 1978) and the Dutch firm of Philips. Each firm began operations in the lamp-making business. Lucas developed from that base by moving heavily into supply of electrical and mechanical components to the automobile industry. The firm became a British leader in the application of production-engineering concepts to the design and manufacture of products. Moreover, the firm's employees made many visits to the USA to learn about their methods of production at the workshop level. That is to say, the focus was intraworkshop rather than the overall coordination of operations and its interfacing with innovation-design. Philips moved from lamps into electronic goods for radios, later developing a very strong inhouse scientific research and design base in Eindhoven. Philips, through a chance conjuncture of wartime events, acquired an American base which provided them with a direct conduit for the transfer of technology and for access to American military contracts. Philips also developed a strong interest in administration, but did so from a more enquiring, science-oriented approach than did Lucas. That difference is reflected in the way in which experiments in administration within Philips were systematically evaluated by highly trained inhouse teams of engineers and psychologists using quasiexperimental designs[18]. In com-

[18] NV Philips, the giant Dutch multinational in electronics began in lamps as did Lucas, but their trajectory of evolvement has been quite different. Philips has been one of the most active and experimental firms in the uses of the social sciences. In a series of careful experiments since the mid-1960s their internal research groups have located the technological, economic and psychological factors which, in their view, affect different forms of work organization. Their approach is paralleled in other Dutch enterprises, both public (e. g., PTT) and private

parison, Lucas approached administration through British concepts of engineering. Lucas, one of the most successful indigenous firms with a place in the electronic industry, emerged through locating themselves in relation to the automobile industry at a time when British auto firms held a significant share of the market. Indeed, relative to many of the small UK carmakers (e.g., Rover, Jaguar, Standard) the Lucas Company was very large and would have been viewed as a reference point for good manufacturing practices in the West Midlands. It was the Lucas Company which facilitated the survival of the Rover Company after its severe losses between 1929-1932. In the post-1945 decades the Lucas Company grew and eventually began both to diversify and to integrate. An important piece of integration occurred with the takeover of the electrical components firm, Smith Industries, which made clocks and more complex devices.

Turning from Lucas to other sectors of the electricals industry it is noticeable that many other successful British firms possess crucial connections to government contracting in relation to the enormous defence industry. However, unlike the American military contracting, that filiere has not been a source of rapid spinoffs into the civilian markets, nor has there been the development of administrative and technical devices which could be widely diffused. Instead the linkages have been loose. That looseness is embodied in the financially successful giant of the postwar period, the British owned GEC, which built up enormous cash reserves. It is a holding company with a small headquarters staff applying rigorous financial criteria to an immense diversity of enterprises. These range from the assembling of simple consumer products like cookerhoods at factories in Birmingham to complex contracts with the Ministry of Defence for the development of sophisticated surveillence systems designed to monitor potential movements of Soviet military aircraft over Western Europe; the latter have not yet been markedly successful.

12 Transatlantic Evolvement II: Britain and the Appropriation Gap

12.1 Introduction

It is clear that after 1890 the British succeeded less well than the USA in acquiring a significant industrial base and that they progressively lost a key source of earnings in shipping whilst just about retaining their position as one of the world's leading financial centres until the 1980s.

The British elected to operate in a different world-economy to that constituted by the Americans and consequently a different filiere developed. The British filiere was largely the outcome of British predispositions interacting with market features, so it might be contended that the British already possessed the capacities for coping in that world-economy. However, staying with a declining world-economy meant the growing loss of economic opportunities over the long term arising from the failure to make a significant penetration of the wealthy mass markets of North America and the more variegated, yet also well-off markets of Europe. Moreover, the British failed to acquire the learning experience relevant to competition through large-scale distribution and large-scale manufacture based on innovation-design.

There was some degree of awareness amongst the British of what foreign competitors were doing, though the analysis of the decline mainly consisted of inaccurate constructions of cause-effect maps coupled with naive injunctions for adaptation. Amongst foreign competitors the position of the Germans and the Americans was the most salient to the British. German templates of organizing had their British admirers from the 1870s onward when German methods of technical education were being promoted in Britain. Later, in the 1920s the creation of the I.G. Farben from smaller German chemical firms was an important additional stimulus to the creation of ICI and was taken as evidence of the German approach to the rationalisation of industries. A similar German influence, albeit with a slightly different template, was noted in the role of AEG in the German electrical industry and in its relationship with major American multinationals like General Electric. From the Germans the British acquired concepts of arranging market conditions rather than of reconfiguring the internal authority and bureaucracy of the firm. In the twentieth century the British continuously attributed an image of efficiency to German methods of manufacture, yet relatively slight attempts were made to undertake the detailed analyses which would have been necessary to facilitate any significant degree of imitation, let alone of appropriation[1].

American influences in Britain were continuous from 1890s onward and their presence went through successions of peaks and troughs. British firms did track American developments and create *fragile and particularistic channels* of learning about products, processes and work organization. For example, Lucas, the large electromechanical firm supplying the auto industry regularly sent engineers to learn from American approaches to production management (Nockolds 1978). Likewise, in the retail sector where the histories of Boots and of Marks and Spencer show that American practices were observed and skilfully integrated with indigenous forms (Tse 1985; Chapman 1974). American supermarket firms gained a clear lead in the 1940s and 1950s through technology, new raw materials, and organization, but were unable to exploit that lead sufficiently in the British context because the supply side of British-owned food and drink firms were anxious to retain their own independence as long as possible. That delay was heightened by the unwillingness of the British consumer to tradeoff lower prices against the loss of the sales assistance. Each factor provided the later British entrants like Cohen of Tesco with sufficient time to learn the new forms of operating. This was likewise the case for the firm of John Sainsbury, who now hold the largest share of the British supermarket sales and are recognised to have the highest throughput per square metre in the UK. It may be argued that Sainsbury's have learned a great deal from the careful scrutiny of American firms, particularly in the batching of goods, the development of profitable lines and the control over the supply side. During the past decade they have used American training methods to anticipate the requirements of the new electronic point-of-sale equipment. Other British firms have shown a strong awareness and interest in American methods. Mention has already been made of ICI who during the 1960s were reputed to be the major employers in Europe of American behavioural consultants and of Pilkington's, the glassmakers, who developed close links with American higher education. In contrast, much less interest was shown by the professional services sector of health and education, where leading employers defined British practices as superior and

---

1 In the British academic context there is a good deal of ad hoc comparative work, but relatively little systematic comparison directed towards policy objectives. Thus the establishment of 'American Studies' in the 1960s was really little more than the juxtaposing of various elements. Likewise the periodic bursts of interest in German and American practices is only slightly underpinned by a filiere linking those comparisons to policy-making. The character of those studies is reflected in Prais, S.J. and K Wagner (1985) Schooling standards in England and Germany: some comparisons bearing on economic performance, *National Institute of Economics Review*, pp. 53–76. The current position has not so far been altered by the rationalisation of higher education which began in the 1980s

therefore as not requiring the stimulus of detailed comparison. It is noticeable that there were few systematic instances of comparison from these areas to place alongside the 1854 report on the American system of manufacture. There was no 'American System of Services'.

There is little indication in the twentieth century that American firms still felt the need to learn from the British and that tendency became stronger throughout the twentieth century. For many Americans, Britain was most like the macho image of "a living museum of the past" as presented by the advertising of the English Tourist Board[2]: Beefeaters, the Tower of London, Horse Guards Parade, Warwick Castle, Shakespeare and the warlike clans of the Scottish glens.

The American presence in Britain was heightened in various ways. First, American firms often chose Britain as a useful base for their European operations, particularly in the 1920s and 1930s before the effects of the European Common Market began to suggest that the borders of France and Germany were becoming the new heartland of European innovation. Those multinational American firms imported, inserted and utilised American practices of administration and succeeded in applying their Americanised standard operating procedures, particularly when these were clearly formalized and were handled at the top by American-trained executives. Second, the British media often utilised *their conception* of American templates as examples of international best practice and as viable alternatives to British practices. For example, in the 1960s the movement in Britain for the removal of piecework systems constantly claimed that leading American firms had already removed piecework. Third, from the mid-1950s onward there was a growing interest in American approaches to management and to administration. Some of the early attempts to develop management education relied on the Harvard case study format only to discover that British managers were rather unwilling to submit themselves to its directorial approach. Also in the early 1960s there was a small burst of growth in management education and American templates were part of the initial set of ingredients, but were modified and rejected in the British context until the mid-1980s when the pressures of the market-economy led to renewed interest in the American approach to large-batch education. That area provides a useful probe into the extent of British self-development and appropriation. Fourth, some American practices were channelled into the British context through specific American agencies; examples of these would include the activities of the consultants McKinsey, who worked with many major British enterprises in

---

[2] The English Tourist Board have been very skilful in selling the British heritage to North Americans; their approach provided a subtle web of linkages to certain groups of professional historians, most obviously those who engaged themselves in industrial archaeology

the late 1960s and through the 1970s. Later, during the massive and sudden growth of management accounting after the mid-1970s came firms like Arthur Anderson. Fifth, a considerable amount of American technology has been brought into Britain throughout the twentieth century. Early examples included shoe machinery, then engineering equipment and transfer lines. Commercial office equipment tended to have a strong American influence (e.g., NCR) and there was a similar influence in new technology for banking. Also computing equipment tended to be American (e.g., IBM) and so were important segments of the new developments in medical technology. In all those areas the Americans were able to export high value added equipment into Britain.

By the mid-1960s the distinctiveness of the British world-economy was collapsing and there was increasing standardization of British tastes (e.g., in beers). These developments shifted the bases of competition towards the economies of scale and so to longer production runs. Also, in the mid-1960s the underlying processes of rationalisation arising from within the European Common Market began to have its impact on British-based consumer goods, especially electricals (e.g., washing machines, fridges) and in the new areas of television sets, where the Dutch-owned Philips acquired a strong market share. A study of job loss across 30 sectors of manufacturing between 1969–1973 shows that the slimming of manpower commenced then and was well underway before the oil cuts of 1974 (Massey/Meegan 1982). Similarly, the shift of innovative activity to the southeast of England and into the European heartland was already becoming evident to industrial geographers.

In the early 1970s the Europeans, particularly the Germans and the Swedes, became the frequently cited templates for British firms. Interest in Swedish experiments in work organization was strongly stimulated by the publicity attending the Volvo plant at Kalmar[3]. However, that interest hardly affected the design of British plants. Instead it was the American reaction to Japanese competition which surfaced at the end of the 1980s and came into the debating arena. In practice the British comprehension of the Japanese approach was strongly filtered through a narrow set of gatekeepers to an important segment of which were American-owned companies like Ford.

In this chapter I examine the extent to which a small sample of innovations may be utilised to reconstruct a strong sketch of the British capacity to im-

[3] During the mid-1970s the British media highlighted the approach to work organization of the Swedish, and the Economic and Social Research Council attempted to provide a channel from that experience into the academic and industrial community. However, there was considerable scepticism from the trade unions and it was far from obvious that the leadership of British enterprises regarded such channels with any degree of seriousness. Since then the Japanese dimension has been more salient, though as yet there is not an organized approach

itate, modify and/or appropriate relevant foreign innovations. The following examples are taken:

1. *Administrative innovations:* work study, Organization Development, the multidivisional form (MDF), plantwide incentive schemes
2. *Technological innovations:* the automobile assembly line, micro-electronics in the service sector, computer-aided production management (CAPM)
3. *Management education:* the MBA

Overall, it will become apparent that each of these was made available in the British context, but that their re-invention was accompanied by the loss of certain of the fundamental core features of the original innovations. What survived sometimes contained a high degree of cosmetic adaptation.

12.2 Administrative Innovations

12.2.1 Taylorism: Work Study

To what extent did the British appropriate Taylorism and the related areas of corporate engineering developed in the USA, or, develop alternative means?

The British approach to the analysis of production techniques and knowledge may be reconstructed by sampling the situation in each decade after 1930. The essential background was given in the previous chapter, particularly the tendency to use quasimarket mechanisms like piece rates. These were calculated from experience rather than by systematic observation[4].

First, the position in 1930 can be ascertained from the archives of the Labour Section of Management Research Group No. 1 based in London for the period 1929-1934[5]. The archives show that the Group consisted of senior members drawn from the leading firms of the period who met to discuss matters of mutual interest, mainly to do with the institutions of work (e.g., payment practices) and to enquire about the content of American methods. The minutes indicate that there was quite a basic problem of interpreting what the Americans actually did. At the meeting in February 1931 there was puzzlement over the methods and role of the foreman in America, which was followed by a talk from a visiting French speaker, Monsieur M. Dubreuill on Taylor's approach[6]. Dubreuill observed that:

---

4 In Britain the American ideas of systematic management were welcomed by some such as E. Cadbury of the famous chocolate firm, but there was considerable opposition, most notably from British psychologists

5 The Minutes of the Management Research Group No. 1 in the pre-1940 period were edited by its secretary and later located at the Business History Unit, London School of Economics where they are known as the Ward Papers

6 Minutes of Management Research Group No. 1. Meeting on 20 March, 1930, see pp. 26-28

> I think the essential part of "Taylorism" is the substitution of measurement for guessing.

He went on in the report to say how he was struck by the practical approach of Americans to the re-organization of production by the use of *statistics* and of *experiments* in shopfloor layout and the use of materials. Despite the efforts of the visitor to explain about method study his audience was primarily interested in the system of payment. In the same period of time the attempt in Britain to replace the guessing of work content by systematic measurement along the lines outlined by Taylorism was not initiated by the British, but by an American, Bedeaux, who devised the Bedeaux unit of work. A key attribute of Bedeaux's approach was that the consultant made two calculations: one measured the amount of activity in time which the operative 'should' be allowed for completing a stated amount of work and the other calculation added onto that figure an additional percentage to allow for rest periods and personal relaxation in the workplace. It was that latter attribute which increased the acceptability of the Bedeaux approach in the otherwise hostile British context. The Bedeaux consultants were active in Britain from the late 1920s into the 1930s during which period it is estimated that they undertook applications for more than 250 clients, many of these were leading firms in engineering, metal work and textiles (Littler 1982). However, the striking features of their presence was that the stress was on measured work as much as on the overall coordinative efficiency of the plant, and that there were few organized attempts to appropriate the innovation. Instead, the British empirical approach to the setting of piece rates through the reliance on the promotion of experienced former operatives continued to dominate.

Second, in the 1940s an initial attempt to create a British professional association for the study of work organization and its practices was held in a London hotel as late as 1941[7]. From then on, and for two decades, several locally based associations were established, such as in Manchester and in Birmingham. Also groupings emerged around British entrepreneurs who had been trained in American methods, like Anne Shaw, and who had returned to establish consultancies in Britain. The development of a national all-embracing institution is depicted in Fig. 12-1, which shows how many small societies were gradually folded into a single body by the early 1960s. The pattern was strongly influenced by events within ICI, whose corporate leadership gave the impetus to a new concept, work study.

Third, the major impetus to the use of the term work study had come from within the leading chemical firm, Imperial Chemical Industries, where, from

[7] The examination of work study was greatly assisted by Barbara Goreham who was the librarian and information officer for the Institute of Management Services, Enfield, Middlesex, England; the Institute kindly provided access to its archives

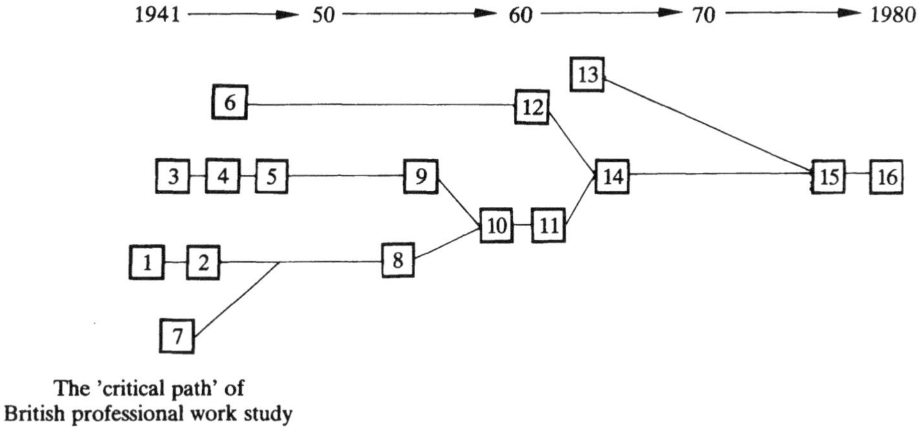

As shown here, there have been over the years no fewer than 16 bodies for professional practitioners in the field, including the Institute of Estimators, Planning and Time Study Engineers — formed at a meeting at the Waldorf Hotel, London on 19 October 1941.

1. Institute of Estimators, Planning and Time Study Engineers
2. Institute of Economic Engineering
3. Shaw Society
4. Motion Study Society
5. Motion Study Society of Great Britain
6. Institute of Industrial Technicians
7. Production and Control Research Group
8. Society of Industrial Engineers
9. Work Study Society
10. Work Study Society
11. Institute of Work Study
12. Institute of Incorporated Work Study Technologies
13. Organisation and Methods Society
14. Institute of Work Study Practitioners
15. Institute of Practitioners in Work Study, Organisation and Methods
16. Institute of Management Services

Figure 12-1. Evolvement of work-study profession in Britain, 1941-1980

the early 1950s, productivity experts had gained strong support amongst the postwar strata of rising management who themselves had witnessed the benefits of systematic approaches to problem-solving and to training within the military. There was also a small infusion of graduate trained men. Within ICI, a small, though rapidly growing group, sought to improve and develop a production knowledge by devising a systematic approach with two prongs, methods study and time study. A core feature of the overall approach became known as critical analysis. The ICI approach was based on a deep scepticism about the rule of thumb approach associated with shopfloor engineering in Britain, whereby a great deal of attention was given to the making of jigs and fixtures. In ICI it was reasoned that all existing practices should be recorded and then submitted to a critical examination to ascertain whether the whole of an existing method could be removed by an alternative, specially designed process. It is relevant to note that such a relatively abstract concept was developed in the large-scale, science-based, chemical industry. Within ICI, work study officers, often new graduates on a generally upward career track, were taught the ICI approach and by 1960 there was one officer for every 100 shopfloor employees. Not surprisingly, ICI personnel came to have a sharp, albeit passing influence, on the development of work study between 1955 and 1975. During those two decades ICI executives reconfigured the existing fragmented associations into a single, composite institute and also came to play a major role in the creation of international groupings for the rest of the world, but excluding North America. In the 1960s the ICI approach to work study was appropriated in the polytechnic sector of higher education from where ten-week full-time courses were used to train cadres of work study personnel.

These three samples indicate very sharply how far British practices lagged behind those of the Americans. Moreover, work study never succeeded in gaining a secure foothold amongst graduates and in the universities, particularly in the engineering schools[8]. Work study did gain control over a roster of shopfloor productivity problems in the 1950s and 1960s, but by the 1970s, when payment systems were increasingly criticised and gradually replaced, a major source of the workload disappeared and attempts to survive by taking on board the new information technologies had been only partially successful. From 1981-1985 the major professional association lost almost one-third of its membership[9].

12.2.2 Organization Development (OD)

OD is defined as a nonstructural approach to organizing in which the identities, conflicts, emotions and attachments of organizational members are the prime targets for systematic change. OD has largely been applied to intramanagement relationships and to white-collar workers rather than to the vertical relationships between capital and labour. Overall the approach of OD is systemic and experiential rather than systemic and cognitive and so little direct

---

[8] The archives show that during the late 1950s and early 1960s work study was only a minor segment in the courses of the Colleges of Advanced Technology, but that even this was reduced when they were upgraded to technological universities in the mid-1960s. The licensing of practitioners in work study was done through the polytechnics and the technical colleges. The Lucas Company in Birmingham were very active in the development of production engineering concepts during the 1950s, but their attempt to fund a bridge into higher education at the prestigious Birmingham University did not have the outcomes they anticipated; see Nockold (1978) – also that aspect is being explored more fully in current ESRC, WORC studies

[9] From the archives of the Institute of Management Services, from interviews and confirmed in the Annual Reports

attention is given to the knowledge bases in organizations or to the content of decisions[10]. However, it should be noted that most major users of OD in North America are actively aware of the distinction and do not expect OD to cover the cognitive aspects because they are the subject matter of other established approaches. Although many of the roots of OD surfaced during the 1930s in Europe (e.g., group psychotherapy) it is the USA which has been the seedbed[11]. In the 1940s American social psychology began to loosen its subordinate position in relation to corporate engineering and to develop its own distinctive claims for relevance and for establishment as a separate internal function within corporations.

Social psychology occupies a special role in the construction and reproduction of American rituals and forms of decision-making. Thus, core concepts like role are not merely descriptive categories of 'what is', but are prescriptive concepts which embody notions of conformity and of 'channelled' agency[12]. The early development of social psychology focused on areas like leadership, two-step influence models and the management of change. The content was taught through classroom methods. Later, attention was directed towards the greater impact of methods of learning which involved the ego of the subject in an active dialogue over the subject matter. Two major lines of involvement soon emerged. One line was in the direct experience of the struggles, traumas and excitements of interpersonal learning through T groups and similar learning situations. The other line made heavy use of individual and collective survey feedback as a trigger to the reflective altering of behaviour. Each of these modes of learning developed in an autonomous, parallel way. The T-group approach became organized through various new institutions like the National Training Laboratory and its method of training and licensing practitioners. Their search for clients, particularly in California, resulted in the evolvement of a whole array of small-group techniques. The survey feedback approach developed widely, yet was given direction and legitimacy through its connection with prestigious practitioners at locations such as the Institute for Social Research on the Ann Arbor campus of the University of Michigan.

American clients of OD, such as General Electric and the military have been active in 'putting social science to work'. In the 1950s a number of influential

---

[10] In Organization Development there has been a strong emphasis upon the experience of organizational relationships, particularly in handling horizontal relations of hierarchised authority between different types of employee. Much less attention has been given to the *content* of managerial decision making; see Clark, P. A. (1972b, Chap. 6) for the distinction between systemic cognitive and systemic experiential

[11] Most standard textbooks on Organization Development now include a brief history of its development

[12] The notion of antagonistic cooperation is highly relevant; see Riesman, Glazer, and Denney (1953)

clients entered the marketplace: food, drink and tobacco firms; petrochemical firms; and the public services. Soon OD acquired an occupational base within firms as well as in the universities. However, that potential only became fully operationalised through a series of experiments and considerable missionary zeal. One particular experiment illustrates the point. In the mid-1950s the senior executives of the Standard Oil Company defined certain new problems of the coordination of manpower in their technologically sophisticated refining plants. These plants were increasingly integrated and were operating continuously. Amongst the problems which had arisen were those of the spatial deployment of operatives and the increasing extent to which problems required action, often anticipatory action, from the cooperative actions of operatives and maintenance crews. The Company invited a mixed bunch of OD practitioners to advise them, including representatives of both the feedback and experiential approaches. From out of that initiative emerged the realisation of the potential for OD and a specific innovation which later became known as Grid Organization Development (GOD). The GOD approach soon became a carefully designed multistage program spread over several years. It possessed most of the five attributes identified by Rogers as desirable in an innovation to promote its utilisation (Rogers 1983). GOD became an exemplar and template for other much more modest packages. As a template, its combination of structured self-learning exercises, and the use of internal tutors, demonstrated the opportunity for the joint relationship between the supplier and the user whilst also providing an economical training method which ensured internal commitment. By the mid-1960s OD approaches were being designed to incorporate a high degree of consumer acceptability.

OD was sold with a missionary zeal during the 1960s and its prestigious early texts proclaimed the alleged weaknesses of rival methods of change. During that period OD was clothed in the optimism of the 1960s and in some of its social ideal, though these became more muted in the 1970s. OD did not become ossified into a single package as did Taylorism and ironically that attribute became a source of criticism from certain academics, including the frequently cited assessment of Robert Kahn[13]. In the universities OD was taught to students, many of whom already possessed some basic discipline in psychology or in micro-OB. Consequently, the initiates were well able to learn the basic practices of OD, and its techniques were conveyed into the users where OD coexisted with the already institutionalised approaches to the design of work organization by corporate engineers. Thus, OD was used to increase the acceptability of just those forms of work organization which

[13] In a frequently cited observation, Robert Kahn of ISR, Ann Arbor suggested that Organization Development was an ad hoc collection of practices which were unrelated to the theories which were being expounded

its early ideology most often ridiculed. In the American context it was taken for granted that OD was about the implementation of corporate designs.

European OD also had its roots in prewar developments of group psychotherapy and these became the focus for attention during the 1940s and 1950s when members of the Tavistock Institute of Human Relations in London applied group experiences to the rehabilitation of returning prisoners of war with severe problems of adjustment[14]. A clinical theory was developed in which it was asserted that members of groups typically engaged in pathological behaviours unless they were able to develop and maintain group mechanisms for rational problem-solving. After 1948 the clinical approach was used to examine industrial work groups, most famously in the newly nationalised coal industry where the clinicians discovered similar pathologies underlying those reported in the clinic. In a skilful development of the original theory, Trist and colleagues asserted that work groups needed to develop the internal mechanisms for self-control of their own task boundaries and what occurred within the boundaries (Trist 1963). In one of the best-known studies using the sociotechnical approach, Trist and colleagues demonstrated that self-control amongst coalminers was impeded by the underlying technical system which had been designed according to the principles associated with Taylorism. In a classic comparison, itself based on the accidental survival of a form of self-organized work group previously known as the butty system, the researchers demonstrated that group autonomy was both more wholesome and more effective. The rediscovered form of work organization was later labelled 'responsible autonomy' and the method of its designing became known as sociotechnical analysis.

It will be recognised that the approach of responsible autonomy actually continued a British tradition of delegated control in a new form by transferring the former control of internal contractors to the group as a whole. In that respect the autonomous work group highlighted responsible autonomy as a modern solution to just the problems of teamwork which were being experienced in American refineries whilst producing a solution to the problem which was largely inimical to American concepts of planning undertaken by corporate technocrats. So what happened to the autonomous work group? For a variety of reasons the sociotechnical approach and the autonomous work group did not gain a significant constituency in Britain. Moreover, the detailed understanding of the design problems for introduction of such groups were not taken seriously and the supply line of practitioners never emerged. Further, in the coal-mining industry where the story began it was the corporate engineers who guided the evolvement of a production technology which progressively facilitated work group autonomy in the 1960s, then

---

[14] Tavistock Institute of Human Relations, Belsize Park, London

reversed that direction in the 1970s by subordinating the coal-face groups to central control.

In practice the influence of the sociotechnical approach in Europe was diffused to admirers in the Netherlands and in Norway, where new institutes sprang up and became centres of application and diffusion. The filiere in both those countries was much more favourable than in Britain. Overall the Netherlands became an active centre for the application of OD in companies. Moreover, the greater length of training of the Dutch graduates coupled to their general orientation to expert applications provided a context for experimentation in the public sector in the Netherlands (e.g., PTT) and in the private sector with giant electrical firms like Philips (see Chap. 11, note 18). Also, a model of organization evolvement based on OD was pioneered from the Netherlands Pedagogische Institute (Lievegoede 1980). In Norway there was a joint application by unions and management of the sociotechnical approaches[15], and OD was developed in areas as diverse as education and the paper industry. Other innovations occurred in France and Denmark (e.g., Hjelholt 1963). Despite these indigenous developments, European corporations tended to look to the USA, and even the leading oil companies employed American consultants to introduce the basic steps in OD applications. Paradoxically, it was Esso in Britain who pioneered domestic developments with the skill-training package of Coverdale and the employment of full-time social science consultants. The more typical pattern occurred where a British firm employed American consultants and/or sent their senior executives to leading American business centres. Thus, in the 1960s ICI imported a great many Americans. However, OD did not become as institutionalised into the power structures of ICI as had work study (Pettigrew 1985). Overall OD in Europe has not so far gained the fragile ascendancy which it enjoys in North America.

In Britain, few sought to appropriate American imports to the British context. Instead American ideas such as job enrichment were inserted into the British settings like ICI (Paul/Robertson 1970). There were two important and interesting exceptions to that pattern which may be mentioned, though neither became the forerunner of a wider movement. First, in the nationalised health service, Revans devised a form of structured-action learning directly based on a modification of American practices of survey feedback and applied these to the less measurement-oriented approach of the senior executives in the hospital and to the medical practioners (Revans 1972). The initial evaluation by Wieland indicated that the ambitious program had largely failed to achieve its objectives, yet a later report from the same evaluation team using fuller data claimed that more than five million sterling (1980 val-

[15] A neat summary of the Norwegian work is provided in Eldon (1978)

ues) had been saved[16]. By then, the program had been forgotten by all except those most centrally involved. Second, a program to promote the general utilisation of social sciences attempted to combine the organization design of the sociotechnical school with the use of OD in novel ways to design new factories[17]. Neither of these attempts survived into the 1980s. The only really substantial appropriation of American concepts occurred with payment systems, and then only with great difficulty. In Britain the American concepts of OD have created as much suspicion as consideration[18]. In the 1960s and early 1970s a number of companies went through the early stages of GOD and some completed the process. Also as indicated, certain leading firms imported American OD consultants, yet what is striking is the low application and the failure to give serious consideration to OD by organizations engaged in major change programs[19].

In contrast to the fragmentary approach to OD of the British and Europeans, certain American initiatives sought to uncover those features most relevant to the development of best practice administration. The most notable initiative was launched by L. E. Davis with Ford Foundation support from America and the support of A. B. Cherns in Britain[20]. The approach, which became known as the Quality of Working Life (QWL) program was based on the idea of pooling knowledge about the design of work organization. The core approach was set by the principal missionary agent, L. E. Davis, as a conscious evolvement of the sociotechnical approach which he had appropriated into the American context by the late 1960s and had synthesised with indigenous approaches to production engineering. Through his concern and his direct experience of working in Britain the sponsors of QWL were drawn into a systematic review of British and European achievements, especially the redesigns of shopfloor work organization of Volvo and the more widespread sociotechnical examples. His focus on the European autonomous work groups (AWG) was initially met with scepticism, yet at the First QWL Conference in 1972 it was the European speakers who attracted

---

[16] The original evaluation study was led by George Wieland and their first reports indicated relatively slight impacts. However, in a report produced almost one decade later it was concluded that savings of several millions of pounds could be attributed to the intervention by Revans. The earlier report was certainly thorough, if restrictive, but the later report totally ignores the possibility that those hospitals most open to change also elected to accept Revans's approach

[17] Albert Cherns directed the five-year program of research on the uses of the social sciences at the Centre for Utilisation of Social Science Research, Loughborough University of Technology, ESRC Grant, HR 86 to Cherns, A. B. and P. A. Clark

[18] In Britain there are widespread suspicions about OD amongst managers and trade union officials. These suspicions are most salient in their humour!

[19] See the forthcoming reports from ESRC, WORC

[20] The QWL label was almost totally unknown before Louis Davis's initiative with Albert Cherns and the support of the Ford Foundation; see Davis, and Cherns (eds) (1975)

most attention and it was American firms like Proctor and Gamble and General Motors whose corporate engineers had discovered that teams were better at handling production problems which occur stochastically (Davis 1971). Their claims were confirmed by the management science theory of teamwork. Hence the interest in the AWG. However, American users, particularly corporate users, were not interested in the devolvement of control to create self-control. Instead their notion of teamwork was more like that of American football. So, like the British employers, though for quite contrary reasons, the AWG as such was largely rejected and its place was taken by teamwork set within an authoritative framework. To be more specific, groups were used and their experiences were shaped by experts from the technostructure who encouraged the members to use their discretion within broadly defined parameters, the so-called tight-loose approach[21]. In the USA the evolvement of OD continued to be eclectic in a way which was remarkably flexible, so even when the Japanese challenge came in the 1980s that flexibility was retained.

Europe generally and Britain in particular lacked the infrastructure for OD and so that approach to adaptation became less salient. OD was treated with suspicion in higher education, and there are few professors of OD in Britain, France and Germany. Many academics in related fields would not wish that situation to change. Moreover, the reluctance of academics is reinforced by the general attitude of managers and of organised labour.

12.2.3 The Multidivisional Form (MDF)

The position of the multidivisional form (MDF) of administration in Britain is the most frequently cited example to support the claim that there has been an extensive diffusion of American practices to Britain and of their imitation by British firms[22]. The claims are made by Chandler and reinforced in studies undertaken in Britain[23], but what are those claims based on and is the MDF in Britain similar to that found in the USA?

In a careful and insightful piece of historical reconstruction coupled to an imaginative analysis, Chandler concluded that by the 1920s several significant American enterprises had encountered top-level organizational problems arising from their policies of corporate diversification. These firms also

---

21 Peters, and Waterman (1982), see the first 100 pages and the use made of the ideas of Weick

22 Clark, P.A. and J Tann (1981) Technological innovation and the division of labour in the 80s: prising open the British case, Conference paper, European Group on Organization Studies

23 Chandler (1962); Chandler (1977); Chandler, and Daems (1980); Channon (1973); Channon (1978)

discovered that the best administrative solution to diversification was the MDF. Chandler suggested that:

> the M-form designation is to be reserved for those firms that combine the appropriate *structural control* and *internal operating attributes* ... (emphasis added) (Chandler 1977: 1)

and he sought to show that once discovered and recognised as a generic solution to a common problem the MDF was then rapidly and selectively diffused to other firms who had diversified. The claims about the position in the USA of the MDF are so widely accepted that debate has narrowed to the explanation. Fligstein and DiMaggio suggest that corporate professionals were the major agency for the diffusion (Fligstein 1985; DiMaggio/Powell 1983).

Chandler claimed that the sequence of evolvement in the USA began with the development of hierarchies in the nineteenth century, by which time the visible hand of American management was challenging the market mechanism over the effective allocation of resources. Further, Chandler sought to demonstrate that the MDF was diffused from the USA to the rest of the industrial world where it was adopted through a process of imitation aided by the channelling activities of consultants. Those claims are best documented for the British case.

There are four principal ways in which the MDF came to be *known about* in Britain. First US firms already using that administrative format continued to do so with respect to their establishments in the UK. By 1966 there were over 1600 US-firms in Britain, and Dunning reported that they accounted for one-tenth of all the goods produced[24]. American firms using the MDF tended to do so through the application of strong internal controls and through the establishment of bargaining at the divisional level rather than at the establishment level. Second, British senior managements did visit the USA and learn about the MDF, though there were very few detailed accounts of these visits. Third, it is more likely that those British firms which were interested decided to employ US consultants, particularly McKinsey, who, during the 1960s were advisers to a wide array of British enterprises: the BBC, ICI, Plessey, Cadbury's, Unigate and Unilever. Fourth, American templates of administration were presented in the leading business schools by academics returning from the USA.

Knowing about the MDF is different from actually adopting its detailed processes of control and coordination. For example, ICI which is frequently cited as an example of a divisionalised firm was largely a functional firm until the 1970s. Personnel and Management Services, for example, were highly

[24] Dunning, JD, cited in Hannah (1983)

centralized at the Millbank headquarters in London. Moreover, in 1950 only four of the MDF-firms in Britain were actually British and during the 1960s the most prevalent forms were the holding company and the functional form of administration (Hannah 1974; 1976; 1983; 1980). The major increase in firms actually tending towards MDF occurred between 1964/1966 and 1970/1972 when the proportion doubled (Hannah 1983). However, the introduction of MDF seems to have arisen from mergers and from a response to declining profits.

Channon has provided one of the most interesting studies of the adoption of MDF in the service sectors and industry (Channon 1973). His analysis was based on a survey which attempted to get at the structural data by using company reports and similar records. However, Clark and Tann conclude that the evidence suggests an only surface similarity between the American format and its British counterpart and they conclude that the fundamental processes were absent (Clark/Tann 1986). Further, they point to the developmental sequence in Britain whereby the holding company preceded the MDF and the latter was often introduced through addition of an extra layer of management. Their conclusion indicates that imitation was difficult and they note that few British executives would be as familiar as their American counterparts with the theoretical debates over centralization/decentralization which attended the American adoptions.

12.2.4 Plantwide Productivity Schemes

American-designed plantwide productivity schemes such as the Rucker Plan and the Scanlon Plan also excited interest in Britain, even amongst trade union officials[26]. They were adopted relatively extensively, particularly during the 1960s and during periods when the government imposed a pay freeze. However, the problems of transatlantic transfer were considerable. The point is illustrated through reference to one agency for their adoption and through reference to an implementation in a specialist firm based in London.

One of the channels through which these administrative systems were introduced into Britain during the 1960s period was the consultancy directed by the late F. R. Bentley[27]. He had been trained and educated in American accountancy practices and was keenly aware of the differences at that time in

[25] Hannah (1974); Hannah (ed) (1976); Hannah (1983); Hannah (1980)

[26] There is evidence to show that some trade union officials were influenced by the origins of plantwide productivity schemes with the former American union official, Joe Scanlon

[27] In interview by the author

the financial data typically held by similar American and British firms. Generally the Americans had more developed data-based systems and were more concerned to routinely record stock figures on a periodic basis. That data were essential as a basis for calculating the monthly costs and for deriving the bonus. Bentley therefore devised a modification of the two American plans to create an admixture from each and in so doing made his scheme, known as the Share of Productivity Plan (SHOPP), highly flexible so that he could advise users on which combination of features would best fit their context (Bentley 1964). Bentley also used the requirements of the plantwide productivity schemes to encourage users to adopt more extensive and detailed monitoring procedures and to involve their employees in the interpretation of the basis for success and failure of the firm.

Unfortunately, there are few detailed evaluation studies of these plans, but one, undertaken in a typically British firm based in London had an interesting life course. Its introduction owed a great deal to the local trade union combine[28]. The tenets of the scheme were explained to the workforce of around 3000 by the consultant and it was evident that employees felt that they understood what might happen. However, the implementation of the scheme stirred up unforeseen antagonisms between blue-collar and white-collar workers. In addition, the monthly bonuses were directly affected by the unusual trading pattern of the company. That might have been an opportunity for management to explain the position of the company to its employees, but that did not occur. Although the scheme lasted four years it was eventually withdrawn. In many respects the problems which arose were similar to those reported in American studies.

It might be expected that British managements would have been highly predisposed to such schemes, yet it should be noted that the British approach was to control through segmented subunits. However, the plantwide scheme actually focused attention on the collective performance of management and of their capacity to take a synoptic view.

12.3 Technological Innovations

12.3.1 Automobile Assembly Lines in the 1920s

In this section the focus is on examples of technological innovations which British-based enterprises sought to acquire for their own usage. Therefore, the situation of a technology transfer through licensing from the USA is ex-

[28] At the Morgan Crucible plants in Battersea, London; reported in Clark, P.A. (1972b, Chap. 9, Sect. 2)

cluded from consideration. The focus is on nonlicensed technologies because these provide the best opportunity to examine British capacities for innovation design. It is useful to take two very different examples of twentieth century technologies at two different periods of the century. The first example will be of the transatlantic transfer of American principles of car assembly into the British industry between 1918 and the late 1920s during which period the British industry was beginning to experience mass distribution and mass production. The second example, is of the transfer of information technology for use in libraries and in the health service, specifically in hospitals. In each mini case the objective is to examine the overall position with respect to British appropriation of these technologies rather than to compare the leading edge of each country.

To what extent was the American approach to automobile production introduced into the British context between 1918 and 1930 and to what extent was that system rejected or appropriated[29]? It is important to start by setting the question in the context of the themes of the book and to observe that the differences in the context of the industry to that prevailing in the USA is relevant. For instance, in France the very successful Renault firm were extremely keen observers of American car production techniques and knowledge, yet their objectives from 1910 to 1938 were to be very selective in their appropriation[30]. Moreover, the relevance of their judgement was reinforced by the case of Citroen who introduced mechanically paced assembly line practices and technology, but found that when the mass market collapsed in the Depression that they were unable to cover the costs of their investment (Boyer 1986).

When referring to the American approach to the production of automobiles it is important to balance the saliency given to the mechanically assembled line introduced by Ford in 1914-1915 with the methods used in General Motors. The Ford assembly for the Model T was exceptional. No other production system since the Model T has had so much volume for a single automobile. Indeed, quite the contrary, the final assembly line actually became the location where seasonal and other variations in the cosmetic styling of cars were introduced. By referring back to the Abernathy analysis and Fig. 4-8 it can be demonstrated that the assembly line was less far down the fluidity-

---

[29] I am very much indebted to the investigations by and discussions with Wayne Lewchuk (McMaster University, Hamilton, Canada). The full thrust of his analysis is contained in Lewchuk (1987)

[30] A summary can be found in Clark and Windebank (1985); see also Coriat (1979); Fridenson (1972); Midler (1980); Monjardet (1974); Poitou (1984). I am indebted to Christophe Midler, Robert Linhart and Patrick Fridenson for their guidance on the French industry and its relations to American developments; clearly there are sharp differences from the position in Britain

specificity dimension than were the engine plant and other production units producing components and subassemblies. Consequently, the final assembly line in American plants was the target for the admixture of a dedicated moving track for carrying heavy weights at a defined pace in combination with relatively flexible power tools utilised to position the components. Moreover, the workforce were continuously redeployed to cope with the varying work content in the different models which proceeded down the line.

In Britain, as already indicated in Chap. 11, the structure and size of the British car market were different from the USA. In the period of the early 1920s two firms, Austin and Morris entered the market for small, cheap cars and relative to their competitors they were highly successful in obtaining market shares.

In looking at the British situation in the 1920s it is important to examine the position of the Engineering Employers Federation (EEF) to which the automobile firms belonged with respect to the principal strategies of managerial control. The EEF promoted the use of management control through the use of piece-rate systems and any firm which removed piece rates would have had to resign from its membership. In that respect an important element of Ford's approach to man-management was precluded, though it should be noted that other American firms continued to use piecework and so did leading European firms like Renault and Peugeot.

The British interest in American production techniques was centred amongst production engineers rather than the boards of British firms. First, at Austin in Longbridge, Birmingham, the senior executives had visited America and the Ford plants where they had been most impressed by the 'spirit of labour' and so on their return they decided to give the heaviest attention to increasing time consciousness amongst employees. They concluded that the assembly line was too inflexible for the requirements and exigencies of the British market because of the frequent changes in style and because the selling season was so uncertain and heavily bunched into a short period. At Austin they made additional introductions of technology after 1924 with elementary conveyors and through a small number of single-purpose machines. On the whole, technological solutions were not so central.

Second, Morris's senior management were also interested in American methods and their leading engineers at the Coventry plant argued in favour of the introduction of the machine pacing of work. By 1923 they had travelled the farthest of any British firm in the direction of flow-line principles; even so their factory was still organized around self-contained departments which produced specialized components in smallish batches. Third, the position at Rolls Royce was rather exceptional because their cars were at the most expensive end of the market. However, as their Derby factories were changing back from wartime armaments to peacetime production there was considerable debate at the most senior levels about the potential for the future. In fact, two conflicting policies emerged. One group maintained that the company should move into mass production and a small group of leading engineers wished to centralize planning, establish control departments, introduce special purpose machines and install assembly lines. However, their vision of a mass-production factory was much more segmented than was the vision of Henry Ford and his advisers; this group were the losers. Their opponents contended that the strengths of the firm lay in their skilled employees. Moreover, these men reasoned the introduction of American practices would stimulate socialism and probably lead to a syndicalist takeover of the company. It was their reasoning which soon triumphed. Consequently, existing forms of work organization were retained.

Ford in Britain was obviously a major channel for the introduction of American concepts of flow production. They were joined by General Motors in 1925 who took over the ailing Vauxhall firm. In fact Vauxhall had, from 1918 onward, been highly interested in introducing flow-line production so that left the EEF to manage without piece rates. Vauxhall purchased a quantity of American machines, yet continued to man them with the existing skilled workers. Then, in 1921 the market collapsed and Vauxhall were more heavily affected than other firms who had invested much less in equipment. The factory was closed for six weeks and method study was applied to create a new layout designed as a high-volume, repetitive plant. Even so the layout stopped short of fully embracing the assembly line principle and the flow of production was semiautomatic rather than continuous. Vauxhall introduced daywork rather than piecework and coupled the change to more direct control from supervision. However, these changes did not increase the competitiveness of the company and in 1925 they were taken over by GM who immediately introduced American advisers. The result of their intervention was the introduction of a system of 23 self-contained departments all feeding subassemblies to an assembly line. Overall there was a flow of materials, but it was combined with a high degree of loose-coupling. In 1928 the company adopted the policy of machine paced work.

Finally, Rover was a typically British small firm who found great difficulty during the 1920s in adjusting their designs and production to the shifting structures and segments of the domestic market (Whipp/Clark 1986). Their purchases of any technology were relatively slight because the income did not permit further investment.

These five examples indicate that during the 1920s the British car firms did not follow very far down the trajectory set by American production systems. In part it may be argued that the British managements made a shrewd assessment of the financial potential of the British market and of their capacity to alter its shape. Also, it may be argued that in the 1920s the senior managements did devise a production knowledge with its own language and recipes for matching the market with production techniques and production institutions. However, it may equally be argued that after the Second World War they failed to revise that knowledge base and to foresee the opportunities in Europe which other firms could envisage: Renault, Fiat and Volkswagon, for example. Also, it should be noted that small firms like Volvo in Sweden had a keen appreciation of the significance of the choice of markets for long-term viability (Nicol 1975).

12.3.2 Information Technology: 1970s and 1980s

Clark and Tann contended that the diffusion and adoption of forms of American information technology in Britain was likely to be inhibited by two nonmarket factors: the power of occupational groups at shopfloor and managerial levels and the extent to which successful examples of implementation can be organically transmitted within the same enterprise (Clark/Tann 1986). The latter point refers to the general absence of linking filiere to facilitate the diffusion of innovations other than by the movement of their originators. These points can be examined with brief reference to five areas: construction, retail, hospitals, libraries and production managers in manufacturing industry.

The construction industry in Britain has the reputation of being opposed to innovation. However, a recent examination has challenged that conclusion by comparing the tendering process as it was in 1965 and as it became in the 1980s[31]. It is clear that in the large firms there has been extensive innovation on a continuing incremental basis in several areas (Clark P.A. 1986b):

1. The use of a whole variety of predictable raw materials whose features have been researched and are subject to design and development
2. The use of new building concepts like steel structures
3. The extensive development of preplanning techniques allied to use of aggregate data sets for calculating costs
4. The transfer of shopfloor labour out of the major firms into subcontracting, so that the major firms mainly employ managers
5. The extensive development of managerial skills in the form of team work and project work on the basis for a temporary organization for the site managers and a stable managerial core

[31] The series of seven proceedings from the Bartlett International Summer School organized from University College, London, provide a thorough coverage of the labour side of the British construction industry. The managerial side will be covered by the research group from Reading University, Biggs, D, Dunning, J, and P Hillebrandt (1987 forthcoming)

6. Extensive development of monitoring and assessment techniques based on financial criteria
7. The realisation that construction firms operate on negative capital and that a core activity is financial management
8. The extensive adoption of small-scale micros and computers to facilitate design and to monitor performance

Relatively speaking the industry has transformed its image of the mid-1960s as an arena of multiple conflicts (Higgin/Jessop 1965) into an industry where the leading firms have imposed a managerial control over the site operation which has inhibited the power of organized labour. That transformation was facilitated by the development of management in the 1960s when firms did not follow the general pattern of highly specialized and segmented functional divisions of management. There are reasons, but the interesting point is the degree of adoption of information technology.

Retail is an area where the power of shopfloor occupational groups is relatively slight and that power has been further reduced by the extensive use of part-time employees. Retail is also an area where there has been extensive change in internal and external organization and raw materials over the past two decades. It is important to observe these sources of change when considering the adoption of electronic point-of-sale equipment (EPOS). The important point to note is that EPOS systems are simply one potential node on a network of information flows which supermarket controllers might wish to utilise. In practice supermarket operation has been changed by the redesign of the raw materials (e.g., tomatoes!) and by central planning as by EPOS systems in specific supermarkets. The striking Anglo-American feature is the closeness of the developments. The Americans are ahead, but by very little and it may well be concluded that the British-based operators are seeking to learn from observation of the American experiences. In their case, several British firms do seem to have developed the means for imitating American practices and for their modification to the British context.

As may be seen there are relatively slight Anglo-American differences in construction and retail. There are much larger gaps for hospitals and libraries.

In American hospitals there has been an extensive adoption of technology, probably because private hospitals have benefitted from the generous insurance policies of their better clientele. Therefore, medical technology in the USA has been more highly developed than in Britain and American companies have looked to the British market for exports. However, the American companies have encountered the curious features of the British medical service where state ownership is still combined with extensive powers for the medical staff in the selection of equipment. The research by the MESS project[32] strongly suggests these medical staff have chosen systems from their personal and departmental perspective. Consequently, the segmented character of decision-making has undermined the potential for systemic data usage. Obviously there are significant exceptions, but the point to be made concerns the overall profile of the hospital service rather than the small number of highly publicised examples of adoption.

Libraries are another area where extensive development in technology was introduced in the USA, but has only slowly penetrated the British context. For example, in the former industrial heartland of Birmingham it is to be noted that the library in the large civic University of Birmingham was using a handwritten system of tickets in 1986 and that in nearby Solihull, a residential dormitory, its prestigious local library used a computer, but did not know how many books its users had removed until 1986! Overall, British libraries are small when compared with American libraries, especially at the level of universities. However, even small North American universities were using more advanced technology in the mid-1970s than were comparable British ones in the early 1980s. As with hospitals there are interesting exceptions. For example, the Technological University of Loughborough introduced new technology in the early 1970s and did so through the agency and missionary zeal of a British librarian returning from the USA with a clear template of American approaches. Although, that library was visited by many interested parties there was not the organic filiere in the professional association of libraries to quicken the diffusion of the new technology. Moreover, the librarian profession was never as strong as the medical profession, yet it was able to activate ideologies of professional control as a protection for established methods.

Production management in Britain has long been described as a low-status occupation (Gill/Lockyer 1979). American techniques have dominated (e.g., PERT), as has American technology, so that leading British texts for students tend to follow American practices. A recent sample survey of the adoption of computer-aided production management (CAPM) revealed the interesting and highly relevant finding that many potential users failed to shift their adoption beyond very elementary stages of buying turnkey operations and then under-using them[33]. Two features were missing from many firms: the development of the software and the development of integrated systems.

---

[32] See the European Micro Electronics in the Service Sector Project, ESRC, WORC

[33] In Britain the Science and Engineering Research Council has initiated investigations of the factors affecting the adoption of production management in a program directed by A.C.M.E.

12.4 Management Education

American concepts of management education have been influential in the British context, yet as indicated earlier, there is a degree of misunderstanding by the British of the shape and uses of the American approach.

Lazonick compares American and British firms in their use of university personnel and contends that American firms have systematically recruited postgraduate, Master's students from both the vertically structured, advanced degrees and from the horizontally structured problem-oriented degrees like the Master's in Business Administration (MBA). Moreover, that admixture of specialists and generalists have been themselves circulated within the typical American corporation. In contrast, British firms were split into those who recruited heavily from the universities (e.g., petrochemicals) and those who recruited infrequently (e.g., automobile industry, textiles). In the British case the recruitment concentrated upon candidates with the Bachelor's rather than the Master's degree and there were classic anecdotes of British Master's students from advanced courses at prestigious universities being paid at the same rate as graduates. Also, in the British case there is a very small supply line of Master's students and very few MBAs.

The influence of American higher education on management education in Britain has come in uneven waves. The events of the early 1960s are revealing. Two examples may be taken.

First, in the development of management education for a postexperience type there was an initial impact from the Harvard case study approach. However, experienced British managers were reluctant to devote the preparation required for analysing the case according to the Harvard system. They were even more reluctant to have the instructor direct the case and force them to make public statements which their peers could evaluate. Also, very few British teachers were sent to learn the method of instruction. Consequently, that case method was soon abandoned – except for the occasional forays by Harvard faculty. In its place there developed the syndicate method by which a course was split into small groups who then discussed the case in private focusing more on evaluation than on analysis. Each syndicate elected a reporter and at the end of the session all syndicates pooled their opinions.

Second, the MBA was introduced into Britain with great difficulty. In Britain it had been taken for granted during the late 1950s and early 1960s that a Master's degree was a vertical, advanced degree taking undergraduate courses in greater depth. Thus, when proposals were advanced by British industrialists to introduce the MBA these proposals met with considerable opposition, especially from the prestigious universities. It seemed that the attempt at importation would fail. However, industry raised sufficient funding to endow two specialist schools, each of which could produce up to 200 Master's students per year. As a result of some extraordinary lobbying, it was agreed that the University of London would permit the creation of a new unit, the London Graduate Business School, but its degree had to be an M.Sc. rather than an MBA! A similar, though more flexible arrangement was developed at Manchester University, whereby the Manchester Business School was created. In fact the growth in postgraduate management education largely occurred in the much less prestigious sector of the small, specialized technical universities, yet even they used the label M.Sc. until the mid-1970s. At that time there was a sudden switch to the MBA label. Even so, the typical British MBA is a one-year course involving 30 weeks of courses and a project-dissertation. To an American it would be very obvious that the stitching together of the MBA is different to their expectations. This is because the British MBA was, until very recently, a timetable into which established disciplinary groups and functional areas poured their expertise according to their definition of its relevance as a subunit. There was little recognition that the MBA should be a designed, horizontal degree in which common managerial problems are addressed from a variety of selected angles. There are other differences. The British MBA will still be given booklists like those received when an undergraduate. There are few set books and that practice is often derided. Put simply, there is a lot of continuity between the British approach to undergraduate learning and the MBA.

12.5 Assessment

In Chap. 6 the flow of two British innovations into the USA was examined and it was shown that these did not fit their new contexts and that the original innovations were transformed by the American users. This chapter has probed into the reverse process of American-led innovations into Britain and has sought to demonstrate that there are patterns in the British approach.

At the macrolevel of society the concept of a filiere can be applied to enquire about the degree to which there are a network of channels through which imitation and/or appropriation might be achieved. In the American case, it has been argued that there are various interconnected ways in which European (and British) practices were ingested and developed as distinctly North American creations. In retrospect it may be seen that the agency of American entrepreneurs attempted to apply certain recipes and discovered some of the major sources of success, especially in the period of mass distribution-mass production in the twentieth century. The American case is replete with examples of learning *through* experiments and from *experiments* as well as of devising means of transmitting the outcomes of learning-by-doing. Obviously there were and are failures, yet there were also significant successes of adopting and adapting indigenous practices to new circumstances. In all respects education and training played a very significant role; the coach or mentor is distinctly part of the typical American repertoire. By comparison the British situation is immensely patchy. Considerable attention has been given in the past to successes of the British and too little sustained analysis has been given to filiere through which success could be analysed and the major lessons disseminated. Also, too much attention has been given by the British to the simplistic analysis of foreign practices. All too often a foreign practice is extracted from its context and recommended for the British situation. That failure of systematic recording has been reinforced by failures in appropriation. The British have given too little attention to the evolvement of innovations and to their matching in the British context. Clearly there are exceptions: firms like ICI were active centres for the development of work study and specific consultants sought means of matching American innovations to the British context. Yet overall that has not occurred in the past.

The examples taken in this section do underplay the extent to which the state and the government has become an agency for innovation. That point may be addressed more fully in the implications and after examining the Japanese situation.

Part IV.
Implications

13 Japan and the Pacific Rim: The New Competition

13.1 Introduction

In the West the Japanese have often been described as good imitators. The arguments and analysis in the preceeding chapters have made it clear that imitation is rather difficult and that appropriation is a major achievement. The Japanese case provides ample evidence of the appropriation of certain distinct competences and of the use of process innovations to maintain and obtain market shares under a situation of competition over diversified segments. It may also be argued that there have been certain features of the Japanese domestic markets, which in conjunction with the strategies for product, process and work organization as devised by leading postwar enterprises, have meant that the domestic market has been a useful developing ground for products which were subsequently marketed in the highly profitable American markets.

The analysis of the Japanese situation has elicited an extraordinary degree of interest and excitement[1]. Much attention has and is being given by American analysts to the issue of whether Japanese practices can be transferred and inserted into the American situation. In the American debate there are two groups, one which considers transfer as relevant and feasible[2], whilst the other is sceptical of relevance and dubious about transfer[3].

This chapter concentrates on three themes. First, I apply the frameworks of analysis set out in Part II and in Chap. 7 to the Japanese case. Of necessity the analysis must be brief and in broad strokes. The objective will be to com-

---

[1] McMillan (1985); Morishima (1985); Nakamura (1985); Hazama (1986); Vogel (1979); Okochi, and Uchida (Eds.) (1980); Koshira (1983); Wilkinson (1984); Smith (1986)

[2] The strongest statement on the feasibility of transferring Japanese innovations is Schonberger (1982). See Fig. 1-1 for an international comparison between the USA and Japan. Another optimistic assessment is that by Ouchi (1984)

[3] A most cautious and sceptical treatment of the capacity of the USA to appropriate Japanese innovations is that by Sethi, Namiki, and Swanson (1984). The position of Japanese firms in the USA and in Europe is under constant review in the media. In Europe there is a sprinkling of Japanese industrial plants; for example, by 1986 there were between forty and fifty in each of France and Britain. The *Financial Times* provides regular assessments for the commercial perspective. For the position of Japanese firms in Britain consult the reviews by Keith Thurley, London School of Economics

plement other analyses, such as that by McMillan, by focusing on the evolvement of innovations (McMillan 1985). Second, it is important to show the Japanese capacity for appropriation. That will be exemplified by examining three waves of appropriation: the 1870s, 1920s and 1950s. The objective will be to sketch the evolvement of a filiere which is highly sensitive to international competition whilst also possessing a strong, though focused, capacity for innovation-design. Third, I will highlight the current strengths of the Japanese in relation to innovation.

13.2 Structuration: Framework

13.2.1 Geopolitical

The Pacific Ocean was once an area of European influence within which the Dutch, French and British each had their spheres. Throughout those events the Japanese retained their independence, whilst undertaking limited trading assignments with the West. That situation began to alter in the nineteenth century when the Japanese began to explore the Western world and to make presentations at international expositions. The pace of interest by the Japanese in Western technologies quickened after 1860. In fact there was a coincidence of timing between the increasing Japanese interest in the wider context and American penetration of the Pacific as an area of influence.

Japan is an island-based society without natural resources and with few good agricultural surfaces; it is located within the Pacific Rim close to the land masses of China and the USSR. Japan is like Britain in its lack of natural resources, only much more so. Moreover there is a similarity in their dependence on water-based logistical systems for the importation of raw materials and for the export of manufactured products. One major difference lies in the size of the population. Japan's population is more than one hundred million and their density is very high, even higher than is the case for Britain.

In the twentieth century, as European influences in the Pacific declined and as American influence rose, the position of Japan became pivotal. Thus, in the key years between 1895 and 1910, afterwhich the North Atlantic region became gradually less of a preoccupation for the USA, Japanese political influence increased. More recently, between the occupation of 1945 and the mid-1970s Japan has held a strategic position in relation to American foreign policy as a base for both intelligence and as a 'frontier fort' in the face of certain possible threats. The Soviet naval ports of Vladivostock and Petrov are nearby. Japan has been a stable political base with governments which are respectful to America's extensive interests in the Pacific region. The military importance of Japan between 1945 and 1975 has been evident in its role as a safe haven and point of reinforcement for American forces, as happened in the Korean and Vietnam Wars. For example, in the Korean conflict Japanese airfields provided the bases from which American airforces were able to cripple the North Korean supply lines as they moved southward on the exposed roads to the Pusan perimeter. Japan's position as an offshore island and the nature of logistical inputs – their bulk – was a focus of interest for American military planners and logistics experts, who soon observed the potential to be achieved by the development of specialized systems of transportation based on standardized units: the approach of containerisation. That development had certain benefits for the Japanese as well as eventually providing the American shipping sector with the rationale for creating political control over freight movements which parallels earlier, seventeenth century, British regulations.

Japan is quite unlike the USA in its population size and density, physical terrain, location and resources. Moreover, there are occasional earthquakes and similar natural disasters.

13.2.2 New World-Economy: JUSA

The current performance of the Japanese economy is regarded with apprehension and admiration by Europeans and North Americans. According to the popular view, the current situation is very exceptional, although the earlier growth of Japan's economy between 1870 and 1940 was at a level which compared very favourably with European states (Nakamura 1985).

Since 1945 the Japanese and American economies have become gradually intertwined. Today the American markets are of central concern to the Japanese manufacturer and the links between Japan and the USA have sometimes been referred to as 'JUSA'. According to the *Economist* that new world-economy now accounts for 30% of world trade and the Pacific Rim surpassed the North Atlantic trading zone in 1985[4]. Today, Japan sends one-third of its exports to the USA and receives one-fifth of America's exports. These figures would be even more striking if America and Canada were treated jointly. Certainly Japan is next to Canada in its trading importance to the USA. American supplies to Japan include high proportions of bulky foodstuffs: 20% of the imported grain, 80% of the imported beef. Japan's exports to the USA include a high proportion of parts, components and pe-

[4] Japan: A Survey, *The Economist*, 7 December, 1985. A recent comparison of the relative strengths of the USA and Japan in high technology concludes that the Japanese have extensive strengths; See High Technology – the clash of the Titans, *The Economist*, 23 August 1986, p. 48 ff

ripheral goods which become incorporated into American assembled goods. In addition there are the highly publicised exports of electrical goods (e.g., televisions, cameras and cars). Japan also has extensive financial investments in California and in the southern United States.

Japan and the USA have become increasingly interdependent, but their economies are not alike. Japanese enterprises, like their European counterparts, view the USA as the major market to be penetrated. It seems that the Japanese target the USA rather than China or the increasingly prosperous, though still poor, countries like India. In 1986 Japan attained a trade surplus with the rest of the world of more than sixty billion dollars of which two-thirds was with the USA. That surplus was roughly equivalent to four percent of Japan's GNP. So it is not surprising that Tokyo, which has become a core zone of the world's processes of capital accumulation, has also become a financial market with major banking enterprises. So far, these banks have invested in heavy loans to the politically safe USA. Surprisingly the world-economy in which Japan has become located is not one in which she is as major an exporter as the United Kingdom or Germany. Each of those countries are proportionately more dependent on exports than is Japan. The significant feature about Japan is the large domestic market with its intense internal competition in areas like car sales and *small* household electrical goods which can be situated in the tiny domestic spaces of the Japanese household. An obvious example is the small television set and it should be noted that the Japanese are hardy watchers of TV. McMillan states that there is strong market demand for all household goods which enhance the quality of life of the Japanese (McMillan 1985). It must therefore be observed that the most salient, and visible, of Japanese exports to the West (e.g., TVs) obscures three features: first that Japan is a major supplier to the USA of high quality components; second, that the Japanese domestic market is so competitive that successful marketing there can provide the *potential* for success in other markets where the consumers' tastes overlap those of the Japanese domestic consumers; and, third, that Japanese exports tend to be very narrowly focused on types of products which contain a high level of value added.

13.2.3 Structuration

When examining the forms of structuration which characterise Japanese society it is important to avoid a static view as found in the account of Abegglen (Abegglen 1958). Instead, particular attention should be given to the changing balance between features which continue and persist and the novel elements which emerge at particular conjunctures and become significant features in the future. There are important respects in which Japanese structuration today is both continuous and also discontinuous with a century ago. It is useful to compare structuration in the period before the Meiji revolution of 1868 and then to track its main features over the past century.

Prior to the Meiji revolution Japanese culture possessed certain mentalities of which five major groupings require emphasis. First, there were certain basic features in the ordering of the members of society into four types of *social identity*. These were the samurai, peasants, artisans and merchants. In each case the identity of a household was indicated by a sign on the doorpost and that sign indicated the rank of all its members as well as the codes of behaviour which they ought to display when meeting one another. Initially the role of the samurai had been to enforce the codes of behaviour through the physical punishment of transgressions, but the most brutal manifestations of that role had been progressively curtailed and application of the rules had increasingly relied on the individuals' conscience to regulate collective behaviour. At the level of the rulers of Japanese society there was both centralization and also the existence of a mock form of government and court somewhat analogous to that which existed in France in the reign of Louis XIV.

Second, Japanese culture possessed an interesting clan-like form of template with three facets:

1. The household was the basic template for organizing, rather than the individual, and a collection of households was the template for large-scale activities.
2. A high degree of homogeneity existed across Japanese culture with central locations acting as both the generators of core values and also monitoring the transmission of the culture. That meant that any new values could be diffused very rapidly.
3. There was a particularly strong conception of the superior/subordinate relationship.

These facets provide a strong contrast with the USA and Britain. Third, a strong form of patrimonial bureaucracy had developed. The male heads of the most powerful households were the employers of the bureaucrats and they expected them to provide an impersonal service based on employment and to devise means whereby the interests of the household could be achieved. Moreover, the new bureaucrats were recruited from the samurai who now became the enforcers of impersonal rules.

Fourth, it is important to give attention to Confucian ideology and to show that its contribution to contemporary Japanese culture reflects a distinctly Japanese variant of Confucian thought and also that its core tenets provide a relevant though very different input to capitalism from the thought which characterised Western European rationalism. Confucian ideology has usually been associated with benevolence and wisdom, as in the Chinese variant, but Japanese Confucianism has a great emphasis upon loyalty to the household, clan, firm and state. It seems likely that the emphasis on loyalty originates from the early seventh century during the Taika reform when Shtoku Taisha imported and subtly adapted Confucian ideology to strengthen obedience to the imperial system. That form was sedimented and persisted for more than a millenia until the later revolution of the Meiji restoration in 1878. In each revolution the changes were superimposed from above. The Japanese ethic was neither strictly capitalist nor Calvinist, yet it was equally frugal and possessed the potential for similar economic outcomes to those associated by Weber with the Western varieties of rationalism. That tendency was further strengthened by Shintoism. The Japanese ethic highlighted dualism as a constant feature of society and one which is found in all spheres: economic, political and religious. That dualism, apparently not well understood in the West, accounts for some of the complexities of understanding Japanese consensual decision-making. The Japanese ethic of Confucianism underpins a variety of Japanese institutions and practices, for example, a high ratio of savings, the relations between government and business, and the notions of human capital accumulation and employment practices.

Fifth, Japanese approaches were strongly characterised by what has been termed *adversity management* (McMillan 1985). That is to say, it was assumed that all key resources, such as land, were scarce and that human resources and values had to cope. Consequently, there was an implicit concern with human resources which manifested itself in almost continuous attention to small-scale social innovations. Adversity management as an underlying principle had some interesting consequences. Work was regarded as an opportunity for training to improve oneself, and discretion at work could be exercised by invoking the claim that one was acting on behalf of the superior. There was a high degree of literacy relative to Western Europe. Planning was an important feature of adversity management, yet its rationality was very different to that found in the West, most notably in its orientation to the time-frame of the past for the interpretation of future events;

> ... planning is seen less as a guide to the future but as an interpretation of the past. In an age of profound discontinuities...planning is seen as a tool to shape current thinking based on interpretations of past assumptions and behaviours (McMillan 1985: 16).

In addition to these indigenous features the shogunate and the Tokugawa had been involved in the fostering of restricted foreign imports, particularly from the French and focusing on education, shipyards and textiles. In the latter case there was extensive use of external and internal contracting.

After the Meiji revolution of 1868 several major developments were introduced in the next three decades: the centralization of initiatives; the importation of selected foreign technologies and organizational innovations which were inserted into the state-sponsored sectors (e.g., French dockyards); the conscious separation of traditional industries from new, modern industries and the giving of positive support to the latter; the development of a dual, multistructured economy in which great power was given to the powerful, core firms who were then in a strong position for the making of contracts with suppliers of components and services; and the introduction of bureaucratic forms of organizing in the two sectors of banking and trading. Even though Western technologies and practices were imported the underlying processes within which they were utilised contained a strong infusion and embedding of Japanese structuration, and there was conscious countering of Western values in a manner somewhat similar to the preoccupations of the early American colonists to create an Americanised culture and set of artifacts. The result was that from 1870 to 1940 the economy grew by rates comparable to those found in the USA and in Western Europe, so that there was slow, yet steady modernisation up to 1914 and then more rapid growth to 1940. In the interwar years agriculture was neglected, but modern industries became established, and the large-scale firms flourished at the expense of small firms. The large firms tended to pay high wages and to provide long-term employment.

The economy experienced pronounced cycles of expansion and downturn, but Japanese growth tended to be the outcome of a configuration of forces involving three large-scale actors: the military, the government and the Zaibatus. Military actions increased after 1895 and their centrality increased with the successful annexation of Manchuria and the development there of a corporatist economy fashioned on European models from Germany and Italy. Defence expenditures increased and in the economic downturns those increases squeezed private consumption thereby heightening the impacts on the economy. The big business was organized by the Zaibatsu who were mainly powerful in the 1920s, when they concentrated financial capital and controlled the highly profitable sectors of commerce and mining. On the whole the Zaibatsu were absent from manufacturing and so their influence declined in the 1930s, although the template of their organization was a target for American restructuring after 1945. In the 1930s the military became influential. The government continued the industrial policies initiated by the Meiji and supported industrialisation.

Two European templates were imported the French and German. The French template of education was grafted onto the pre-existing system which already possessed a relatively high degree of literacy, and the general direction of the influence was to create a high minimum level of education. The German influence was more evident in the sphere of government where the Prussian Diet was selected as a template, quite possibly because it was associated with a relatively quiescent form of politics and was consistent with the previously existing institutions. The British influences were relatively slight, though Britain became a relevant political ally and British-built ships were used by the Japanese navy in their defeat of Russia in 1895. Ten years later China was defeated and by then the ambitions of Japan in the Pacific arena were becoming more apparent. American influences on Japan in this period were slight, though the Japanese had been impressed by American technology even in the 1860s.

The next 40 years culminating in the defeat of 1945 were ones of increasing Japanese adaptation and appropriation of Western technology. From 1910 onward attempts were made to import Taylorism and in the 1920s those efforts made a dramatic impact in a limited sphere. Also, in the 1920s the textile industry purchased redundant British technology which they operated more efficiently and also introduced innovations in the work roles. Those innovations coincided with the importing of Taylorism, and it is important to note that the Japanese retained the techniques of work analysis, but instead of regarding the corporate engineer as the designer of the operatives' tasks, the operative was given the job of improving their own performance. In effect the outcome was similar to the notion of work simplification utilised by Hewlett Packard in the 1960s, yet it was very different from the formal American practices and the more informal British methods. It may be concluded that the Japanese succeeded in appropriating Taylorism. The outcome was the surfacing of several efficiency movements followed by visits to the USA, specialized societies, translations of Taylor's work and that of similar writers, and the application of payment schemes. However, these approaches were not adopted in the chemical industry. The movement was extended in the 1930s with the importation of German principles of rationalism. During this period American concepts were largely rejected because of the antagonism between the USA and Japan. However, there were European imports: the mutual-aid program of Krupps in Germany; the welfare concepts of certain British firms; the approach to communication developed by National Cash Register in the USA.

The period 1905–1945 was one in which the Zaibatsu were powerful in creating economic empires. At the level of the enterprise the corporate style was one of paternalism and managerial familism in which the template of the household was rescripted so that the labourer was treated as the child of the employer and was therefore someone who had to obey instructions, be trained and cared for in a manner defined by the head of the firm. During this period labour organizations were systematically discredited.

The state became very powerful in the legitimating of rules and in promoting meritocratic structures and careers, whilst management as a new occupational strata was able to present themselves as a Mandarin type of elite applying the vision of adversity management. Consequently, managements became associated with the best use of resources and with continuous small-scale innovations and they were able to avoid presenting themselves as either an interest grouping or as the agents of another interest group such as capital.

In the 1930s there was a general shift towards militarism.

13.3 Appropriation Rather than Imitation/Rejection

The long-held view that the Japanese are good at imitation underestimates the possibility that the Japanese may turn out to have been (and continue to be) highly effective at the selective appropriation of foreign technologies and foreign forms of organizing: production technique, institutions and knowledge. The case for appropriation requires further scrutiny. Three examples spanning the past century may be taken.

First, it is useful to commence the examination by referring to Dore's analysis of the introduction of a French-designed shipyard in the 1860s during the reign of the Tokugawa (Dore 1973). The dockyard was established through the state bureaucracy. Dore estimates the extent of continuity of past practices at 25%, the proportion of borrowing and adoption at 35% and the proportion of novel invention at 40%. Continuity was evident in several ways: the use of total commitment and moralism through exhortation, role models and structured guidance; the application of compulsory savings; the retention of craftsmen; long service and the relating of wages to the length of service. Borrowing and adoption were found in two areas: the rational legal approach to defining procedures and setting boundaries of space and time to activities, and in the use of classrooms in the yard for training of older skills and the introduction of training for new skills. Invention of new forms occurred in five areas:

1. The introduction of minute regulations
2. The introduction of the work-gang concept
3. The regular intake of new workers
4. The use of continuous hierarchisations
5. The new ideology of authority

The outcome became a prototype template which influenced the establishment of private dockyards because its organizational innovations were perceived to be relevant to the handling of the problems of size.

Second, in the period between 1910 and 1930 the American approach of Management Science associated with Taylorism, Church and Gantt was the focus of intense interest in areas like textiles (Okochi/Uchida 1980). Cole states that Japanese interest began in 1908 through a joint venture between the Westinghouse Electric Company and Nippon Electric which led to visits by Japanese engineers to the American plants where they observed the application of time and motion studies coupled to incentive schemes. By 1911 Taylor's *Principles of Scientific Management* had been translated into Japanese and eventually sold nearly two million copies. The basic ideas of Taylor on standard motions were applied in the railways and the textile industry. Also synthesised standards were soon created as a database from which individual task profiles could be constructed. That development paralleled American evolvement and was considerably ahead of British practice. In Japan the reception given to time and motion study was highly enthusiastic. Japanese management, in this period, felt the need to reform its operating practices of personnel control in order to address the perceptions of relative inefficiency compared to foreign rivals as well as to handle the possibility of conflict between the owners, the Zaibatsu, and labour. Unlike the British management (as reported by Child 1969), Japanese management took an active, collective role in the diffusion of Taylorist principles, emphasising their role as an impartial body of knowledge about human work. Time and motion study approaches were grafted onto adversity management and the general orientation of Japanese management to the running of production systems. The principles of time and motion study were sedimented as a series of section criteria through which production problems were identified, and time and motion study became an integral aspect of what Rosenberg terms a general focusing device. As a result Japanese management built on that base. An important modification of American time and motion study occurred in that the transfer of the responsibility for action in the removal of inefficient practices was located at the shopfloor level. That may be compared with the American tendency to create a sharp separation between the corporate engineers from the technostructure and the operatives. In the Japanese case *analysis and action* were concentrated at the shopfloor level.

In Japan time and motion study was diffused to a number of sectors – excluding the chemical industry – through the establishment of influential institutions during the 1920s (cf. Britain). It became influential in the direct organization of work in the electrical industry and that might account for the deep differences in shopfloor management noticed by Dore in his comparison between a Japanese and a British plant assembling television sets.

Third, in the 1950s two American experts in the application of statistical quality control, Deming and Juran, also made a strong impact on Japanese management (Schonberger 1982). These two men lectured extensively in Japan on the achievements of American approaches to the establishment of quality control standards and to the correction of low performance. As with Taylorism the American visitors emphasised the central role of corporate engineers and the responsibility of staff rather than line management for quality control standards. However, the Japanese noted that the American approach seemed to be more concerned with the distant application of a standard than with the removal of the causes of defects and so in their applications they reversed the emphasis and placed the strongest responsibilities on the employees and the immediate shopfloor groupings of workers and managers. The principle behind that was, quite simply, that it was easier to remove faults at their source than take corrective actions later. Following the visits of Deming and Juran the Japanese Union of Scientists and Engineers (JUSE) was established as a central forum for learning about the new techniques and as a key agency for their diffusion to its members throughout Japan. JUSE is an active, influential and recognised agency for increasing the consciousness of management about quality control. From that beginning, quality control management became the focus for an array of problem-solving practices which covered a diversity of areas: inventory, safety, the use of energy, and operator training. Gradually these became more formalized into quality circles.

These three brief examples illustrate a general point and provide an important contrast to the transatlantic diffusion of innovations from America into Britain. In the Japanese case there appears to be a high degree of cumulative evolvement of production techniques, production knowledge and production institutions. That evolvement contained both a degree of continuity in Japanese structuration combined with transformations which took the form of appropriating external ideas and practices and incorporating them into newly constituted practices. Those developments took place, initially, within the Japanese domestic markets where they were simply part of an overall filiere which was amply suited to the nature of international competition after 1950.

13.4 Markets and Knowledge

13.4.1 Current Strengths

Decoding the roots of contemporary Japanese success has become a major and significant preoccupation within American managements and amongst business school academics. There has been a marked shift in the content of the American interest from the sociological inquiry into Japanese culture undertaken by Ruth Benedict in *The Chrysanthemum and the Sword* in the 1940s to the rash of books and articles from the late 1970s onward (Benedict 1946). It would be unfair and inaccurate to highlight any particular interpretations, yet the two crisp articles in the *Harvard Business Review* by the late Bill Abernathy and colleagues may be noted because of the extent of their impacts (Abernathy et al. 1980, 1981; Abernathy 1983). Essentially those articles concluded that the bases of Japanese success consisted of a network of highly interrelated channels and that forms of structuration were coupled with the specific structure of the domestic market and with forms of shopfloor production techniques, production knowledge and production institutions which embodied previously undetected strengths. The Japanese filiere turned out to have areas of excellence in many areas which Americans, who previously compared themselves with the British, had considered to be their main strengths. Debate continues over which channel of the filiere is the most remarkable, whether it is the shopfloor forms of work organization and technological innovation or the macrolevel cultural and economic institutions. Yet what is clear is that there is a Japanese configuration which has contributed to the current success. It is also clear that a simplistic examination of Japanese culture will be insufficient because too many accounts neglect the dynamic evolvement of Japanese culture and its cumulative transformations as new problems become a target of focusing devices.

13.4.2 Markets

Japanese success in the international marketplace is, at the moment, based on a narrow range of products and components, the latter being heavily supplied to the USA rather than to Europe. Particular attention should be given to the structure of the Japanese markets and to a consideration of their performance and potential in terms of the perspective of flexible specialization. Piore and Sabel concluded that from 1870 to 1970 the most successful route to capital accumulation was one which created mass markets and supplied them with mass-produced consumer goods: the Fordist template of capital regulation (Piore/Sabel 1985). I have shown that in the American case mass markets were created and that they provided the context in which those corporations who successfully used their *design capacity* to link mass distribution with mass production had a head start in achieving the economies of scale which would eliminate rivals. The USA proved to be a series of world-economies in which those features surfaced and it was able to capture the strategic heights to create a filiere which non-American corporations found impossible to counter in the American markets (up to the 1960s) and often too difficult to resist in their home markets unless assisted by the actions of the state. Piore and Sabel consider that the USA was enabled to use its domestic market as a very satisfactory learning ground for obtaining overseas markets.

If I apply a similar analysis to the Japanese situation then two points become evident. First, the Japanese have been able to produce the very products which their competitors believed were in a mature state of production and they have done so by introducing unforeseen (by the non-Japanese) changes in *process innovation* which were later coupled to product innovation with products from the fourth longwave. Second, the structure of the Japanese market has provided intense competition in several areas of consumer production which later became core parts of the Japanese success in the international markets. These included automobiles, cameras and consumer electronics like radios from the third longwave.

The case of automobiles provides a neat illustration. At one stage, in the early 1960s, Japanese policy groups considered that the automobile industry ought to be rationalised because there were too many competing firms for the size of the market. In practice the advice was ignored and the Japanese began to compete on the basis of satisfying segments of the consumer demand with cars whose features were the target of more exacting incremental innovation than had been the pattern in the USA. The cars which were small and as fuel efficient as those produced in Europe were as heavily styled as the American cars. Moreover, in order to minimise the costs of production and heighten production flexibility the Japanese car firms began to innovate throughout the whole production process. The commodification of time was *the* focus of attention, especially the costs of all raw materials including those components still with the supplier and those components passing through the firm in various stages of assemply. Whereas the industrial engineers within American car firms had used computer-based technologies to create material resource planning (MRP), which used buffer stocks to smooth the interdependencies between stages in the production process, the Japanese firms sought to apply the template of the supermarket to the assembly firm. To do so the assemblers reshaped their relationship with component suppliers so that faults were eliminated and the required quantity of supplies arrived at several designated time points during the working day. In that way, inventory from suppliers was minimised and responsibilities for interruptions were pushed downstream.

Within the car plants simple systems were introduced to increase production flexibility. Two areas may be noted: changeover times for equipment and inventory levels. In the several key industries (e.g., plastics) Japanese management uncovered ways of reducing the time taken to change over equipment from one type of production tool to another quite different type. The whole process of the changeover was the focus for detailed time and motion analyses coupled to considerable scrutiny of the work organization. The result was that changeover times were reduced to levels which North American and European experts had previously considered unattainable. In addition certain firms began to reduce their inventory levels so that production was only undertaken when a consumer had made a purchase. Starting with a consumer purchase a whole sequence of triggering events, based on a system of tickets now called the Kanban system, was introduced – in Toyota and then diffused to other firms. That is the Just-in-Time (JIT) system which many Western firms are now attempting to ingest (Schonberger 1982). Once JIT had emerged as a set of practices it became evident to certain careful observers that JIT was ideally suited to the exposure of all kinds of inadequacy in production techniques and in production knowledge. Fortunately, for the Japanese they had developed the production institutions, for example, quality circles, which made JIT feasible. JIT *evolved* as the solution to a highly competitive market situation in which firms were faced by competitors who were prepared to be highly flexible in their consumer products.

Japanese cars had been exported to the USA in the early 1960s, but had failed to obtain sales and were even less successful than European firms. After the 'oil shock' of 1973-1974 the American consumer became more interested in cars which were economical in their fuel consumption. So, when the new wave of superior, well built, small and efficient Japanese cars began to appear in North America after 1974 they encountered a receptive market and were able to obtain a segment which American car firms found difficult to service.

13.4.3 Production Knowledge and Techniques: Process Innovations

Current research on Japan, both indigenous and that undertaken by foreign analysts, has concluded that Japanese industrial enterprises have made two major and related innovations. First, they have overturned North American concepts of production knowledge and production technique. Second, they have gained an expertise in the continuous incremental innovation of the production process. Each of these will be examined and the section will give attention to the role of knowledge intensification in Japan.

The Japanese approach to production knowledge has been labelled flexible Fordism and Toyotoism, so it is important to clarify the basic distinction of the new flexible approach from the earlier one (Dohse et al. 1984). Fordism has been the generic label applied by exponents of the French regulation school to the balance achieved by American capital in having the state provide Keynesian mechanisms to balance the consumption and production facets of the economy. In Fordism the producer is also the consumer, and so the two sides of the same role provide the basis for high rates of capital accumulation. By Fordism is also meant the adoption of fairly rigid and highly dedicated forms of technology and work organization in which the minute division of labour is coupled to centralized, and hence inflexible steering of the enterprise.

The Japanese variant of Fordism represents an extension and also a different route of development. In Japan the paradigms for technology and the production techniques possess certain distinctive features:

1. The average working time per annum of the employee is almost 2,000 hours whilst that of the American worker is under 1,800
2. The connection between the payment of wages and the content of work is loose; therefore, the potential for a struggle over the details of labour time is more problematic for the Japanese trade unionist than for his/her British counterpart.
3. There is a very high level of flexibility in the deployment of staff. There are no rigid job demarcations so that workers can be quickly deployed amongst a multiplicity of tasks. There is a self-regulating distribution of tasks with responsibility decentralized to the shopfloor.
4. There are fewer direct workers in the functions of quality control, production control, maintenance and rectification. Quality control is pushed down to the shopfloor level into Quality Circles.
5. The workforce contains the core of protected workshops which are supplemented by seasonal and part-time workers.
6. The emergence of JIT systems has removed temporal buffers and increased the visibility of workers without work. Moreover, first-level supervision has been given a highly active role in applying the techniques of work simplification to everyday problems.

These six features have also been labelled by some European observers as Toyotoism.

By developing a more flexible approach the Japanese have actually rejected many of the taken-for-granted assumptions of post-1945 American production management. Most obviously they have questioned the longer term utility of the American system by which corporate engineering took over the main tasks of steering the firm through centralizing the setting of both strategic and tactical parameters in the technostructure rather than on the shopfloor. By implication the Japanese approach raises certain questions about the use of a planning philosophy based on 'push' rather than the Japanese emphasis on 'pull'. In the classic American push systems such as material resource planning and master production scheduling very fixed plans are constructed, yet these plans tend to decay quite quickly in the face of everyday contingencies on the shopfloor as well as those arising from suppliers and customers. For that reason American and British firms have been experimenting with a mixture of the established push approaches and the Japanese pull.

The foregoing points also illustrate the claim of Japanese analysts that the Japanese have acquired a distinctive competence in process innovations and that has led some Japanese to suggest that their strength has been to take Western ideas of organizing and to reshape those ideas and practices. Put simply, the strength is organizational as much as technological. Abernathy argued that the Japanese had created a new paradigm for combining technology and work organization (Abernathy et al. 1981). An important feature of that new paradigm has been in the relationships between the core, assembly organization and its suppliers. As indicated earlier the manufacturers are largely assemblers and that degree of focusing permits them to benefit from the effects of the learning curve. Though it should be noted that their approach to process innovation permits that learning to occur. American management scientists have often assumed that the operation of a learning curve effect is virtually automatic, but the British case shows that the advantages have to be won. In Japan the assembler accepts the responsibility for the quality of the product, yet – in the system of sector dualism – he is bound tightly to a fragmented constellation of suppliers. To an important degee the assemblers set the framework within which the suppliers have to operate, particularly the tight coupling of commodified time which reduces work in progress to a minimum.

Schonberger, in his widely cited analysis of how JIT approaches can be appropriated by American managements has claimed that the current Japanese strength is confined to high volume, repetitive production (Schonberger 1982).

The Japanese approach to process innovation has resulted in product innovations. For example, the use of automated production in consumer electronics has been combined with a sharp reduction in the number of components inside a particular end product. The *Economist* reports that there are 30% less components and so consequently it is feasible to automate as much as 80% of the production tasks (Economist 1985). In the case of television sets the Japanese required only 1.9 man hours for assembly per set compared with 3.9 hours in Germany and 6.1 hours in Britain.

The development of production knowledge and technique rests on channels for the dissemination of scientific knowledge and on the creation of an extensive design and developmental activity (D and D). The overall Japanese approach has been referred to as *knowledge intensification* (McMillan 1985). This initially involved heavy purchases of patents from the USA and the Japanese seem to have been able to obtain a high proportion of licensing in the sunrise industries for about one-sixth of the cost of their development to the American research-economy. More recently that earlier policy has been extended as Japan's gross investment in R and D rose from well under two percent of gross domestic product in 1970 to more than two and a half percent by 1983. According to OECD Japan's share of international R and D has increased relative to its competitors and the benefits of that investment are reflected in the six-fold increase of patents registered in the USA between 1965 and 1985. Knowledge intensification consists of the heavy D and D on both indigenous products and on foreign products. It is undertaken within the firms so it is interesting to note that Japan has the equivalent number of scientists to the USA, but they are concentrated inside the firms rather than as in the USA where they are distributed between the universities, independent institutes and corporations. That latter policy can lead to considerable confusion as is illustrated by Noble's analysis of the evolvement of the concept of the automated factory (Noble 1984).

The policy of knowledge intensification is a filiere of considerable scope and extensiveness. First, the trading firms of which there are 8,000, are regular monitors of international developments in technology. They produce regular reports which are diffused to the inhouse scientists and executives. Cole contends that there is a highly sophisticated structure for the diffusion of information which supplements the formal reports with extensive personal linkages (Cole 1971). Second, there has been a policy of establishing creative technological policies which are designed to provide a focus stretching forward into the future, whilst being less utopian and more pragmatic than equivalent American policies. In Japan there have been three areas of development for knowledge intensification on technology: for city-related needs and the quality of life of the consumer; for high-technology sectors; and for energy, including the use of ceramics in engine design. Fourth, at the level of the firm there are elaborate devices for circulating information and for obtaining agreements at shopfloor and managerial levels. In Japan engineering is central, yet it is much more embedded in the steering of the firm through the continuous development of process technology and innovations, sometimes in quite dramatic ways. Moreover, the Japanese have not really required the M-form of corporate organization in the same way as have many American corporations (Ouchi 1984).

13.4.4 Production Institutions: Education

In the immediate post-1945 period there was a heavy initial overlayer of American influence during the occupation. This created a curious admixture of influences and consequences. On the one hand, America was respected as the "determined victor". The Americans attempted to breakup the influence of prewar groups, particularly the Zaibatsu and to promote the growth of trade unionism. However, these American overlayers were soon absorbed and redirected, largely with American cognisance. Thus the potential power

---

[5] Ouchi, op. cit.

of labour organization was defeated, especially in the automobile industry. Moreover, new institutions emerged like the MITI who played a contingent role in supplying the newly emerging trading consortium with market intelligence and some vision of industrial policies. There was renewed interest in American concepts of work organization and or production technique which led to the introduction of quality control (see Sect. 13.3).

A distinctly non-American development was the further growth of education. The basic principle was to continue the generally high level of minimum education and at the level of higher education 1,000 colleges and universities were created. These provided generalised university courses with a very strong impetus to engineering courses. Unlike American universities, those in Japan left the development of industrial research to the corporations because specialized knowledge was regarded as the province of the corporations. Thus, there was no attempt to imitate the specialized American MBA (cf. Britain). The Japanese approach in higher education is to prepare management for the handling of events which are presumed to be largely outside their control (cf. American assumptions) and therefore there is strong focusing on the collection of all kinds of information. The focus on information collecting occurs at all levels from the shopfloor to the gate-keeping agencies like MITI which collate international information.

There are close links between higher education and capital just as there are in the USA, but the substance of the linkages is different. Japanese education is highly competitive and Japanese students are constantly compared with each other and with an abstract notion of standards. There is extensive use of examinations and tests which create an awareness of performance-related hierarchies. Yet these do not have the divisive outcomes sometimes encountered in the American corporation.

14 Summary and Implications

14.1 The Core Problems

Macro-organization studies as a *process* perspective should make two related alterations in its definition of the important problems requiring systematic research and theorizing during the remainder of this century. First, serious attention has to be given to the organization as a user and a generator of innovations of all kinds: in knowledge, raw materials, technology and in social institutions. In order to undertake that task a useful theory of innovation has to be integrated into macro-organization studies. Second, innovation occurs in a network of channels which reach outside the narrow sector of industrial and commercial organizations into the institutions of society to form a filiere which requires detailed attention through cross-cultural comparisons of the extent to which the filiere of a specific society permits the individuals and organizations within that society to monitor and to appropriate relevant innovations and then to promote their further evolvement. This final chapter draws together the main themes and interim conclusions.

The objective of the book has been to connect otherwise fragmented perspectives so that their strengths and directions can be assessed in a more exacting manner. The prime objective has been to show that one particular direction is a useful route to understanding and to developing best practice methodologies.

Macro-organization studies is at an interesting turning point in its own evolvement because the critics of its earlier preoccupation with structure and with cross-cultural similarities have been so effective in their critiques and attacks that the image of a dominant orthodoxy is no longer viable.

For much more than a decade the critics have alleged that the approaches to the analysis of organization such as the early Aston Programme (Pugh/Hickson 1976) dominated the definition of problems and provided the exemplars for solutions. The critics have argued for a processual approach which took into account the historical perspective and which examined organizations within a societal framework and with reference to the uneven development of capitalism as an economic form (Clegg/Dunkerley 1980). Today the critics are so powerful that the remaining exponents of the former orthodoxy are now labelling the critics as the new orthodoxy and are attempting to defend themselves against the avalanche of criticism (Donaldson 1985). Thus, with the demise of the former orthodoxy it becomes necessary to consider the directions to be taken by its successors. In practice the exponents of new directions have given scant attention to two arrays of problems: those concerned with how organizations achieve a transition from one configuration of variables to another configuration (Clark/Starkey 1987) and those concerned with the economic and societal relationships within which organizations are embedded[1]. Those two sets of concerns have been addressed in Part II and Part III, so this final chapter will seek to draw the major points together into a summary which will then form the basis for drawing certain implications for the debates surrounding the issues of renaissance and retardation as applied to the USA and to Britain.

Three core problems were identified in Chap. 1 with reference to: the fragmentary approach to innovation; the over-emphasis upon alikeness in cross-cultural studies, with neglect of the transfer of innovations; and the weakness of processual approaches in organization studies. First, with respect to the study of innovation, a serious attempt has been made to provide a coherent perspective which counter-balances and complements the approach from neoclassical economics to comparative factor prices. Innovations have been conceptualised along the four dimensions of knowledge, raw material, artifacts and societal embedding. Particular attention has been given to the dimension of knowledge and to the requirements for its continual alteration. In addition a balance has been introduced between the analysis of the supply side and the user side. Overall, attention has been given to the evolvement in the shape and uses of innovations. Second, the transatlantic transfer of innovations has been used as a model both to explore the evolvement of innovations and to examine how three different contextual facets have strong impacts on evolvement, namely, culture, market structure and geophysical resources. By examining societal culture and market structure it has been possible to examine the tendencies of the Americans and the British to adopt different patterns towards innovation.

The case has been made that the British and the Americans possess different tendencies in innovation and that any apparent convergence may well be explained by the better survival of American innovations and their importation into the British situation by American-dominated channels and agencies. The specific case of certain types of innovations without agencies has been used to reveal important underlying differences between America and Britain. Third, the objective has been to highlight the deep neglect by current approaches in organization studies to the problems of innovation-design and of the ways in which the evolvement of different kinds of innovation and

[1] Compare the approach of Clegg and Dunkerley (1980) with the approach taken by the economists, Freeman and Perez; for example: Freeman, and Jones (1985), esp. Introduction; Perez (1983)

their trajectories impinges on definitions of best practice. Orthodox approaches to organization change through Organization Development (and the like) have been preoccupied with simplistic and mechanistic multistage prescriptive schemas, but these fail to adequately audit the contextual requirements for adaptation and moreover their theory of organizational transitions fails to handle momentum and inertia and underplays the tremendous cognitive changes required for successful transitions. The synthesis achieved in Part II points to the needs for new thinking about organizational transitions.

The approach taken to these core problems will be summarised in the following sections which will provide the background for reconsidering the current thinking in organization studies about the degree to which British and American organizations have the innovation-design capacity to achieve organizational renewal.

14.2 Innovation Evolvement

The process perspective in macro-organization studies requires a useful theory of innovation. The roster of problems to be addressed was presented in Chap. 2. The key starting point is the definition of an innovation in terms of four related dimensions: the knowledge base and the embodiment of that knowledge in specific raw materials and artifacts, as well as the embedding of the knowledge, raw materials and artifacts in a socioeconomic context. The defintion of an innovation by these four dimensions very much prises open the range of innovations which can be considered. Previous studies tended to focus on well-shaped artifacts and to study their spread through a given population. Consequently, many previous studies treated the embodied knowledge as unproblematic and neglected the shaping role on the innovation of the specific socioeconomic context.

The dominance in the setting of problems by the discipline of neoclassical economics was both a strength and a source of problems. Until their recent revision, the neoclassical models had implied that market signals objectively shaped the cognitions of decision-makers so that only rational decision paths were followed. Moreover, the original rational model of relative factor prices contained a preconceived framework, but economists have already commenced the task of revising that theory to incorporate behavioural assumptions. In particular the knowledge dimension of innovation was already being investigated. Also, the theory of transaction costs had been extended by economic historians to show how the knowledge of production technique was itself shaped by the production institutions linking capital and labour.

Chapter 3 scrutinised the state of theory about the diffusing of innovations with respect to three perspectives of economics, sociology and geography and demonstrated that the dominance of the supply side models was under revision. However, it was also demonstrated that the revisions were still largely from the point of view of the supplier rather than the user. The supply models had originally assumed that analysis should start with an innovation which was already fully developed, had a stable shape and was recognised by the users to have defined uses. In other words the innovation was a tight package of elements arranged in a coherent manner. It was also assumed that the diffusion of innovations followed an S-shaped pattern based on the initial spread to key centres and its subsequent cascading down a hierarchy of centres and through a neighbourhood effect. Those very deterministic models of the supply side have been revised in the following ways:

1. Innovations are now treated as bundles of elements whose linkages may be tight or loose. The bundle can be prised open and the elements used in differentiated and novel combinations.
2. The shape of the innovation is now treated as problematic and subject to continuing evolvement.
3. The uses of the innovation are now presumed to be uncertain and also open to further and discontinuous evolvement.
4. It is now recognised that the users of innovations vary in their activity. Some users merely seek to ingest the original shape and uses and to retain those features in a stable manner, whilst other users seem to appropriate the innovation and to shape its subsequent shape and uses themselves.
5. The recognition that innovations evolve in their shape and uses became a major preoccupation of the supply side because evolvement has many practical consequences for the supplier. Hence the development by the supply side of models which take consumer activity into account.
6. It was recognised that the user might perceive an innovation as highly uncertain and so complex that the innovation might be rejected. Therefore, attention was given by the supply side to the processes by which an innovation ought to be ingested and institutionalised.
7. It was recognised that there are important innovations whose diffusion is not undertaken by an active agency.
8. The objectivity of the information provided by the supplier was revealed to be problematic, because suppliers are concerned to produce data which show their innovation in the best light.

These developments represent a considerable advance on the previous paradigm, yet they neglect very basic features of the position of the user.

Macro-organization studies has a particular interest in the technological dimension of innovations and in how the process perspective on the evolvement of technology might be constructed. Various established definitions of technology were rejected as unsuited to the present purposes. Instead, the approach taken in Chap. 4 starts by recognising the multidimensional character of innovations as a configuration which is evolving unevenly through time. Specific attention was given to the knowledge dimension, especially to the notion of paradigms and of their influence on the trajectories of evolvement followed by innovations. The notion of paradigms was taken in conjunction with the notion of technology trajectories and the two ideas were carefully delineated and their linkages were specified.

Trajectories of technology may be discerned through the very abstract and extremely fruitful typology of transformation machines, transfer machines and control machines (Kaplinsky 1984). Moreover the typology can be utilised to examine information technology and its impacts in the service sector and amongst white collar and managerial staffs. I have indicated its impacts on the areas of innovation-design. Trajectories of technology are constituted from embodied knowledge of the suppliers and the users. Technology may be embodied in either or both the product (or service) and the process by which it is created. The theory of the sector life course provides a fruitful and relevant perspective which is still being developed. In particular, it is necessary to recognise that several state-of-the-art profiles can coexist in different market segments and in different world-economies. Also, much more attention has to be given to the occurrence of discontinuities when the paradigms which underpin the old profiles are rigidly adhered to by existing power groups of engineers, whilst new paradigms are shaping the profiles of the future. The notion of a life course is preferred to that of a life cycle. It was noted that the same technological inputs can have quite contrary impacts in different settings, even within the same sector and the same country. The new technologies may be utilised to reinforce existing organizational structures or to trigger their creative destruction. The cross-cultural analysis in Part III illustrated that point.

Economists have given saliency to the construction of theories which predict how different combinations of the factor prices of technology and labour will lead to specific outcomes. Such theories have been used to explain the propensities of the British and the Americans to utilise technology. However, the earlier versions of the theory made a too simplistic translation of the outcomes into the inputs. They assumed that the decision structures of the decision-maker could be reconstructed from the outcomes of the decision. The process perspective of organization studies contends that the paradigms of the user may have unintended outcomes, either beneficial or harmful. In other words it is entirely possible for American decision-makers to choose technologically biased solutions on the basis of neo-rationality and to encounter unforeseen gains.

Part II gave particular attention to the user, especially to four features: politics, multiple innovations, innovation-design capacity and appropriation. The political nature of decision-making was emphasised to highlight the conflicts within management as well as the conflicts between management and other strata. Political struggles continuously impact on the capability of the members of the organization to cope with the multiplicity of potential innovations with which they are faced. The user is typically faced with innovations from many very different directions and for that reason ought to attempt to develop a capacity for innovation-design which is capable of recognising when transitions are required and how they might be undertaken. However, because of the influence of the political factors already noted, that adaptation may be entirely inhibited by organizational inertia. A full-blown capacity for innovation-design implies a capability at appropriating core innovations and being active in their matching to the user's specific needs. That requires the development of organizational paradigms and as the comparisons between America and Britain illustrate, those paradigms may require major investiments in education und training.

14.3 Anglo-American Patterns and Transfers

In Part III the perspective formulated in Part II was extended to fully encompass the societal dimension and then applied to Anglo-American patterns of innovation and to transfers between the two societies. The nature of the network of channels and of the historical cultural requirements for the successful appropriation of innovations was presented in terms of the dynamic interactions between the filiere and other factors. It was argued that national patterns of innovation do exist so that a typical variety can be identified. The national filiere for innovation may, or may not, have the enabling capacities to appropriate innovations from other settings. A distinction may be drawn between societal filieres which are capable of identifying significant innovations, particularly innovations in knowledge and in work-organization, in such a way that useful innovations can be retained, edited and reproduced. The possession of that capacity may be termed robust appropriation. The robustness of the American filiere was illustrated through two different case studies in both of which loosely formulated British practices were appropriated (see Chap. 6). The case studies of 'teamwork' were used to exemplify differences between the executive strata in the two countries.

The notion of a filiere was then amplified with reference to three intertwined facets: the economy, structuration and region. Economic pulsations occur, yet their patterning cannot be discerned with any certainty. It is a matter of judgement and debate whether there are longwaves. The theory of longwaves was adopted in its most cautious format. Generic innovations and longwaves were connected. The theory of transaction costs was used to analyse the role of the state and of production institutions in enabling some societies to benefit more from economic circumstance than their neighbours. The theory highlighted the differences in production institutions between the USA and Britain and the consequences of these differences for the usage of production techniques and the continuing, incremental evolvement of production knowledge. It was argued that the production institutions in Britain were less favourable to generic, radical and incremental innovation (see Sect. 7.3.4).

Structuration refers to the sedimented repertoires of cultural frameworks. Specific attention was given to three areas: time, thinking practices and education. It was argued that innovation depends on adopting certain approaches to time, particularly in the commodification of time and in the development of future-oriented programs. Thinking practices shape the core of any filiere and influence the role of education. Together these three areas influence the extent to which the cultural frameworks of the executive strata adjust to new challenges and to transitional periods. It was suggested that the cultural frameworks of American society possess a degree of malleability and flexibility which is not currently possessed by British frameworks. The regional dimension was introduced and then developed in conjunction with the longwaves and the structuration dimensions in four chapters which compared Britain and the USA. Of necessity these took the form of an analytic sketch to illustrate the main themes and to provide a scaffolding from which subsequent studies can be developed. The underlying thrust to these four chapters was the claim that the patterns of innovation in the two countries do differ and the earlier suggestion of robustness in the American approach to the appropriation of innovations was strongly supported.

Chapter 12 examined the fate of various American innovations in the British context. It was demonstrated that American innovations and templates do have an important impact in Britain, but it is not evident that the British typically appropriate relevant innovations, though there have been some notable areas of success (e.g., advertising). In large areas of British work organization there has been the continuity of previous predispositions. Also the adoption of American innovations is at a cosmetic level rather than being a substantial appropriation.

The position of Japan was examined in Chap. 13. The Japanese have developed a filiere which has been able to appropriate significant Western innovations and to ally these to indigenous forms of structuration. In particular, certain sectors of the Japanese economy have developed forms of organization which minimise transaction costs in the mature sectors of automobiles and of electrical goods. Their success is in striking contrast to the position of the British whose weakness in those two areas has considerable consequences for the importation of expensive consumer goods. However, it is not suggested here that the current Japanese success inevitably means that a new world-economy centred on Japan must necessarily arise.

It is relevant to emphasise that the principles which have been elaborated in Part II should be developed through further elaboration of the basic perspective and by systematic analysis of the transatlantic evolvement of innovations.

14.4 New Divides

An analysis of the extent to which America and Britain are heading into a state of retardation or renaissance ought to consider whether the selection criteria which have influenced innovation in the twentieth century are likely to continue into the future. There is a growing consensus amongst exponents of the political economy perspective which claims that a transition has been underway since the mid 1970s. However, their claims are difficult to separate from the analysis of longwaves.

Political economists examining the contemporary scene claim to detect a transition in the means by which capital is accumulated. It is convenient to take the second divide thesis as illustrative of the arguments which are being made (Piore/Sabel 1985). As noted earlier the thesis contends that between 1850 and 1880 there was a struggle between two quite different systems of capitalism. One system relied on craft- and community-based clans to achieve a highly developed division of labour for the production of goods in small batches from labour-intensive workshops operating within a complex system of market linkages. The other system relied on large-scale distribution allied to large-scale manufacture based on the substitution of high-speed machinery and applied science for craft labour. The former mode was represented by the regions of Birmingham in England and Lyons in France, the latter by the rise of the American system of manufacturers. During the struggle between the two systems the latter system tended to be the most prolific at capital accumulation, hence the economic decline of regions of Britain and the rise of new regions in the USA during the twentieth century. After the first divide there came almost a century containing two longwaves during which large-scale homogeneous markets and standardized production were predominant.

The second divide thesis claims that today there is a new struggle between alternative systems of capitalist organization, similar to the struggle more than a century earlier. The contenders today are the already established system of mass consumption/production and flexible manufacturing. The former system has sometimes been labelled Fordism and its limitations and contradictions have been under discussion for some time. It is argued that the new system is based on segmentation of the former mass consumption markets of the wealthier regions. Segmentation has occurred in the provision of goods designed for consumer needs and provided in a faster, tighter time scale. Satisfying these varied tastes requires that suppliers learn about the economies of scope as well as the economies of scale. Thus, the second divide consists of a struggle between Fordism and flexible production. Sabel implies that flexible production requires a return to the earlier forms of community-based production and he cites certain regions of northern Italy as examples of a new wave of social and work organization (Sabel/Zeitlin 1986).

The thesis is an interesting one, yet it is problematic on three counts. First, the trajectory of evolvement of modern technologies has been highly significant in creating an organizational capacity of flexible specialization. Production specialists have proclaimed the flexibility of the new technologies and have argued that manufacturing will be increasingly capable of high flexibility and shorter production runs: the economies of scope. For example, the Italian clothing firm of Benetton (not cited by Sabel) has applied modern information technology to its European-wide selling operations and uses a collection of small local firms to supply garments in a made-up but undyed state. The information from sales is analysed on a daily basis to schedule the dyeing activity. The result is a very tight time framework in which yesterday's sales figures are used to decide today's colours. The goods are then transported directly to the consumers. Benetton are able to retain their flexibility through the use of many small local firms to supply the raw, undyed garments and so there is a small-scale filiere. However, that arrangement of design activity and coordination is very different from the midnineteenth century manufacture of firearms in Birmingham, England. Second, the analysis of flexible specialization in Japan overstates the degree to which the Japanese work community is similar to those of the last century. Third, the second divide thesis does not satisfactorily examine the possibility that a new longwave is underway.

Boyer contends that what is happening today is better described as the replacement of Fordism by neo-Fordism (Boyer/Coriat 1986). That is to say, the new forms of flexible technology are being used by certain innovative firms to slightly alter the structure of the market by responding more flexibly and quickly to consumer-demand patterns. Neo-Fordism would replace the rigid compartmentalised and hierarchised bureaucracies of the early twentieth century with tightly bounded clan-like relations. The result would be the introduction of teamwork at shopfloor levels and amongst management. In fact, neo-Fordism expresses the direction which the Japanese are already well into and which the best-selling American texts of Peters, Waterman and Kanter have been recommending since the start of the decade (Peters/Waterman 1982; Kanter 1984).

Longwave theorists could interpret the contemporary situation as one of structural crisis in which the articulation between the economic and social systems becomes severed and after which there is a political search by governments and by corporations for a new articulation (Freeman/Jones 1985; Perez 1983). Each wave is characterised by the discovery in the upswing of a mingling of technological styles and organizational forms which are highly successful at raising the overall levels of profits. The peak of any longwave tends to be an economic frenzy of relatively short duration which is highly deceptive because all the older sectors are still joining the techno-organizational paradigm which characterised the upswing. In the late 1960s Fordism seemed to be at its most successful and its templates of best practice were gaining adherence around the world, yet by the early 1970s car plants like those of Renault, which had been designed along Fordist lines, were proving inflexible and expensive, whilst those plants like Kalmar, which anticipated the future paradigm, provided the innovative Volvo with an opportunity to monitor the experiment and engage in the learning of a new paradigm. Once the peak is passed there is a falling away of the investment opportunities for embodying the old technology in new products and processes. Consequently the downswing becomes a period of experimentation in new forms of techno-organizational systems at all levels of society. There *ought to be* a proliferation of re-assessments of existing best practice, but that requires a firm understanding of what best practice was, and some agreement on the alternatives. It is easier to identify a definite form of best practice in the USA than in Britain, and the American scene has been a dramatic case of re-assessments. Also, there ought to be a proliferation of new solutions and considerable trial-and-error learning in which existing forms of occupational structuring are replaced. These searches will be driven by the profit motive and those who are not impelled by the profit motive face the possibility of extinction. However, the steering mechanism of the new longwave becomes the new best practice techno-organizational paradigm. The recent work of Wheelwright and Hayes is an excellent *example* of the formulation of a new paradigm (Hayes/Wheelwright 1985).

In each longwave there is a distinctive occupational structure which becomes best practice. During the third and fourth longwaves, successful firms tended to be those which applied the administrative science perspective to their organizational structures and ingested its paradigm (e.g., ICI). Thus, there was a growth of white-collar and managerial tasks in the areas of coordination and of design and those changes reduced the overall costs whilst shifting the distribution of income away from shopfloor levels into the middle strata. The paradigm was based in the USA and it surfaced in various sectors, e.g., oil production, electricity, automobiles and radios. In a similar way those two longwaves introduced major changes in the social institutional contexts of credit systems, advertising and planned obsolescence.

What of the fifth longwave? Perez suggests three central features will be embodied in the new paradigm which is still in the process of being created (Perez 1983). First, the premium will be placed on the use of information-intensive products and processes because these are much cheaper than those based on energy consumption. The stress will be on designs which utilise electronic technology whilst economising on materials, components, moving parts and energy. The result will be both a major shift in the product mix, a transformation in process technologies and in the relationship between the service sector and manufacturing. Second, systems will change from inflexible, dedicated manufacturing of a small range of standard products to plants which are flexible and can produce a range of products: the economies of scope. Third, a filiere will be developed between suppliers, assemblers and sellers based on the systematization of communication and the development of common languages. That tendency is already evident in the recent acquisitions by GM of electronic firms and the similar strategy by Mercedes in Germany. The new systemic linkages will probably be based on information technology rather than on the just-in-time social systems developed by the Japanese.

14.5 Renaissance or Retardation?

The issue of whether the USA is about to follow the pathway to retardation cannot be resolved easily, though certain lines of analysis can be constructed. With respect to the USA the case for renaissance has been made, but that requires substantial alterations in the structure of the economy and in the paradigms held by schools of business and by corporations (Abernathy 1978; Lawrence/Dyer 1983). The issue is whether that transformation can take place. What are the signs? The signs can be observed most keenly by comparing American and British response sets to the current structural crisis. There is a sharp difference. In the American case the fall in competitiveness produced a changing sequence of responses. At first in the early 1970s the rise of Japanese sectors evoked interest, but by the mid-1970s the search was on by business and research institutes to audit the problem and to discover whether the repertoire of existing solutions would be sufficient. Within a decade there had been a sustained investigation and analysis of the Japanese visits, observations and new texts. The early observations highlighted the gaps in performance between Japan and the USA in a very detailed manner, thereby providing a potential for reflective action. Many writers debated whether Japanese practices should and could be adopted. Their debates were accompanied by new forms of experimentation and by even more-searching critiques of American performance. For example, investigations delved into asking just how good are American managements in the use of computer-aided methods for design and production[2]? Contrary to the image of the business schools as analytic ivory tower, the leading schools produced a diversity of responses. Thus, the image is one of renewal.

But how sustained is that image at the level of structuration and at the level of paradigms which routinely guide decision-making? That question is weighty because any examination of Britain in the 1880s might well have drawn the conclusion that dramatic and impending structural reforms were underway and that Britain had the time to adjust. However, Britain subsequently followed political and economic pathways which did not give a salience to innovation as a problem. In contrast to Britain, American society has a much greater capacity to rewrite and recast the roles and scripts which underpin successful performance, but that established capacity rested on the recipes and structural repertoires of Fordism rather than neo-Fordism and its successors.

In the case of the British the issue is almost one of avoiding further retardation, yet it might be argued that there are signs of renaissance. Under a radical Conservative government, there has been a period when the most deeply held consensus values and prestige systems of patronage have been fiercely challenged. The onslaught was so sustained that the personage of the Queen was invoked[3]. Also, in the area of education, the Manpower Services Commission, which was created to introduce change, became a huge financial wedge for introducing short educational and training episodes which differ sharply from what was previously being offered by the educational system. Likewise there has been a sharp increase in privatization and in the number of people on part-time employment. Yet there are few substantial signs of positive change. The point can be made very clearly with respect to the claims made about new forms of work organization and new employment contracts. It has been claimed that British organizations have become more flexible. However, closer examination reveals that this does *not* mean achieving success at economies of scope through greater organizing skills, but refers to the reapplication of a market model to employment relationships. Indeed the much-vaunted flexible firm looks like one typified by very lean design capabilities. There are few signs of additional development of robust design capacities[4].

---

[2] See, for example, the study by Harvey Brooks and Maryellen Kelley, Technical, Economic and Organizational Factors Explaining the Diffusion of Programmable Automation in Machining at the John F. Kennedy School of Government, Harvard University

[3] See *Sunday Times*, 20 September 1986, p 1 and the special insert in *The Economist*, 18-25 July 1986. Also a fascinating article in *Paris Match*, 12 August 1986, pp 48-49 on the relations between the Queen and Prime Minister Thatcher

[4] See the ESRC, WORC program at Aston University

References

Abegglen, J. (1958): *The Japanese Factory*. New York: Free Press.

Abernathy, W.J. (1978): *The Productivity Dilemma. Roadblock to Innovation in the Automobile Industry*. Baltimore: Johns Hopkins.

Abernathy, W.J. (1983): Technological Ferment and the Auto Industry in Japan, USA and Britain. Recorded lecture, ESRC WORC, Aston University.

Abernathy, W.J., Clark, K.B., and A.M. Kantrow (1983): *Industrial Renaissance: Producing a Positive Future for America*. Boston: MIT Press.

Abernathy, W.J., and K.B. Clark (1985): Innovation: Mapping the Winds of Creative Destruction. *Research Policy* 14, 3-22.

Abernathy, W.J., and R.H. Hayes (1980): Managing our Way to Economic Decline. *Harvard Business Review* October, 69-77.

Abernathy, W.J., Clark, K.B., and A.M. Kantrow (1981): The New Industrial Competition. *Harvard Business Review* October, 68-81.

Abramovitz, M. (1966): Resources and Output in the US Since 1870. *American Economic Review* 46, 5-23.

Aglietta, M. (1976): *A Theory of Capitalist Regulation*. London: New Left Books.

Aguren, S. (1984): *Volvo Kalmar Revisited. Ten Years of Experience*. Stockholm: Swedish Employers Federation.

Albrow, M. (1970): *Bureaucracy*. London: Pall Mall.

Aldcroft, D.H. (1968): *The Development of British Industry and Foreign Competition, 1875-1914*. London: Methuen.

Aldrich, H.E. (1979): *Organisations and Environments*. Englewood Cliffs, N.J.: Prentice-Hall.

Aldrich, H.E., McKelvey, W., and D. Ulrich (1984): Design Strategy from the Population Perspective. *Journal of Management* 10 (1), 67-86.

Alford, B.W.E. (1973): *W.D. & H.O. Wills and the Development of the U.K. Tobacco Industry, 1786-1965*. London: Methuen.

Alford, B.W.E. (1976a): Strategy and Structure in the U.K. Tobacco Industry. *Management Strategy and Business Development*. Hannah, L. (Ed.) London: Macmillan, Chapter 3.

Alford, B.W.E. (1976b): Penny Cigarettes, Oligopoly and Entrepreneurship in the U.K. Tobacco Industry in the Late Nineteenth Century. *Essays in British Business History*. Supple, B. (Ed.) Oxford: Oxford University Press, 49-68.

Alpers, S. (1983) *The Art of Describing: Dutch Art in the Seventeenth Century*. Chicago: Chicago University Press.

Allport, F.H. (1962): A Structuronomic Conception of Behaviour: Individual and Collective. *Journal of Abnormal and Social Psychology* 64 (1), 3-30.

Arieas, P. (1986): *Le Temps d'Historie*. Paris: Seuil.

Argyris, C., and D.A. Schon (1978): *Organisation Learning. A Theory of Action Perspective*. Reading: Addison-Wesley.

Arrow, K. (1962): The Economic Implications of Learning-by-Doing. *Review of Economic Studies* 29, 155-173.

Baker, W.J. (1984): *Sports in the Western World.* River City: Rowan and Littlefield.

Baker, W.J., and J.M.Carroll (1983): *Sports in American Society.* River City: Rowan and Littlefield.

Baldamus, W. (1961): *Efficiency and Effort. An Analysis of Industrial Administration.* London: Tavistock.

Bale, J.R. (1980): The Adoption of Football in Europe - a Geographic Perspective. *Canadian Journal of the History of Sport and Physical Education* XI (1), 56-86.

Bale, J.R. (1984): International Sports History as Innovation Diffusion. *Canadian Journal of History of Sport* XV (1), 38-63.

Bandura, A. (1977): *Social Learning Theory.* Englewood Cliffs, N.J.: Prentice-Hall.

Bandura, A. (1986): *Social Foundations of Thought and Action: A Social Cognitive Theory.* Englewood Cliffs, N.J.: Prentice-Hall.

Bardou, J.P., Chanaron, J-J., Fridenson, P., and J.M.Laux (1982): *The Automobile Revolution. The Impact of an Industry.* North Carolina: North Carolina University Press.

Barker, T.C. (1977): *The Glassmakers Pilkington: The Rise of an International Company.* London: Weidenfeld and Nicolson.

Barnett, C. (1976): *The Human Factor and British Industrial Decline. An Historical Perspective.* London: Working Together Campaign.

Bayley, S. (1983): *Harley Earl and the Dream Machine.* New York: Knopf.

Bell, R.M. (1972): *Changing Technology and Manpower Requirements in the Engineering Industry.* London: Sussex University Press.

Benedict, R. (1946): *The Chrysanthemum and the Sword.* Boston: Houghton Mifflin.

Bentley, F.R. (1964): *Relating Wages to Productivity.* London: Bentley.

Bergson, H. (1903): *Introduction a la Metaphysique.* Paris: Seuil.

Blake, R.R., and J.S.Mouton (1974): *Grid Organisation Development.* Houston: Gulf Publishing.

Boyer, R. (1985): New Technologies and Employment in the Eighties. Working Paper No.8526, Paris: CEPREMAP.

Boyer, R., and B.Coriat (1986): Technical Flexibility and Macro Stabilisation. Working Paper to Conference on Innovation Diffusion, Venice, March.

Braudel, F. (1958): Time, History and the Social Sciences. *Varieties of History: From Voltaire to the Present.* Stern, F. (Ed.) (North American edition) New York: Free Press.

Braudel, F. (1972): *The Mediterranean and the Mediterranean World.* London: Collins.

Braudel, F. (1983): *The Structure of Everyday Life.* London: Collins.

Braudel, F. (1984): *The Perspective of the World.* London: Collins.

Braverman, H. (1974): *Labor and Monopoly Capital. The Degradation of Work in the Twentieth Century.* New York: Monthly Review Press.

Briggs, A. (1963): *Victorian Cities.* London: Odhams.

Bright, J.R. (1958): *Automation and Management.* Boston: Irwin.

Brittain, J.H., and J.H.Freeman (1980): Organisational Proliferation and Density Dependent Selection. *Organisation Life Cycles.* Kimberley, J.R., and R.H.Miles (Eds.) San Francisco: Jossey Bass.

Broeg, R. (1983): The Impact of Sport on Americans. *Sports in American Society.* Baker, W.J., and J.M.Carroll (Eds.) River City: Rowan and Littlefield.

Brookman, C. (1984): *America and Culture and Society.* London: Macmillan.

Brooks, H., and M.-E.Kelley (1986): Technical, Economic and Organisational Factors Explaining the Diffusion of Programmable Automation of Machining. Working Paper, J.F.Kennedy School of Government, Harvard University.

Brown, L.A. (1981): *Innovation Diffusion. A New Perspective.* London: Methuen.

Brunsson, N. (1985): *The Irrational Organisation: Irrationality as a Basis for Organisation Action and Change.* London: Wiley.

Burawoy, M. (1984): *The Politics of Work.* London: Heinemann.

Burnett, J.M. (1969): Ceremony, Rites and Economy in the Student System of an American High School. *Human Organisation* 28 (1), 1-10.

Burns, T., and G.M.Stalker (1961): *The Management of Innovation.* London: Tavistock.

Campinos-Dubernet, M. (1986): The Rationalisation of Labour in the (French) Construction Industry: The Limits of Orthodox Taylorism. *Proceedings of the 7th Bartlett Summer School.* London: University College.

Carroll, P.N., and D.W. Noble (1980): *The Free and the Unfree. A New History of the United States.* London: Pelican.

Caves, R.E. (1980): Productivity Differences Among Industries. *Britain's Economic Performance.* Caves, R.E., and L.B.Krause (Eds.) Washington: Brookings Institution.

Chandler, A.D. (1962): *Strategy and Structure.* Boston: MIT Press.

Chandler, A.D. (1977): *The Visible Hand. The Managerial Revolution in America.* Cambridge: Belknap.

Chandler, A.D. (1980): The Growth of the Transnational Industrial Firm in the US and the UK: A Comparative Analysis. *The Economic History Review* 33 (3), 396-410.

Chandler, A.D. (1984): The Emergence of Managerial Capitalism. *Business History Review* 58 (4), 473-503.

Chandler, A.D., and H.Daems (1980): *Managerial Hierarchies. Comparative Perspectives on the Rise of the Modern Industrial Enterprise.* Cambridge: Harvard University Press.

Channon, D.F. (1973): *The Strategy and Structure of British Enterprises.* London: Macmillan.

Channon, D.F. (1978): *The Service Industries: Strategy, Structure and Financial Performance.* London: Macmillan.

Chapman, S.D. (1974): *Jesse Boot of Boots the Chemists. A Study in Business History.* London: Hodder and Stoughton.

Child, J. (1969): *British Management Thought.* London: Allen and Unwin.

Chirot, D. (1985): The Rise of the West. *American Sociological Review* 50 (2), 181-194.

Church, R. (1968): The Effect of the American Export Invasion on the British Boot and Shoe Industry. *The Journal of Economic History* 2, 223-254.

Churchman, C.W. (1968): *Challenge to Reason.* New York: McGraw Hill.

Clark, K.B. (1985): The Interaction of Design Hierarchies and Market Concepts in Technological Evolution. *Research Policy* 14, 235-251.

Clark, N. (1985): *The Political Economy of Science and Technology.* Oxford: Blackwell.

Clark, P. A. (1972a): *Organisational Design: Theory and Practice.* London: Tavistock.

Clark, P. A. (1972b): *Action Research and Organisation Change.* London: Harper Row.

Clark, P. A. (1975a): Intervention Theory: Matching Role, Focus and Context. *The Quality of Working Life Volume One.* Davis, L. E., and A. B. Cherns (Eds.) New York: Free Press.

Clark, P. A. (1975b): Key Problems in Organisational Design. *Administration and Society* 7 (2), 213-256.

Clark, P. A. (1976): Some Analytic Requirements of an Applied Organisational Science. *The Management of Organisation Design: Strategies and Implementation.* Kilmann, R., Pondy, L. R., and D. P. Slevin (Eds.) New York: North Holland.

Clark, P. A. (1979): Cultural Context as a Determinant of Organisational Rationality: A Comparison of the Tobacco Industries in Britain and France. *Organisations Alike and Unlike: International and Inter-Institutional Studies in the Sociology of Organisations.* Lammers, C. J., and D. J. Hickson (Eds.) London: Routledge and Kegan Paul.

Clark, P. A. (1981): *The Administrative Structure of Canadian Textile Establishments.* Ottawa: Economic Council of Canada.

Clark, P. A. (1984): Leadership Theory: The Search for a Reformulation. *Leaders and Managers.* Hunt, J. G., Hosking, D-M., Schreishiem, C. A., and R. Stewart (Eds.) New York: Pergamon Press.

Clark, P. A. (1985): A Review of Theories of Time and Structure for Organisational Sociology. *Research in the Sociology of Organisations, vol. 4.* Bachrach, S. B., and Mitchell, S. M. (Eds.) Connecticut: JAI Press.

Clark, P. A. (1986a): Le Capitalism et la Reglementation du Temps de Travail: Une Critique de la These d'E. P. Thompson. *Temps Libre* March, 27-32.

Clark, P. A. (1986b): The Economy of Time and the Managerial Division of Labour in the British Construction Industry 1965-1985. *Proceedings of the 7th Bartlett International Summer School.* London: University College London.

Clark, P. A. (1987): Social Technology and the Construction Firm: Structure, Innovation and Process. *Management Strategy of Large UK Construction Firms in the Domestic and International Operations.* Biggs, W. D., Dunning, J., and P. G. Hillebrandt (Eds.) London: Macmillan.

Clark, P. A., and J. R. Ford (1970): Methodological and Theoretical Problems in the Investigation of Planned Organisational Change. *Sociological Review* 18 (1), 29-32.

Clark, P. A., and K. P. Starkey (1987): *Organisation Transitions and Innovation-Design.* London: Frances Pinter.

Clark, P. A., and J. Tann (1986): Cultures and Corporations: The M-form in the USA and in Britain. Paper to the International Academy of Business, ESRC, WORC Aston University.

Clark, P. A., and R. Whipp (1985): Industrial Change in Britain: A Comparative Historical Investigation Into Innovation. *ESRC WORK Working Paper No. 5,* Aston University.

Clark, P. A., and J. Windebank (1985): Innovation and Renault 1900-1982. Product, Process and Work Organisation. *ESRC WORC Working Paper No. 13,* Aston University.

Clegg, S., and D. Dunkerley (1980): *Organisation, Class and Control.* London: Routledge and Kegan Paul.
Coakley, J., and L. Harris (1983): *The City of Capital. London's Role as a Financial Centre.* Oxford: Blackwell.
Coase, R. (1937): The Nature of the Firm. *Economics* 4, 386-405.
Cochran, T.C. (1981): *Frontiers of Change: Early Industrialism in America.* Oxford: Oxford University Press.
Cohen, M.D., March, J.G., and J.P. Olsen (1972): A Garbage Can Model of Organisation Choice. *Administrative Science Quaerterly* 17, 1-25.
Cole, R. (1971): *Japanese Blue Collar: The Changing Tradition.* Berkeley: University of California Press.
Coleman, D.C. (1973): Gentlemen and Players. *Economic History Review* 2 (26), 99-112.
Coleman, D.C. (1969): *Courtaulds: An Economic and Social History.* Oxford: Oxford University Press.
Coleman, D.C. (1977): Courtaulds and the Beginning of Rayon. *Essays in British Business History.* Supple, B. (Ed.) Oxford: Oxford University Press.
Coleman, J.S., Katz, E., and H. Menzel (1966): *Medical Innovation. A Diffusion Study.* New York: Bobbs Merrill.
Conley, P. (1970): *Experience Curves as a Planning Tool: A Special Commentary.* Boston: Boston Consulting Group.
Coombs, R. (1983): Automation, Management Strategies and Labour Process Change. Working Paper, UMIST, Manchester.
Coriat, B. (1976): *L'Atelier et le Chronometre.* Paris: Seuil.
Crafts, N.F.R. (1985): *British Economic Growth During the Industrial Revolution.* Oxford: Clarendon Press.
Crane, D. (1972): *Invisible Colleges. Diffusion of Knowledge in Scientific Communities.* Chicago: University of Chicago Press.
Crouzet, F. (1982): *The Victorian Economy.* London: Methuen.
Crozier, M. (1964): *The Bureaucratic Phenomenon.* London: Tavistock.
Crozier, M. (1973): *The Stalled Society.* New York: Viking Press.
Crozier, M. (1984): *The Trouble with America.* Berkeley: University of California Press.
Cullen, J.C. (1972): Space, Time and the Disruption of Behaviour in Cities. *Environment and Planning* 4, 459-470.
Davis, L.E. (1971): The Coming Crisis for Production Management: Technology and Organisation. *International Journal of Production Research* 9, 65-82.
Davis, L.E., and A.B. Cherns (Eds.) (1975): *The Quality of Working Life.* (two volumes) New York: Free Press.
David, P.A. (1975): *Technological Choice, Innovation and Economic Growth. Essays on American and British Experience in the 19th Century.* Cambridge: Cambridge University Press.
DeBresson, C., and J. Lampel (1985a): Beyond the Life Cycle II. An Illustration. *Journal of Product Innovation Management* 3, 170-187.
DeBresson, C., and J. Lampel (1985b): Beyond the Life Cycle: Organisational and Technological Design. An Alternative Perspective. *Journal of Product Innovation Management* 3, 188-195.

Denzig, A. (1956): *The History of American Football: Its Great Teams, Players and Coaches*. Englewood: Prentice-Hall.

DiMaggio, B., and W.W. Powell (1983): The Iron Cage Revisited: Institutional Isomorphism and Collective Rationality in Organisational Fields. *American Sociological Review* 82, 147-160.

Dohse, K., Jürgens, U., and T. Malsch (1984): From "Fordism" to "Toyotism"? The Social Organisation of the Labour Process in the Japanese Automobile Industry. *IIGB Discussion Paper* International Institute for Comparative Social Research, Wissenschaftszentrum, Berlin.

Donaldson, L. (1985): *In Defence of Organisation Theory: A Reply to the Critics:* Cambridge: Cambridge University Press.

Dore, R.P. (1973): *British Factory, Japanese Factory*. London: Allen and Unwin.

Dosi, G. (1982): Technological Paradigms and Technological Trajectories. *Research Policy* 11 (3), 147-162.

Dubois, P., and D. Monjardet (1979): *La Division du Travail dans l'Industrie. Etude de Cas Anglais et Francais*. Paris: Groupe de Sociologie du Travail.

Duchesneau, T.D., Cohn, S.F., and J.E. Dutton (1979): *A Study of Innovation in Manufacturing: Determinants, Processes and Methodological Issues*. Report to the National Science Foundation, No. RDA-7423652, University of Maine.

Duncan, R.B. (1972): Characteristics of Organisation Environments and Perceived Environmental Uncertainty. *Administrative Science Quarterly* 17, 313-327.

Duncan, R.D. (1974): A Contingency Approach to Designing Organisations for Innovation. Paper given at Pittsburgh Seminar on Organisation Design, and available from North Western University.

Dunning, E., and K. Sheard (1979): *Barbarians, Gentlemen and Players*. London: Martin Robertson.

Durkheim, E. (1915): *The Elementary Forms of the Religious Life*. London: Allen and Unwin.

Economist, The (1985): Japan: A Survey. *The Economist*, 7 December.

Economist, The (1986): High Technology – The Clash of the Titans. *The Economist*, 23 August, 48f.

Edwards, R. (1979): *Contested Terrain. The Transformation of the Workplace in the Twentieth Century*. London: Heinemann.

Eisenstadt, S.N. (1966): *Modernisation: Protest and Change*. Englewood Cliffs, N.J.: Prentice-Hall.

Elbaum, B., and W. Lazonick (Eds.) (1986): *The Decline of the British Economy*. Oxford: Clarendon Press.

Eldon, M. (1978): *Three Generations of Work Democracy Experiments in Norway – Beyond Classical Socio Technical Analysis*. Trondheim: Technical University of Trondheim.

Elias, N. (1978): *The Civilising Process*. Oxford: Blackwell.

Elias, N., and E. Dunning (1971): Folk Football in Medieval and Early Modern Britain. *The Sociology of Sport*. Dunning, E. (Ed.) London: Frank Cass, 313-327.

Eveland, J.D. (1979): Issues in Using the Concept of 'Adoption' in Innovations. *Journal of Technology Transfer* 4 (1), 1-14.

Eveland, J.D., and E.M. Rogers (1980): Measuring the Innovation Process in Public Organisations. Working Paper, National Science Foundation, Washington, USA.

Fligstein, N. (1985): The Spread of the Multi-Divisional Form Amongst Large Firms, 1949-1979. *American Sociological Review* 50, 377-391.

Foster, J. (1974): *Class Struggle and the Industrial Revolution. Early Capitalism in Three English Towns.* London: Weidenfeld and Nicolson.

Foucault, M. (1977): *Discipline and Punish. The Birth of the Prison.* London: Penguin.

Fox, A. (1984): *History and Heritage. The Social Origins of British Industrial Relations.* London: Allen and Unwin.

Foxall, G., and J. Tierney (1984): From CAP 1 to CAP 2: User-Initiated Innovation From the User's Point of View. *Management Decision* 22 (5), 3-15.

Freeman, C. (1974): *The Economics of Industrial Innovation.* (second edition) London: Frances Pinter.

Freeman, C. (1982): *Unemployment and Technical Innovation.* London: Frances Pinter.

Freeman, C. (Ed.) (1983): *Longwaves in the World Economy.* London: Butterworth.

Freeman, C., and D.T. Jones (1985): *Technological Trends and Employment 4. Engineering and Vehicles.* London: Gower.

Freeman, C., and L. Soete (Eds.) (1985): *Technical Change and Full Employment.* Oxford: Blackwell.

Fridenson, P. (1972): *Histoire des Usines Renault.* Paris: Seuil.

Friedman, A. L. (1977): *Industry and Labour. Class Struggle at Work and Monopoly Capitalism.* London: Macmillan.

Friedman, M. (1962): *Capitalism and Freedom.* Chicago: University of Chicago Press.

Galbraith, J. R. (1977): *Organisation Design.* Reading: Addison Wesley.

Gallie, D. (1978): *In Search of the New Working Class.* Cambridge: Cambridge University Press.

Gallie, D. (1983): *Social Inequalities and Class Radicalism in France and Britain.* Cambridge: Cambridge University Press.

Gardiner, J. P. (1983): Design Trajectories for Airplanes and Automobiles During the Past Fifty Years. Working Paper, SPRU, University of Sussex.

Gearing, E. (1958): The Structural Poses of the Eighteenth Century Cherokee Villages. *American Anthropologist* 60.

Gent, P. (1974): *North Dallas Forty.* London: Michael Joseph.

Gershuny, J. I. (1983): *Social Innovation and the Division of Labour.* Oxford: Oxford University Press.

Gershuny, J. I., and I. D. Miles (1983): *The New Service Economy. The Transformation of Employment in Industrial Societies.* London: Frances Pinter.

Geyl, P. (1964): *History of the Low Countries. Episodes and Problems.* New York: Macmillan.

Giddens, A. (1979): *Central Problems in Social Theory: Action, Structure and Contradiction in Social Analysis.* London: Macmillan.

Giddens, A. (1981): *A Contemporary Critique of Historical Materialism.* London: Macmillan.

Giddens, A. (1985): *The Constitution of Society.* Oxford: Polity Press.

Gideon, S. (1967): *Space, Time and Architecture.* (fifth edition) New York: Oxford University Press.

Gill, R. W. T., and K. G. Lockyer (1979): *The Career Development of the Production Manager in British Industry.* London: British Institute for Production Management.

Gold, B. (1975): *Technical Change Economics, Management and Environment.* Oxford: Pergamon.
Gold, B. (1979): *Productivity, Technology and Capital.* Massachusetts: Lexington.
Gold, B. (1980): *Productivity, Technology and Capital. Economics Analysis, Managerial Strategies and Government Policies.* Lexington: Lexington Press.
Gold, B. (1981): *Evaluating Technological Innovations. Methods, Expectations and Findings.* Lexington: Lexington Press.
Goodrich, C. L. (1975): *The Frontier of Control. A Study in British Workshop Politics.* London: Pluto Press.
Goudsblom, J. (1967): *Dutch Society.* New York: Random House.
Granovetter, M. S. (1973): The Strength of Weak Ties. *American Sociological Review* 78, 1360-1380.
Greenberg, D. S. (1966): "Hindsight": DoD Study Examines Return on Investment in Research. *Science* 18, November.
Gregory, D. (1982): *Regional Transformation and Industrial Revolution: A Geography of the Yorkshire Woollen Industry.* London: Macmillan.
Gross, N. G., Giaquinta, J. B., and M. Bernstein (1971): *Implementing Organisational Innovations.* New York: Basic Books.
Gruneau, R. S., and J. G. Albinson (1976): *Canadian Sport: Sociological Perspectives.* Toronto: Addison-Wesley.
Gurvitch, G. (1964): *The Spectrum of Social Time.* Dordrecht: Reidel.
Hage, J. (1980): *Theories of Organisation: Form, Process and Transformation.* New York: Wiley.
Hagerstrand, T. (1952): The Propagation of Innovation Waves. *Human Geography* 4, 3-19.
Hall, E. T. (1959): *The Silent Language.* Garden City: Doubleday.
Hall, R. H. (1972): *Organisations: Structure and Process.* (first edition) Englewood Cliffs, N. J.: Prentice-Hall.
Hall, R. I. (1984): The Natural Logic of Management Policy Making: Its Implications For the Survival of an Organisation. *Management Science* 30 (8), 905-927.
Hannah, L. (1974): Managerial Innovation and the Rise of the Large Scale Company in Inter-War Britain. *Economic History Review* 27, 252-270.
Hannah, L. (Ed.) (1976): *Management Strategy and Business Development.* London: Macmillan.
Hannah, L. (1980): Visible and Invisible Hands in Great Britain. *Managerial Hierarchies. Comparative Perspectives on the Rise of the Modern Industrial Enterprise.* Chandler, A. D., and H. Daems (Eds.) Boston: Harvard University Press.
Hannah, L. (1983): *The Rise of the Corporate Economy.* London: Methuen.
Hannan, M., and J. Freeman (1977): The Population Ecology Model of Organisations. *American Journal of Sociology* 82, 929-964.
Harley, C. K. (1971): The Shift From Sailing Ships to Steamships: A Study in Technological Change and Its Diffusion. *Essays on a Mature Economy: Britain After 1840.* McCloskey, D. N. (Ed.) London: Methuen.
Harley, C. K. (1974): Skilled Labour and Choice of Technique in Edwardian Industry. *Explorations in Economic History* 11, 391-414.
Harre, R. (1979): *Social Being.* Oxford: Blackwell.
Hart, M. (1972): *Sport of the Socio-Cultural Process.* New York: Dubeque.

Hayes, R.H., and S.Wheelwright (1985): *Restoring our Competitive Edge. Competing Through Manufacturing*. New York: Wiley.
Hazama, H. (1986): Japanese Industrialisation and Labour-Management Relations. Working Paper, Waseda University, Japan.
Heerding, A. (1986): *The History of N. V. Philips' Gloeilampenfabrieken. Volume 1. The Origin of the Dutch Incandescent Lamp Industry*. Cambridge: Cambridge University Press.
Hexter, H.J. (1972): Fernand Braudel and Monde Braudellian. *Journal of Modern History* XI (3), 1-23.
Hickson, D.J., Pugh, D.S., and D.C.Pheysey (1969): Operations Technology and Organisation Structure: An Empirical Re-Appraisal. *Administrative Science Quarterly* 14 (3), 378-397.
Hickson, D.J., Butler, R.J., Cray, D., Mallory, G.R., and D.C.Wilson (1986): *Top Decisions. Strategic Decision Making in Organisations*. Oxford: Blackwell.
Higgin, G.W., and W.N.Jessop (1965): *Communications in the Building Industry: Uncertainty and Interdependence*. London: Tavistock.
Hinkle, R.P. (1979): The Mirassou Generations. *The Friends of Wine* Nov/Dec, 60-63.
Hinton, J. (1983): *Labour and Socialism. A History of the British Labour Movement 1867-1974*. Boston: University of Massachusetts Press.
Hjelholt, G. (1963): Training for Reality. Working Paper No.5, Industrial Management Division, University of Leeds.
Hobsbawm, E.J. (1983): *The Invention of Tradition*. Cambridge: Cambridge University Press.
Hoch, R. (1972): *Rip Off! The Big Game*. New York: Doubleday.
Hofstadter, R. (1955): *Social Darwinism in American Thought*. Boston: Beacon Press.
Hofstede, G. (1980): *Cultures Consequences. International Differences in Work Related Values*. London: Sage.
Hounshell, D.A. (1984): *From the American System of Mass Production, 1800-1932: The Development of Manufacturing Technology in the United States*. Baltimore: Johns Hopkins University Press.
Howell, N., and M.L.Howell (1969): *Sports and Games in Canadian Life*. Toronto: Hunter Rose.
I.D.E. (1980): *Industrial Democracy in Europe*. Oxford: Oxford University Press.
Ingham, G. (1984): *Capitalism Divided? The City and Industry in British Social Development*. London: Macmillan.
Janssen, J. (1986): Disparities Constituting the Dynamics of Modes of Production. Middlesex and London Building Production in the Transition From Feudalism to Capitalism. *Proceedings of the 7th Bartlett International Summer School*. London: University College, London.
Jeremy, D.J. (1981): *Transatlantic Industrial Revolution*. Oxford: Blackwell.
Jones, E. (1982): *Accountancy and the British Economy 1840-1980*. London: Allen and Unwin.
Jones, G.S. (1972): *Outcast London*. Oxford: Oxford University Press.
Jonsson, B. (1981): Corporative Strategy for People at Work - The Volvo Experience. Company Report, *AB Volvo*.
Kaldor, M. (1980): Technical Change in the Defence Industry. *Technical Innovation and British Economic Performance*. Pavitt, K. (Ed.) London: Macmillan.

Kamien, M.I., and N.L. Schwartz (1982): *Market Structure and Innovation*. Cambridge: Cambridge University Press.
Kanter, R.M. (1984): *The Change Masters. Corporate Entrepreneurs at Work*. London: Allen and Unwin.
Kaplinsky, R. (1984): *Automation: The Technology and Society*. London: Longmans.
Karpik, L. (1978): Organisations, Institutions and History. *Organisation and Environment: Theory, Issues and Reality*. Karpik, L. (Ed.) Beverley Hills: Sage.
Katz, D., and R. Kahn (1978): *Social Psychology of Organisations*. New York: John Wiley.
Keeble, D., Owens, P.L., and D. Thompson (1982): *Centrality, Peripherality and the EEC Regional Development*. Cambridge: Cambridge University Department of Geography.
Keidel, R.W. (1984): Baseball, Football and Basketball: Models for Business. *Organisation Dynamics* Winter, 5-18.
Kondratiev, N.D. (1976): The Long Waves in Economic Life. *Lloyds Bank Review*, 41-60.
Koshira, K. (1983): Personnel Planning, Technological Change and Outsourcing in the Japanese Automobile Industry. Working Paper, Department of Economics, Yokohama National University, Japan.
Kranzberg, M. (1970): Transferring a Dynamic Technology During Socio-Political Change: The American Experience 1607-1850. Paris: Colloques Internationaux, CNRS No. 538.
Kuhn, T.S. (1970): *The Structure of Scientific Revolutions*. (second edition) London: Chicago University Press.
Kuznets, N. (1953): Schumpeter's Business Cycles. *Economic Change*. Kuznets, N. (Ed.) New York: Wiley.
Ladurie E. Leroy (1979): The "Event" and the "Long Term" in Social History: The Case of the Chouan Uprising. *The Territory of the Historian*. Ladurie, E. Leroy (Ed.) London: Harvester Press.
Lall, S. (1980): Developing Countries as Exporters of Industrial Technology. *Research Policy* 9 (1), 24-53.
Lammers, C.J., and D.J. Hickson (1979): *Organisations Alike and Unlike: International and Inter-institutional Studies in the Sociology of Organisations*. London: Routledge and Kegan Paul.
Landes, D.J. (1969): *The Unbound Prometheus: Technological Change and Industrial Development in Western Europe from 1750 to the Present*. London: Cambridge University Press.
Landes, D.J. (1983): *Revolution in Time*. Boston: Belknap.
Lane, C. (1981): *The Rites of Rulers. Ritual in Society. The Soviet Case*. Cambridge: Cambridge University Press.
Langrish, J., Gibbons, M., Evans, W.G., and F.R. Jevons (1972): *Wealth from Knowledge*. New York: John Wiley.
Larsen, J., and R.A. Rogers (1977): Reinvention of Innovative Ideas: Modified? Adapted? Or None of the Above? *Evaluation* 6, 136-140.
Lawrence, P.R., and R. Dyer (1983): *Renewing American Industry*. New York: Free Press.
Lazonick, W. (1983): Industrial Organisation and Technological Change. The Decline of the British Cotton Industry. *British History Review* 57 (2), 195-236.

Lazonick, W. (1987): Strategy, Structure and Management Development in the United States and Britain. *The Development of Managerial Enterprises*. Kobayashi, K. (Ed.) Tokyo: University of Tokyo Press.
Lazonick, W., and W. Mass (1983): The Performance of the British Cotton Industry 1870-1913. Working Paper to the ESRC Summer School on Work Organisation, Warwick University.
Learned, P.E., and A.T. Sproat (1966): *Organisation Theory and Policy. Notes for Analysis*. Illinois: Irwin.
Leontief, W. (1985): The Choice of Technology. *Scientific American* 252 (6), 25-33.
Lewchuk, W. (1987): *The British Motor Vehicle Industry. The Roots of Decline 1895-1983*. (pre-publication draft) McMaster University, Canada.
Lievegoede, B.C.J. (1980): *The Developing Organisation*. London: Celestial Arts.
Linton, R., and A. Kardiner (1952): The Change from dry to wet Rice Cultivation in Tanala-Betsileo. *Readings in Social Psychology*. Swanson, G. (Ed.) New York: Holt.
Littler, C.R. (1982): *The Development of the Labour Process in Comparative Societies*. London: Macmillan.
Loveridge, R., Child, J., and J. Harvey (1985): New Technologies in Banking, Retailing and Health Services, the British Case. Working Paper, ESRC WORC MESS Project, Aston University.
Lupton, T., and I. Tanner (1980): Work Organisation and Design in Europe. *Changes in Working Life*. Duncan, K.D., Gruneberg, M.M., and D. Wallis (Eds.) London: Wiley.
Macklin, K. (1974): *The History of Rugby League Football*. London: Stanley Paul.
Maddison, A. (1982): *Phases of Capitalist Development*. Oxford: Oxford University Press.
Mandel, E. (1978): *Late Capitalism*. London: Verso.
Mandel, E. (1980): *Longwaves of Capitalist Development: The Marxist Interpretation*. Cambridge: Cambridge University Press.
Manning, P.K. (1979): Metaphors of the Field: Varieties of Organisational Discourse. *Administrative Science Quarterly* 24, 660-671.
Mansfield, E. (1961): Technical Change and the Rate of Innovation. *Econometrics* 29, 741-761.
Mansfield, E. (1968a): *Industrial Research and Technological Innovation: An Economic Analysis*. New York: Norton.
Mansfield, E. (1968b): *The Economics of Technological Change*. New York: Norton.
Mansfield, E. (1977): The Diffusion of Eight Major Industrial Innovations. *The State of Science and Research: Some New Indicators*. Terleckjy N.E., Boulder: Westview Press.
March, J.G., and J.P. Cohen (1976): *Ambiguity and Choice in Organisations*. Oslo: Universitetsforlaget.
March, J.G., and H.A. Simon (1958): *Organisations*. New York: Wiley.
Martin, F., Swan, N., Banks, I., Barket, R., and R. Beadry (1979): *The Inter-Regional Diffusion of Innovations in Canada*. Ottawa: Economic Council of Canada.
Marx, K. (1954): *Capital*. London: Lawrence and Wishart.
Mason, R.O., and I.I. Mitroff (1981): *Challenging Strategic Planning Assumptions. Theory, Cases and Technique*. New York: Wiley.

Mason, T. (1980): *Association Football in English Society 1863-1915.* London: Harvester Press.

Massey, D., and R. Meegan (1982): *The Anatomy of Job Loss. The How, Why and Where of Employment Decline.* London: Methuen.

Mathias, P. (1983): *The First Industrial Nation: An Economic History of Britain 1700-1914.* London: Methuen.

Mauss, M. (1904): Essai sur les Variations Saisonnierres les Societies Eskimaux. *L'Annee Sociologique* IX, 39-102.

Mayr, O., and R. C. Post (Eds.) (1981): *Yankee Enterprise. The Rise of the American System of Manufacture.* Washington D.C.: Smithsonian Institution.

Mensch, G. (1979): *Stalemates in Technology: Innovations Overcome the Depression.* Massachusetts: Balinger.

Meyer, M.W., Stevenson, W., and S. Webster (1985): *The Limits to Bureaucratic Growth.* Berlin: de Gruyter.

Midler, C. (1980): L'Organisation du Travail et ses Determinants, Successful Doctoral Thesis, Universite de Paris.

Miles, R.E., and C.C. Snow (1978): *Organisational Strategy, Structure and Process.* Tokyo: McGraw-Hill.

Miles, R.H. (1980): *Macro-Organisational Behaviour.* Santa Monica: Goodyear.

Miller, D., and P.H. Friesen (1982): *Organisations. A Quantum View.* Englewood Cliffs, N.J.: Prentice-Hall.

Miller, E.J., and A.K. Rice (1967): *Systems of Organisation. The Control of Task and Sentient Boundaries.* London: Tavistock.

Mintzberg, H., Raisinghani, D., and A. Thoret (1976): The Structure of Unstructured Decision Processes. *Administrative Science Quarterly* 21, 246-275.

Mitchell, C. (1970): *Networks.* Manchester: Manchester University Press.

Mitroff, I.I. (1981): *Creating Dialectical Social Science. Concepts, Methods and Models.* Dordrecht: Reidel.

Mohr, L.B. (1982): *Explaining Organisational Behaviour: The Limits and Possibilities of Theory and Research.* San Francisco: Jossey Bass.

Monjardet, R. (1974): *Rapports entre Organisation du Travail et Development Technologie dans l'Industrie Automobile.* Paris: ACT.

Monopolies Commission (1961): *Report on the Supply of Cigarettes and Tobacco and of Cigarette and Tobacco Machinery.* London: HMSO.

Montmollin, M., and O. Pastre (Eds.) (1984): *Le Taylorisme.* Paris: La Decouverte.

Morishima, M. (1985): *Why has Japan "Succeeded"? Western Technology and the Japanese Ethos.* Cambridge: Cambridge University Press.

Mowery, D.C., and N. Rosenberg (1979): The Influence of Market Demand Upon Innovation: A Critical Review of Some Empirical Studies. *Research Policy* 8, 103-153.

Musson, A.E. (Ed.) (1969): *Science and Technology in the Industrial Revolution.* Manchester: Manchester University Press.

Myers, S., and D.G. Marquis (1969): *Successful Industrial Innovation.* Washington D.C.: National Science Foundation.

McCabe, W.L., Smith, J.C., and P. Harriott (1984): *Unit Operations of Chemical Engineering.* (4th edition) London: McGraw Hill.

McCloskey, D. (Ed.) (1971): *Essays on a Mature Economy: Britain After 1840.* Boston: Harvard University Press.

McFarlane, A. (1978): *The Origins of English Individualism*. Oxford: Blackwell.

McMillan, C.J. (1985): *The Japanese Industrial System*. Berlin: de Gruyter.

Nadler, G. (1981): *The Planning and Design Approach*. New York: Wiley.

Nakamura, T. (1985): *Economic Growth in Pre-War Japan*. New Haven, Conn.: Yale University Press.

Nelson, D. (1983): Taylorism in American Society 1900-1930. *Le Taylorisme*. Montmollin, M., and O. Pastre (Eds.) Paris: La Decouverte.

Nelson, R., and S. Winter (1977): In Search of Useful Theory of Innovation. *Research Policy* 6 (1), 36-77.

Nelson, R., and S. Winter (1982): *An Evolutionary Theory of Economic Change*. Massachusetts: Harvard University Press.

Newby, H. (1978): *The Deferential Worker*. London: Penguin.

Newsome, J.R. (1982): Automation, Rationalisation and Concentration: A Case Study. Doctoral Thesis, Technology Policy Unit, Aston University.

Nicol, G. (1975): *Volvo*. London: Methuen.

Noble, D. (1977): *America by Design. Science, Technology and the Rise of Corporate Capitalism*. New York: Knopf.

Noble, D. (1984): *Forces of Production. A Social History of Industrial Automation*. New York: Knopf.

Nockolds, H. (1978): *Lucas. The First 100 Years*. London: David and Charles.

North, D.C. (1981): *Structure and Change in Economic History*. New York: Norton.

North, D.C. (1985): Transaction Costs in History. *European Journal of Economic History* 14 (3), 557-576.

O'Connor, J.J. (1984): *Accumulation Crisis*. Oxford: Blackwell.

Okochi, A., and H. Uchida (Eds.) (1980): *Development and Diffusion of Technology in Electrical and Chemical Industry*. Tokyo: Tokyo University Press.

Olleros, F-J. (1986): Emerging Industries and the Burnout of Pioneers. *Journal of Product Innovation Management* 3 (1), 5-18.

Ouchi, W.G. (1980): Markets, Bureaucracies and Clans. *Administrative Science Quarterly* 25, 129-142.

Ouchi, W.G. (1984): *The M-Form: How America can Recapture the Competitive Edge*. Reading: Addison Wesley.

Parkes, D.N., and N.J. Thrift (1980): *Times, Spaces and Places. A Chronographic Perspective*. London: Wiley.

Parsons, T. (1951): *The Social System*. London: Routledge and Kegan Paul.

Parsons, T., Bales, R.F., and E.A. Shils (1953): *Working Papers in the Theory of Action*. New York: Free Press.

Paul, W.J., and K.B. Robertson (1970): *Job Enrichment and Employee Motivation*. London: Gower.

Pavitt, K. (1985): Sectoral Patterns of Technical Change: Towards a Taxonomy and a Theory. *Research Policy* 13, 343-373.

Peel, J.D.Y. (1971): *Herbert Spencer. The Evolution of a Sociologist*. London: Heinemann.

Pelz, D.C. (1981): Quantitative Case Studies of Urban Innovations. Working Paper, CRUSK, University of Michigan at Ann Arbor.

Pelz, D.C. (1982): Use of Information in Innovating. Working Paper, CRUSK, University of Michigan at Ann Arbor.

Pelz, D.C., and F.C. Munson (1980): A Framework of Organisational Innovating. Working Paper, CRUSK, University of Michigan at Ann Arbor.

Penrose, E.T. (1959): *The Theory of the Growth of the Firm.* Oxford: Blackwell.

Perez, C. (1983): Structural Change and Assimilation of New Technologies in the Economic and Social Systems. *Futures* October, 357-375.

Perrow, C.A. (1967): A Framework for the Comparative Analysis of Organisations. *American Sociological Review* 32, 194-208.

Peters, T.J., and R.H. Waterman (1982): *In Search of Excellence.* New York: Harper and Row.

Pettigrew, A.M. (1985): *The Awakening Giant. Continuity and Change in ICI.* Oxford: Blackwell.

Piore, M.J., and C.A. Sabel (1985): *The Second Industrial divide: Possibilities for Prosperity.* New York: Basic Books.

Poitou, J.-P. (1984): *Le Cerveau de l'Usine.* Report to the CNRS, Paris.

Pollard, S. (1965): *The Genesis of Modern Management.* London: Pelican.

Prais, S.J., and K. Wagner (1985): Schooling Standards in England and Germany: Some Comparisons Bearing on Economic Performance. *National Institute of Economics Review,* 53-76.

Price, R. (1980): *Masters and Men: Work Control in Building and the Rise of Labour 1830-1914.* Cambridge: Cambridge University Press.

Pugh, D.S., and D.J. Hickson (1976): *Aston Programme: Volume One.* London: Saxon House.

Pulos, A.J. (1983): *American Design Ethic. A History of Industrial Design to 1940.* Cambridge: MIT Press.

Quinn, J.B. (1980): *Strategies for Change - Logical Incrementalism.* New York: Irwin.

Rader, B.G. (1983): The Impact of Social Elites on Sports. *Sports in American Society.* Baker, W.J., and J.M. Carroll (Eds.) River City: Rowan and Littlefield.

Reader, W.J. (1975): *Imperial Chemical Industries. A History.* Oxford: Oxford University Press.

Reader, W.J. (1977): The Imperial Chemical Industries and the State. *Essays in British Business History.* Supple, B. (Ed.) Oxford: Oxford University Press.

Revans, R.W. (1972): *Hospitals: Communication, Choice and Change.* London: Tavistock.

Riesman, D. (1954): *Individualism Reconsidered.* New York: Doubleday.

Riesman, D., and R. Denney (1954): Football in America: A Case Study in Culture Diffusion. *Individualism Reconsidered.* Riesman, D. (Ed.) New York: Doubleday.

Riesman, D., Glaser, N., and R. Denney (1975): *The Lonely Crown. A Study of Changing American Character.* New York: Doubleday.

Rogers, E.M. (1962): *The Diffusion of Innovations.* New York: Free Press (first edition).

Rogers, E.M. (1983): *Diffusion of Innovations.* New York: Free Press.

Rogers, E.M. (1986): Three Decades of Research on the Diffusion of Innovations: Progress, Problems and Prospects. Working Paper for the 1986 Venice Symposium on Innovation Diffusion: University of Southern California.

Rogers, E.M., and R.K. Rogers (1976): *Communication in Organisations.* New York: Free Press.

Rogers, E.M., and F.S.Shoemaker (1971): *Communication of Innovations: A Cross-Cultural Perspective.* New York: Free Press.

Rosenberg, N. (1963): Technological Change in the Machine Tool Industry 1840-1910. *Journal of Economic History* 23, 414-443.

Rosenberg, N. (1969): *The American System of Manufacturing.* Edinburgh: Edinburgh University Press.

Rosenberg, N. (1975): America's Rise to Woodworking Leadership. *Perspectives on Technology.* Rosenberg, N., Cambridge: Cambridge University Press.

Rosenberg, N. (1976): *Perspectives on Technology.* Cambridge: Cambridge University Press.

Rosenberg, N. (1981): Why in America? *Yankee Enterprise. The Rise of the American System of Manufacture.* Mayr, O., and R.C.Post, Washington D.C.: Smithsonian Institution.

Rosenberg, N. (1982): *Inside the Black Box. Technology and Economics.* New York: Cambridge University Press.

Rosenberg, N., and W.Vincenti (1978): *The Britannia Bridge. The Generation and Diffusion of Technical Knowledge.* Boston: MIT Press.

Rosenbloom, R.S., and W.J.Abernathy (1982): The Climate for Innovation in Industry. The Role of Managerial Attitudes and Practices in Consumer Electronics. *Research Policy* 11, 209-225.

Rothwell, R., and C.Freeman (1974): Sapho Updated - Project Sapho II, *Research Policy* 3, 258-291.

Rothwell, R., C.Freeman, A.Horsley, V.T.P.Jervis, A.B.Robertson, and J.Townsend (1974): Sapho Updated: Sapho Phase II, *Research Policy* 3, 258-291.

Sabel, C.A. (1982): *The Politics of Work.* Cambridge: Cambridge University Press.

Sabel, C.A., and J.Zeitlin (1986): Historical Alternatives to Mass Production. *Past and Present* 108, 133 177.

Samual, R. (Ed.) (1977): *Miners, Quarrymen and Saltworkers.* London: Routledge and Kegan Paul.

Sandberg, L. (1974): *Lancashire in Decline.* Ohio: Ohio University Press.

Saul, S.B. (1967): The Market and Development of the Mechanical Engineering Industries in Britain 1860-1914. *Economic History Review* 20, 111-130.

Sawyer, J.E. (1954): The Social Basis of the American System of Manufacturing. *Journal of Economic History* 14, 361-379.

Schein, E.D. (1985): *Corporate Culture.* London: Sage.

Scherer, F. (1982): Inter-Industry Technology Flows in the US. *Research Policy* 11 (4), 227-246.

Schonberger, R.J. (1982): *Japanese Manufacturing Techniques. Nine Hidden Lessons in Simplicity.* New York: Free Press.

Schumpeter, J. (1934): *Theory of Economic Development.* Harvard Economic Studies, Volume 46.

Schumpeter, J. (1939): *Business Cycles.* New York: Oxford University Press.

Schumpeter, J. (1954): *History of Economic Analysis.* New York: Oxford University Press.

Schumpeter, J. (1961): *The Theory of Economic Development.* New York: Oxford University Press.

Scott, W. H., Halsey, A. H., Lupton, T., and J. Banks (1956): *Technical Change and Industrial Relations*. Liverpool: Liverpool University Press.

Scott-Kempis, D. (1983): The Nature and Significance of Incremental Technical Change: A Case Study of a UK Paper mill, Working Paper Prepared for the Technical Change Centre, London.

Segal, H. P. (1984): *Technological Utopianism in American Culture*. Chicago: University of Chicago Press.

Sethi, S. P., Namiki, N., and C. L. Swanson (1984): *The False Promise of the Japanese Miracle*. Boston: Pitman.

Shaw, B. (1986): Appropriation and Transfer of Innovation Benefit in the UK Medical Equipment Industry. *Technovation* 4, 45-65.

Skinner, W. (1984): Re-Inventing the Factory. *Issues* 1 (2), 3-11.

Smelser, N. J. (1959): *Social Change in the Industrial Revolution. An Application of Theory to the Lancashire Cotton Industry 1770-1840*. London: Routledge and Kegan Paul.

Smith, C., Child, J., and M. Rowlinson (1987): *Innovation in Work Organisation - The Cadbury experience*. Cambridge: Cambridge University Press.

Smith, J. C. (1972): The Native American Ball Games. *Sport of the Socio-Cultural Process*. Hart, M. (Ed.) New York: Dubeque.

Smith, R. S. (1981): Military Entrepreneurship. *Yankee Enterprise. The Rise of the American System of Manufacture*. Mayr, O., and R. C. Post (Ed.) Washington D.C.: Smithsonian Institution.

Smith, T. C. (1986): Peasant Time and Factory Time in Japan. *Past and Present* 113, 165-197.

Solow, R. (1957): Technical Change and the Aggregate Production Function. *Review of Economics and Statistics* 39, 312-330.

Sorge, A. (1983): Cultured Organisation. *International Journal of Management and Organisation* 13, 106-128.

Stoneman, P. (1983): *The Economic Analysis of Technical Change*. Oxford: Oxford University Press.

Supple, B. (Ed.) (1976): *Essays in British Business History*. Oxford: Oxford University Press.

Szalai, A. (1971): *The Use of Time*. Amsterdam: Mouton.

Tann, J. (1981): *The Selected Papers of Boulton and Watt: The Engine Partnership*. London: Diploma Press.

Temin, P. (1975): *Causal Factors in American Economic Growth in the Nineteenth Century*. London: Macmillan.

Terleckjy, N. E. (1975): *Improvements in the Quality of Life: Estimates of Possibilities in the US, 1974-83*. Washington D.C.: National Planning Association.

Terleckjy, N. E. (1977): *A Study of Science and Research: Some New Indicators*. Boulder: Westview Press.

Thompson, E. P. (1967): Time, Work Discipline and Industrial Capitalism. *Past and Present* 38, 56-97.

Titley, U. A., and N. McWhirter (1970): *Centenary History of Rugby Football Union*. London: Rugby Football Union.

Touraine, A. (1955): *L'Evolution du Travail Ouvrier aux Usines Renault*. Paris: CNRS.

Townsend, J. (1980): Innovation in Coal-Mining Machinery. The Case of the Ander-

ton Shearer Loader. *Technical Innovation and Economic Performance.* Pavitt, K. (Ed.) London: Macmillan.
Trist, E.L., Higgin, G.W., Murray, H., and A.B.Pollock (1963): *Organisational Choice: The Loss, Rediscovery and Transformation of a Work Tradition.* London: Tavistock.
Tse, K.K. (1985): *Marks and Spencer.* London: Pergamon.
Turner, H.A. (1962): *Trade Union Growth, Structure and Policy. A Comparative Study of the Cotton Unions.* London: Allen and Unwin.
Uselding, P. (1980): Diffusion of Technology in America. *Development and Diffusion of Technology in Electrical and Chemical Industry.* Okochi, A. (Ed.) Tokyo: University of Tokyo Press.
Utterbeck, J.M., and W.J.Abernathy (1975): A Dynamic Model of Process and Product Innovation. *Omega* 3 (6), 630-656.
Utterbeck, J.M., and W.J.Abernathy (1978): Patterns of Industrial Innovation. *Technology Review* 7, 41-47.
van der Pijl, K. (1984): *The Making of the North Atlantic Ruling Class.* London: Allen and Unwin.
Vogel, E.M. (1979): *Japan as Number One. Lessons for America.* Boston: Harvard University Press.
von Hippel, E. (1976): The Dominant Role of the Users in Scientific instrument Innovation Processes. *Research Policy* 5, 212-239.
von Hippel, E. (1978): Transferring Process Equipment Innovations From User-Innovators to Equipment Manufacturing Firms. *R & D Management* Vol. I, 13-22.
von Hippel, E. (1982a): The Appropriability of Innovation Benefit as a Predictor of the Source of an Innovation. *Research Policy* 11, 95-115.
von Hippel, E. (1982b): The Customer-Active Paradigm for Industrial Product Generation. *Research Policy* 7, 240-266.
von Hippel, E. (1983): Novel Product Concepts From Lead Users: Segmenting Users by Experience. Working Paper No. 1476-83. Cambridge: MIT Press.
Vonnegut, K. (1955): *Player Piano.* London: MacMillan.
Walford, G. (1987): *Restructuring Universities. Political Power in the Management of Change.* Beckenham: Croom Helm.
Wallerstein, I. (1974): *The Modern World-System. Capitalist Agriculture and the Origins of the European World-Economy in the Sixteenth Century.* New York: Academic Press.
Wallerstein, I. (1980): *The Modern World-System II. Mercantilism and the Consolidation of the European World-Economy 1600-1750.* New York: Academic Press.
Weber, M. (1947): *Max Weber. The Theory of Social and Economic Organization.* New York: Free Press.
Weick, K. (1969): *The Social Psychology of Organising.* Reading: Addison-Wesley.
Weick, K. (1977): Enactment Processes in Organisations. *New Directions in Organisational Behaviour.* Staw, B.M., and G.R.Salancik (Eds.) Chicago: St.Clair Press.
Weyand, A.M. (1955): *The Saga of American Football.* Englewood Cliffs, N.J.: Prentice-Hall.
Whipp, R., and P.A.Clark (1986): *Innovation and the Auto Industry: Product, Process and Work Organisation.* London: Frances Pinter.

Wiener, M.J. (1981): *English Culture and the Decline of the Industrial Spirit 1850–1980.* Cambridge: Cambridge University Press.

Wild, R. (1971): *Mass-Production Management.* New York: John Wiley.

Wilkins, M. (1974): The Role of Private Business in the International Diffusion of Technology. *Journal of Economic History* 34, 167–188.

Wilkinson, E. (1984): *Misunderstanding Japan.* Tokyo: Chuokoron-Sha.

Williams, J.J. (1985): Working Papers on the Compatability of Zero-Based Budgeting Systems with Other Forms of Organisational Control, University of Alberta.

Williams, K., Williams, J., and D.Thomas (1983): *Why are the British so Bad at Manufacturing?* London: Routledge and Kegan Paul.

Williamson, O.E. (1975): *Markets and Hierarchy: Analysis and Antitrust Implications.* New York: Free Press.

Williamson, O.E., and W.G.Ouchi (1981): The Markets and Hierarchies Program of Research: Origins, Implications and Prospects. *Perspectives on Organisation Design and Behaviour.* Van de Ven, A., and W.F.Joyce (Eds.) New York: Wiley.

Willman, P., and G.Winch (1985): *Innovation and Management Control: Labour Relations at BL Cars.* Cambridge: Cambridge University Press.

Wittfogel, K.A. (1957): *Oriental Despotism. A Comparative Study of Total Power.* New Haven, Conn.: Yale University Press.

Woodward, J. (1965): *Industrial Organisation: Theory and Practice.* London: Oxford University Press.

Wrege, C.D., and A.G.Perroni (1974): Taylor's Pig-Tale: A Historical Analysis of Frederick W.Taylor's Pig-Iron Experiments. *Work Study and Management Services* November, 564–575.

Zaltman, G., and R.B.Duncan (1977): *Strategies for Planned Chance.* New York: Wiley.

Zerubavel, E. (1979): *Patterns of Time in Hospital Life.* Chicago: Chicago University Press.

16 Author Index

Abernathy, W.J. 14, 20, 35-36, 48, 82, 98, 103-113
Abromovitch, M. 43
Aldrich, H.A. 4
Allport, F.H. 12
Arrow, K. 43

Bell, R.M. 96
Boyer, R. 90-91
Braudel, F. 41
Braverman, H. 91
Bright, J.R. 91-92
Brown, L.A. 58, 82-84

Chandler, A.D. 5, 7, 18, 39, 84, 98
Child, J.C. 148
Clark, K.B. 14, 35-36, 82, 104
Clark, N. 6, 99
Clark, P.A. 4, 45, 46, 60, 61, 77, 81, 110, 133-135, 144, 148, 151
Cohen, M.O. 13
Coombs, R. 96-98, 119
Coriat, B. 90, 91
Crafts, N.F.R. 19
Crozier, M. 6
Cullen, D. 63

David, P.A. 14-16, 44
De Bresson, C. 109-111
DiMaggio, B. 7, 129
Donaldson, L. 9
Dosi, G. 101, 102
Duncan, R.D. 8
Dunning, E. 170-171

Elbaum, B. 6
Elias, N. 129, 166
Eveland, J.D. 58, 75, 79-80, 125

Freeman, C. 48, 84

Gardiner, P. 142-144
Gearing, E. 9-12, 139
Gershuny, J. 119
Giddens, A. 48, 130
Gold, B. 45, 124, 149
Granovetter, M.S. 72-73
Grieco, M. 127
Gross, N. 70-71, 74

Hagerstrand, T. 59-63, 67
Hall, E.T. 103
Hall, R.H. 8
Hall, R. 139-140
Hickson, D.J. 8, 12, 126
Hofstede, G. 138
Hounshell, D.A. 18

Jeremy, D. 161-165

Kantor, R.M. 142
Kaplinsky, R. 113, 141
Karpik, L. 137
Kondratiev, N.D. 41-43, 201

Lammers, C.J. 5
Lampel, J. 109-111
Landes, D. 6, 43
Lawrence, P.R. 20, 21-24
Lazonick, W.T. 6, 163
Littler, C.R. 19
Loveridge, R. 148

Maddison, A. 19
Mansfield, E. 45, 63-64
March, J.G. 41, 131
Marx, K. 41, 200
Mason, T. 172
Mathias, P. 17
McMillan, C. 18
Miles, R. 8
Miles, R.H. 131-132
Miller, D. 4
Miller, E.J. 151

Mintzberg, H. 125
Mohr, L.B. 5, 6, 32, 74, 169

Nadler, G. 150
Nelson, R. 48-50, 99, 100
Noble, D. 48, 55, 101, 183
North, D.C. 44

Olleras, F.-J. 110

Parssons, T. 154
Pavitt, K. 113-121
Penrose, E.T. 17
Pettigrew, A.M. 36, 121, 149
Piore, M. 19, 24, 45
Pugh, D.S. 8
Pulos, A.T. 185

Riesman, D. 166-174, 179, 181
Rogers, E.M. 5, 35, 47, 57, 65-75
Rosenberg, N. 18, 24, 44, 102
Rothwell, R. 74

Sabel, C.A. 19, 24, 45
Scherer, F. 114
Schonberger, R.T. 128
Schumpeter, J. 17, 41, 58
Shaw, B. 84
Simon, H.A. 47
Skinner, W. 124
Smith, C.J. 150
Snow, C.C. 181-182
Spender, J.C. 131-132
Starkey, K.P. 45, 149

Touraine, A. 112-113
Townsend, J. 96-98
Trist, E.L. 98

von Hippel, E. 58, 86-89

Wallerstein, I. 41, 74
Weick, K. 82, 130-131, 138-140
Whipp, R. 45, 48, 107, 144-148
Wiener, M.J. 17, 170
Wild, R. 91
Wilkins, M. 155-156, 162
Williamson, O.E. 44
Willmann, P. 113
Winch, G. 113
Winter, S. 48-50, 99, 100
Woodward, J. 93, 107, 109

Zaltman, G. 8

17 Subject Index

aero industry 13, 87, 142
agricultural innovation 15-16, 65-66
agriculture 38, 65-70, 261-262, 263, 287
American administrative sciences 279-286
- agribusiness 287-288
- corporate education 286-287
- designing 257, 275-279
- system of manufactures 24-25, 254-256, 264-268, 270
- systems of organizing 253-268, 269-296
analytically structured narrative 197, 211, 253
Anglo-American appropriation gap 325-350
armaments 24, 254-256, 264, 268
Aston Programme 8, 93, 95
asymmetrical power relations 219
automobile industry 20, 35, 104, 109-110, 120, 124, 128, 129, 141, 142, 144-148, 150, 154, 229, 292-294, 320-321, 341-345, 365 See references to Abernathy, W.T.

bandwagon effect 207
best practice 65
Birmingham (England) 247
boot and shoe industry 165-166
British systems of organising 233-251, 297-323, 325-350, 371-382

Cherokee Indians 9-12, 131
cigarette industry 288-290, 305, 318-320
The City (of London) 232, 245, 312-315
coal-mining 87, 98-99, 247, 307-308, 335
computer 47
construction 38-39, 249, 308, 345-346
core problem areas 3, 371-373
cross cultural innovation 6-7, 13-16
customer active paradigm (CAP) 86-88

decision making states and strategies 72, 81-82
deep structural crises 206, 380-381
design grid 11, 12
designing 226, 291
diffusion: curve 60
-: hierarchy effect 60-61
-: neighbourhood effect 60-61
duality of structure 216

economic chivalry 215
economy: allocative mechanism 197-212
education 63, 136-137, 152-153, 223-224
electrical 52, 101, 291, 322-323
engineering 308

France 195-196
French regulation school 200

garbage can models 13
geopolitical see structuration framework

health/hospital sector 81, 127, 129, 149, 302-303, 346
historical contingency 211

imitation model 58-59, 63-64
information technology 345-347
innovation: adoption models 66f.
-: areas of
-: dimensions 38, 40
-: economics of 40-46, 59-64, 68
-: entrenching/disrupting 36-37
-: evolvement of 31, 70, 373-376
-: facilitating features 68-70
-: first user 64
-: implementation phase model 79-83
-: individual trajectories 62-63
-: initiation phase model 77-78
-: intersection with organization 70-72

innovation: marketing model 83-88
-: role of agencies 84-85
-: selection environment 49-55
-: shape and uses 31
-: supply oriented models 57-88
-: transnational transfer 55-56
-: users position 32
-: varieties and type 32, 34-35
innovation studies: changing assumptions 46-49
- -: criticisms of established theories 5-6, 45-49
- -: dominance of economics 40-41
- -: fragmentation 31
institutions of work 227-228

Japan 353-370
just-in-time 193

learning by doing 43-44
longwave theories 197-212
loose, devolved organization 299-303

Master of Business Administration 39-40, 224, 348-349
multi divisional form 338-340

Netherlands 23, 236-239
new generic sectors 205
newspapers 139
North Atlantic Economy 263-264
occupational knowledge 225-226
organization development 332-338
organization sciences 52-55

payment systems 308-312, 340-341
photography 49-50
processual perspectives 7-13, 193-195

railways 274-275
recipe knowledge 193-232
recurrence event model 12, 217
regions: core/periphery 210, 230-232
Renaissance/retardation debate 17, 19-24, 381-382
retail 93-96, 326, 346

science/chemicals 36, 101, 121, 291, 313-314, 321-322
sport 39-40, 161-162, 167-168, 170-191
structural perspectives 8-9
structural repertoire 8-12, 217
structuration framework 197-198, 211, 216-226, 228-230, 334-335
subcontracting: internal 247
supplier-user interactions 58-59

Taylorism 282-286, 329-332, 362-363
technique: choice of 12-16
technological trajectories 92-100
technology 89-121
- as process 89-121
-: central position 89
technology ferment 82-83
technology paradigms 99-102
technostructure 11
temporal dimension 4, 11, 79-80, 211, 219-221
textiles 19, 133-135, 162-165, 241-244, 316-318, 379
thinking practices 217-219, 221-226
transaction costs 212-216
transitions 4, 13, 228-230
typical variety thesis 24-25, 294, 296

unintended outcomes 193
urban centres: significance of 215, 220, 235-236
USA: founding cultures 258-262

Venice Conference 73

western European capitalism 233-236
work study 223
world-economy 231, 245, 272-273, 313, 355

Bei Fragen zur Produktsicherheit wenden Sie sich bitte an:
If you have any questions regarding product safety,
please contact:

Walter de Gruyter GmbH
Genthiner Straße 13
10785 Berlin
productsafety@degruyterbrill.com